SCIENCE AND TECHNICAL WRITING

A MANUAL OF STYLE

PHILIP RUBENS
GENERAL EDITOR

AN OWL BOOK
HENRY HOLT AND COMPANY
NEW YORK

Henry Holt and Company, Inc.
Publishers since 1866
115 West 18th Street
New York, New York 10011

Henry Holt® is a registered trademark
of Henry Holt and Company, Inc.

Published in Canada by Fitzhenry & Whiteside Ltd.,
195 Allstate Parkway, Markham, Ontario L3R 4T8.

Library of Congress Cataloging-in-Publication Data
Science and technical writing : a manual of style /
Philip Rubens, general editor.—1st ed.
p. cm.
Includes bibliographical references and index.
1. Technical writing. I. Rubens, Philip. II. Series.
T11.S378 1992 91-36422
808'.0666—dc20 CIP

ISBN 0-8050-3091-3

Henry Holt books are available for special promotions
and premiums. For details contact:
Director, Special Markets.

First published in hardcover in 1992
by Henry Holt Reference Books.

First Owl Book Edition—1994

Designed by Katy Riegel

Printed in the United States of America
All first editions are printed on acid-free paper.∞

5 7 9 10 8 6 4

The use of illustrations and text from various companies in this work constitutes neither approval nor disapproval by these contributing companies of the concepts illustrated by their appearance.

This publication follows the National Council of Teachers of English Guidelines for Non-Sexist Language.

Contents

Exhibits

Preface

This text will help writers plan consistent and useful scientific and technical documents. It includes advice on all aspects of information design, from audience analysis to indexing to document design. Every aspect of a document, regardless of its length, requires advance planning. Thus, the first 11 chapters describe how to create useful text; the next two chapters describe how to create usable illustrations and data displays; and the final chapter describes how to create useful document designs.

Three chapters—Chapter 5, "Abbreviations," Chapter 6, "Specialized Terminology," and Chapter 7 "Numbers and Symbols"—contain closely related information useful in preparing technical and scientific documents. Here readers can discover how to use typographic conventions for mathematics and science.

Three other chapters—Chapter 12, "Creating Useful Illustrations," Chapter 13, "Creating Usable Data Displays: Tables, Charts, and Diagrams," and Chapter 14, "Designing Useful Documents"—provide both traditional and novel insights into the creation of nontextual elements for scientific and technical information. These chapters also discuss the ways in which text and graphics are interrelated.

We have been unafraid to discuss basic communication issues that influence any piece of information. Thus, we proclaim clarity, consistency, and conciseness as the basis of good scientific and technical writing. Writers must recognize the relationships among words, phrases, clauses, sentences, paragraphs, sections, chapters, and manuals. Successful communication also requires the author to develop an accurate profile of an audience and then write for that audience. Audience analysis has long been a problem in technical writing; however, we have unabashedly offered our guidelines.

Punctuation also receives attention throughout this text. Writers have considerable flexibility in choosing the appropriate punctuation mark (or lack thereof). Most punctuation rules have one or two qualifications, and few can be applied inflexibly in all situations. However, most are sensible, and all of them foster ease in reading and help clarify the text.

While we have tried not to duplicate information found in other style guides, we are not adverse to offering general guidelines that seem sensible. Thus, we advise that everyone who writes have a suitable college dictionary handy. Writers working in a field that uses specialized words will benefit from having an appropriate technical dictionary available.

Spelling is also closely related to the use of specialized terms in scientific and technical information. Abbreviated terms—whether abbreviations, initialisms, acronyms, or symbols—save time and avoid repetition. They also require less space in tables and illustrations. Abbreviated terms are easily understood in context. However, since different organizations and agencies have adopted their own abbreviated terms, authors collaborating on a single text often have difficulty agreeing on abbreviating conventions.

Terminology relates to the substantive content of documentation in that it deals with the nouns and verbs that convey meaning. Often, a very specific technical meaning or concept must be communicated. Scientific and technical communication relies on writers and readers sharing a common language set. If a writer operates outside the accepted language set for a particular discipline or subject, poor communication will result.

Science writing, and technical writing in general—as compared with other expository writing—depends more heavily on numbers and symbols, on measure and mathematics. This is necessarily so because of the quantitative nature of the ideas discussed. Definitions must be accurate and complete, operations mathematically rigorous, and measurements precise.

Precision in both symbols and terms has become increasingly important during the past few years as American industry has recognized the need to address an international market. While a variety of measures have been suggested, communications between two or more linguistic groups has always been a problem. Sign language, lingua franca, and pidgin are the traditional solutions. These options are still useful in many instances, but are limited for the communication of exact scientific and technical information.

Translation, as an alternative, presents other problems. The shortage of qualified translators further extends the delays from research to publication. Developing multilingual scientists, engineers, and technicians would require a revolution in education worldwide. The solution offered in this text is to develop reduced language lexicons that can be used across a variety of languages with ease.

A major component of scientific and technical information is the presence of graphics: illustrations, photographs, tables, charts, and diagrams. Graphics, when used properly, will increase the usability of a technical publication. Yet in modern scientific and technical publications, most graphics fail to realize this potential. Editors who would not tolerate an ungrammatical sentence or a misplaced comma ignore confusing, ugly, unprofessional graphics. The problem will only grow worse as writers without graphics training or experience use convenient computer programs to create illustrations, tables, charts, and diagrams.

A technical document represents a point in a continuum of human knowledge. Authors of technical documents draw on works that precede their own and, in turn, provide material for future readers. For a variety of reasons, technical authors must document how their work relates to other research in their field. Acknowledging these references is a vital part of writing for scientific and technical audiences.

Properly citing sources satisfies legal requirements, strengthens an author's argument, maintains intellectual integrity, and provides a valuable resource for the reader. Obviously, when other authors are quoted directly, it is necessary to credit them by identifying the source of the quotation. However, even when direct quotations are not used, it is important to give proper credit when the ideas and conclusions of others are used.

The contributors would like to acknowledge the help of Robert Berlo and Peter Murphy (Lawrence Livermore National Laboratory) and Robert Waite (IBM). The editor would like to acknowledge the secretarial and computing assistance of Laura Garrison, Pat Marra, and Myra Williams. He would also like to thank his family—all of them (especially Brenda)—for having the good sense to leave him alone for the many months of this project. Finally, he regrets that Clea could not have been here to leave her much missed, and always appreciated, mark on all these pages.

—The Contributors and Editor

ONE

Audience Analysis and Document Planning

Preparing a Formal Outline
Preparing an Informal Outline
Using the Nonlinear Outline

ANALYZING THE AUDIENCE

1.1 Before writing anything, describe an audience by

- Identifying audience characteristics,
- Assessing its objectives and needs,
- Planning for subgroups within the audience.

Conducting the Audience Analysis

1.2 Conduct either a formal—based on surveys and questionnaires—or an informal—based on discussions—analysis to create an audience profile.

Formal Audience Analysis

1.3 During formal analysis

- Conduct surveys,
- Use structured interviews,
- Gather questionnaires.

Some organizations often do formal analyses as part of their marketing plan.

Informal Audience Analysis

1.4 Gather information about the audience by talking with people who will read the final document. For example when writing

- Product documentation, talk to people who use the product (or a similar product).
- An article for a periodical or journal, talk to people who read that publication. Especially talk to those who have published in that or similar periodicals.

1.5 Interview marketing, development, and other staff. These specialists have market research results as well as access to customers.

1.6 When interviewing marketing and development staff

- Ask open-ended questions and follow up on incomplete answers.
- Ask about the users' backgrounds (how they work, why they will read the document, and what they need from it).
- Attend meetings at which the product or service will be discussed.

1.7 Find out about the audience by reading

- Notes and reports by product trainers or maintenance personnel who have had contact with the audience,
- Previous issues of a specialized periodical.

Identifying Audience Characteristics

1.8 Identify the audience characteristics and remember them while writing. Before you begin writing consider such important audience characteristics as

- Educational and professional background,
- English-language ability,
- Knowledge and experience levels,
- Reading situation.

Educational Background

1.9 Ask for information about educational background to assess the audience's reading ability and its willingness to read. A college-educated audience should be able to read more difficult texts than a high school– or grade school–educated audience.

In most cases, simple language—common words or technical terms appropriate to a particular readership—and a direct style—typical sentences without unusual structures—offer the best approach for all audiences.

Professional Background

1.10 Know the basic requirements of the jobs the readers perform. Do not confuse a job title with professional functions. For example, readers of technical and science writing could perform many professional roles at the same time:

- Scientists can be doctors, engineers, programmers, or technicians.
- Legislators can be judges or lawyers.
- Any professional could be a manager.

1.11 Job functions can imply different levels of knowledge. Compare, for instance, the difference between a design engineer's and a technician's knowledge of engineering theory.

1.12 Consider how a document will help readers do their jobs. Maintenance documents, for instance, must have less text and, perhaps, more illustrations to help readers complete their work quickly.

Knowledge and Experience Levels

1.13 Use professional and educational background to determine the audience's knowledge and experience on a subject. Use this information to evaluate what readers know and what information they need.

1.14 Categorize readers as a single-level audience if they are members of a specific group. While it is difficult to assign readers to such exclusive groups, a useful distinction can be novice, intermediate, and expert.

Novices have minimal knowledge or experience, and may even fear the product or subject. In reference information, they want basic concepts and procedures. In instructional materials, novices need to see quick results; successful experiences reassure them.

Intermediate audiences have some knowledge or experience. For example, if a document shows how to operate a drill press, an intermediate audience may have experience with similar equipment.

Experts are typically very knowledgeable. For example, an expert using software documentation may be a programmer who uses many of the software's applications.

However, a document that describes not only a particular subject matter (such as chemistry) but also the use of a particular tool (such as a computer) or technique (such as spectroscopy) complicates this view of the single-level audience.

1.15 Categorize readers as a multiple-level audience if they include technical experts (programmers, engineers, scientists) who are unfamiliar with certain tools or techniques. For example, the reader may have general knowledge and experience with mathematics, physics, electronics, and spectrometry. However, she may be an inexperienced computer user and may lack specific knowledge about emission spectroscopy. A document that describes how to use a software package to obtain emission data, and how to interpret that data using specialized mathematics, must address various levels of audience knowledge and experience.

1.16 Consider a document's implied as well as explicit audiences. For example, a technical manual prepared for novices may also be read by financial managers. This same manual may also have to support product maintenance. Hidden audiences affect a document's organization and style. In Section 1.15, for instance, the document may have a benefits summary for sales purposes or provide a reference table for expert readers.

English-Language Ability

1.17 Consider the audience's English-language ability. Many people employed in technical disciplines have graduated from American universities but come from other countries; English may be their second or even third language.

1.18 Consider, too, that a second technical language may be quite different from a second conversational language. The technical author has advantages over other writers, because technical English uses a small subset of the English language. (See also Chapter 8.)

Reading Situation

1.19 Consider the physical and psychological conditions under which the audience reads the document:

- A scientific article, for instance, may be read in a relaxed atmosphere at home or in an office.
- A spreadsheet software tutorial may be read on the job, at the keyboard, while dealing with interruptions.
- A heavy-equipment maintenance manual may be read while repairing the equipment in the field.

Identifying Audience Objectives and Needs

1.20 Use audience objectives and needs to shape how you approach the document:

- Objectives reflect what the audience wants to do after reading the document; for example, install a videotape recorder.
- Needs indicate questions the audience will have that the document should answer. Readers may not even know they will ask these questions, but the writer must anticipate them—and supply answers.

Audience objectives may be long term, short term, personal, or job related. They may or may not be directly related to the document.

1.21 Note that most technical documentation is written for readers with job-related objectives. Identify those objectives. Find out whether the audience will read the document to do a task or to expand its knowledge.

Addressing Diverse Audiences

1.22 To satisfy a diverse audience's needs, address both different experience levels and different goals. Follow these general guidelines when writing for multiple audiences:

- Rank goals in terms of the questions the document must answer first, second, third, and so on.
- Write for one audience group at a time, and indicate which group you are addressing. Expect that any other audiences may need the same information.
- Produce one document for all groups, or divide the information into more than one document.
- Include navigation aids—tables of contents and lists of figures and tables, page headers and footers, headings within the text, appendixes, tab dividers, and the like—to make information easy to find.

Creating the Profile

1.23 Use the audience characteristics, objectives, and needs to develop an audience profile or a profile of each subgroup of a diverse audience. To create the profile

- Group related features in a written sketch of the typical reader,
- For a diverse audience, write a profile for each kind of reader,
- Form mental images of these composite people,
- Get to know the profiles before writing anything,
- Plan the document for typical readers and write to them,
- Provide the kind of information and presentation the readers need to achieve their goals.

DETERMINING THE APPROPRIATE MEDIUM

Technical and scientific communication employs a variety of media forms—brochures, booklets, newsletters, articles, and technical manuals—each with specific characteristics and goals.

Exhibit 1–1:
Communication Media Characteristics

Medium	Purpose	Style
Brochure	Asks for action. Convinces readers to do or decide something.	Catchy, easy to read, lots of graphics.
Booklet	Offers overviews or introductions. Also short-term presentations for one-time or reference use.	Usually conversational, but not sales oriented.
Newsletter	Conveys information at regular intervals about related topics.	Usually journalistic, like a newspaper.
Article	Discusses a single topic in depth, usually for publication in a magazine or book.	Depends on the periodical or book editor, may be formal or conversational.
Technical Manual	Describes a product or process in detail.	Depends on the audience, but usually less formal than an article.

Brochures

1.24 Use a brochure to convince readers to take action or make a decision. In technical or business disciplines, readers need information before they can take action. To achieve its primary goal, a brochure must offer this information.

A brochure typically serves as a marketing or promotional tool. Although we tend to think of brochures as selling products or services, a brochure also might describe a company's pension plan or a professional trade show.

Brochures are rarely more than 16 pages, regardless of the page size. To attract readers, a brochure often contains photographs, illustrations, and clever headings, and is printed in color. Because a brochure is generally read quickly, write it using short sentences and paragraphs, and insert frequent headings to mark major topic divisions.

Booklets

1.25 Use a booklet to

- Convey information about a product or service, or to describe a new or technical subject readers need to know about before they can take action.

- Provide introductory, overview, or "one-group only" information. A university, for instance, could publish a booklet that details all campus services and features; the booklet introduces the university to new students and serves as a reference for continuing students and staff.
- Present both information that readers need only once or occasionally, such as troubleshooting and installation instructions, and regularly used information, such as computer command summaries.

Depending on its communication goal, a booklet may resemble a brochure or a technical manual, ranging in length from a few pages up to 50 pages.

Newsletters

1.26 Use a newsletter to address one major topic, either in a long article or several shorter articles, or to address a variety of topics in a collection of articles.

1.27 There are four types of newsletters: informational, public relations or promotional, internal (family), and hybrids (any combination of the other types).

1.28 Use an informational newsletter to present valuable information for which readers may be willing to pay. Independent research or publishing firms prepare such subscription newsletters.

1.29 Use a public relations newsletter to present and promote an organization's image. For example, a health-care facility may produce a newsletter for the general public that describes recent medical advances, features of its health-maintenance plan, and its staff's qualifications.

1.30 Use an internal newsletter to report an organization's activities and events. A corporate newsletter, for example, may contain articles on regional sales, carpooling, and employee achievements.

1.31 Some newsletters are hybrids. Consider the newsletter for customers that describes shortcuts for using a company's products and previews product changes; this kind of newsletter both informs and promotes.

Like a brochure, a newsletter tends to be short—between 4 and 16 pages long—and designed for fast reading. Multiple text columns similar to those used in general-circulation newspapers and magazines are a popular newsletter format.

Articles

1.32 Use an article to provide an in-depth discussion of a single topic, usually intended for publication in a magazine or book. Most readers expect to read an article in one sitting. There are three article types: professional journal articles, trade journal articles, and articles in the commercial or popular press.

1.33 Use a professional journal article to report results of scientific or engineering research to others in the same field.

Use an academic writing style and assume that the audience consists of other subject-area experts.

1.34 Use a trade journal article to report current news about a specific industry or discipline, case histories about the experiences of practitioners in the field, and product reviews.

1.35 Use articles for the commercial or popular press to report news about products or advances of interest to the general public. Such articles should be written without specialized vocabulary (acronyms and abbreviations) or detailed technical support (complicated data displays or specialized symbologies). The information should focus on the significance of the content for the reader's daily life.

Technical Manuals

1.36 Use a technical manual to describe in detail one system (or product) or a group of related systems. A technical manual frequently contains many internal divisions and numerous figures and tables.

Few readers need to understand every detail of a system's installation, application, operation, and repair. For example, a technician who repairs a heart monitor does not need to know when and why a doctor orders a heart test, or how to interpret one. Technical manual types fill specific reader needs. They can function as

- Tutorial/training guide,
- User's guide/operator's manual,
- User reference manual,
- Reference manual.

1.37 When an audience differs in levels of experience and responsibility, divide the necessary information among different manual types. Multiple manuals may be needed even for one audience, if readers have different needs at different times.

1.38 When organizing information into manuals or sections of manuals, consider both when the information will be used and by whom. Consider what information will be used

- Only once, such as installation information,
- Only by first-time or novice users, such as introductions or tutorials,
- Regularly, but by specialized audiences, such as preventive maintenance procedures,
- Only if something goes wrong, such as troubleshooting instructions.

Several smaller manuals are not only more convenient for readers, but they also offer advantages for the writer:

- Manuals can be produced one at a time, as information becomes available.
- Smaller manuals are easier to update than large manuals.
- Supplementary manuals can be added without reorganizing and reprinting a single, large manual.

Tutorial/Training Guide

1.39 Use a tutorial or training guide to provide skills. A tutorial uses explanation, repetition, hands-on practice, and other motivating methods to help the reader attain a desired skill level. There are two types of tutorials: stand-alone and classroom (or instructional).

1.40 Use a stand-alone tutorial to help readers learn without an instructor. This tutorial often comes with products as different as personal computers and telescopes. Stand-alone tutorials must be as complete as necessary for the audience to gain the skill without expert assistance.

1.41 Use a classroom tutorial to support an instructor, who will supply additional information, answer questions, and help solve problems. For example, many medical technicians attend instructor-aided sessions to learn how to use a new instrument. Materials for classroom tutorials are usually incomplete; that is, the audience cannot use them to gain the desired skill without the instructor's assistance.

1.42 In a tutorial, the sequence of procedures is important and should be designed to aid learning. For example, simple skills lead to more complex skills and general principles to particular applications.

1.43 Divide tutorials into short sections or training modules; assume that the audience will read sequentially, from beginning to end of a section or module.

User's Guide/Operator's Manual

1.44 Use a user's guide or operator's manual to offer procedures for using a product or system. Include procedures needed to perform specific tasks or groups of related tasks. Omit procedures that do not address these tasks; for the sake of brevity, also omit procedures that present advanced (and perhaps more efficient) ways of completing the tasks.

1.45 Organize user's guides or operator's manuals based on the tasks or groups of related tasks needed to complete daily work, not based on the product's or system's structure. Base this task order on chronology, frequency, or importance.

1.46 Depending on the audience profile, assume that experts have some system experience and do not need basic concepts explained. For other audiences, user's guides and operator's manuals include discussions of the basic concepts and/or theory of operation. Place such discussions at the beginning of the document or as an appendix, depending on the audience profile.

1.47 In the procedural chapters of user's guides and operator's manuals, begin each procedure by describing what it accomplishes and why it is efficient. Write short procedures and arrange as a list of numbered steps.

1.48 Include detailed tables and lists, as well as information (such as conceptual background) that not all audience members need, in appendixes. Include an index to help readers find information quickly. (See also Chapter 11.)

User Reference Manual

1.49 Use a user reference manual as a compromise between a user's guide and a reference manual; choose it when resources or schedule dictates that a product will have only one manual.

The user reference manual tries to be both procedural and task-oriented. However, because it must be a complete information source, this approach is not always possible. For example, if a product has 100 commands, time and space constraints may not allow them to be organized by audience and application; an alphabetical presentation may be necessary.

While most user's guides and operator's manuals address only primary tasks, a user reference manual describes everything users can do with a system. Effective headings and a good index help readers, who often have varied technical backgrounds and experience, find the information.

Reference Manual

1.50 Use a reference manual to provide encyclopedic information about a product; include complete technical details about its operation. A reference manual shows how a system is designed and operates. For example, the reference manual for a multiple-line telephone system may contain wiring schematics and switch-setting tables of interest to technicians, but not to telephone users. Many reference manuals are multipurpose and serve the needs of installation, operations, and maintenance personnel. However, a complicated system may require a reference manual for each task or component.

PLANNING A DOCUMENT'S CONTENT

1.51 After analyzing the audience and choosing the appropriate medium for the document, plan its content. The document's structure—the information order and its level of detail—is important when describing a complicated concept or a technical task.

Document planning involves at least three activities:

- Collecting information about the subject,
- Preparing an outline, and
- Selecting an organizational method.

Collecting the Information

1.52 Base information collection procedures on the document type being prepared and the source material available. For example, product specifications may be a good information source when writing a technical manual, but less useful when writing a brochure. Laboratory notes may supply important source information for a technical journal article, but not for a newsletter article about the same research.

1.53 When written source material does not exist, rely on experience, interviews, and product access to collect information:

- If the author is also the researcher or developer of the product to be documented, experience and product knowledge are the major information sources.
- If the author is not the developer, the author usually must interview the developer to obtain information.

1.54 Be aware that someone can know too much about a topic to write about it effectively. If a researcher writes her own material, for instance, she may easily make inappropriate assumptions about audience knowledge, unless she remembers the audience characteristics.

Analyzing Written Source Material

1.55 Find extensive written source material, including marketing materials, earlier versions of documentation, data sheets and specifications, and notes. By analyzing this material, the author can

- Consider where and how to find additional source information,
- Determine whether it contains adequate source information, and
- Formulate questions for interviews and/or subsequent research.

1.56 To prepare for the analysis, inventory the source materials and arrange them in an order that supports the analysis (for example, order of presentation).

1.57 When assessing source materials, consider these questions:

- What is the purpose of each source document? What does it help users do?
- How is each source document organized? What is the organizational pattern? Does this pattern support the document's purpose?
- What assumptions does each source document make about its readers?
- Does each source document include special terms? Does it define these terms? Should it?
- Imagine a question that a reader might have about a topic in each source document. Does the document contain information that can answer questions the reader might have about a topic in each source document?

1.58 During this assessment, keep in mind audience needs and the document's goals. This focus reduces the risk of missing important information or of spending too much time on unimportant information.

Interviewing Sources

1.59 If a document's author is not the primary researcher or product developer, obtain information from other people.

1.60 To prepare for an interview

- Bring a tape recorder to the session.
- Develop a list of questions.
- Explore with the interviewee whether additional sessions are possible.
- Give the interviewee an advance copy of the questions, if possible, to prompt ideas before the discussion.
- Schedule the interview for a convenient time, place, and duration.

1.61 Keep in mind that some interviewees will be comfortable and speak easily, while others will have trouble communicating. Participate in and control the interview according to the personality of the interviewee.

1.62 If the interviewee seems uncomfortable, start the interview with a question based on his or her expertise. Ask for additional explanation as needed, or pose open-ended questions.

Conducting a Hands-on Evaluation

1.63 When evaluating a product before writing about it, pretend to be a target user. Think of tasks the user must perform and then try to do them. Try to use all the basic product functions. Observe whether they operate consistently; note any inconsistencies.

Organizing the Document

1.64 Whatever the medium chosen for the document, make the organization consistent and rational, so readers can follow it easily and understand the information it presents. The type of medium—brochure or article, newsletter or technical manual—plays a major role in document organization. A brochure, for instance, might include specifications at the end; a technical reference manual is more likely to begin with those specifications.

Understanding the Challenge

1.65 Document organization requires thinking about the readers' expected usage patterns and needs. By thinking through the document, the author can begin to focus the information and identify missing information. The author can also decide whether any information is unnecessary, or whether there is too much information for one document.

1.66 Two factors make organizing a document a challenge. First, the organization must be obvious. A reader must be able to understand this plan simply by reading the document without further explanation.

1.67 Second, to be sure that readers understand the organization, the author must decide how to guide readers through the document. Guideposts can be as simple as headings and cross-references or as elaborate as graphic icons and color coding incorporated in the page design.

Understanding Typical Organizational Approaches

1.68 Be sure that a document's organization explains and logically arranges all necessary ideas. Although there is no perfect organization for any technical or scientific document, some organizational methods will be more appropriate than others, depending on the document's audience and purpose.

1.69 Typical organizational approaches include chronological, spatial, climactic, and task-oriented. Some documents may need an organization that combines more than one of these approaches.

1.70 Use chronological organization when time is the organizing principle. It orders information according to when concepts develop, events occur, or actions happen, generally from the earliest to the latest.

1.71 Use spatial organization when placement or geography is the organizing principle. Consider, for instance, describing equipment by parts or components, or presenting product sales figures or health statistics by geographic region.

1.72 Use climactic organization to progress from least to greatest impact, concluding with the most interesting or forceful concepts, ideas, or facts. Brochures, when convincing readers to take action or make a decision, use this organization.

1.73 Use task-oriented organization when readers will use information to do something. Thus, an instruction manual organizes its information into procedures or activities users perform.

1.74 Task-oriented organization depends on two principles:

- Organizing the information into tasks (procedures or activities), and
- Ordering or arranging the tasks in the document.

1.75 Task-oriented organization emphasizes the actual activities someone must perform. These activities include

- Physical performance—the movements necessary to cause changes in states. For example, setting switches, moving levers, and entering specific characters all require physical movement.
- Cognitive performance—the ability to apply information to novel situations. For example, selecting switches to configure a machine for local conditions or constructing a unique command string based on a generic example both require readers to understand the idiosyncratic nature of their tasks.

1.76 Four common ways to order tasks are

- Most important to least important—a user's guide for a computer operating system with a graphic interface might first describe how to interpret the interface and interact with the system. Next, it might describe using files.
- Most frequent to least frequent—maintenance schedules often list first the procedures to be performed every 5,000 miles, followed by those performed every 12,000 miles, and so on.
- Simple to complex—a user's guide for a computer graphics program might begin by describing how to copy shapes and then enlarge, reduce, or stretch them. Finally, it might explain how to create original shapes.
- Required to optional—instructions for assembling an exercise bicycle might conclude by explaining how to install the pulsimeter available as an optional feature.

Outlining the Document

1.77 Outline a document to organize it and define its content before writing. The outline provides a baseline for the document; parts of the outline may change during writing. Along with this flexibility, an outline

- Allows authors to try various organizing principles,
- Controls the level of detail,
- Ensures planning and thinking before writing,
- Establishes the relative importance of topics,
- Focuses on the document's purpose,
- Provides the potential wording for headings, and
- Reveals missing information and other obstacles.

1.78 Use a formal, detailed outline for reviewers who must approve a document.

1.79 Use an informal outline when the outline will be only a tool for the writer's use.

Preparing a Formal Outline

1.80 Prepare a formal outline to show a document's hierarchical structure. Depending on the level of detail, the outline may present chapter titles, section headings, one or two levels of subheadings, and even groups under subheadings. The formal outline's logical structure makes clear the relationships among topics.

1.81 Most formal outlines

- Catalog the background information needed to write the document,
- Evaluate the scope of the headings, and
- Identify the number of headings in a document.

1.82 A formal outline should help reviewers judge a document's organization as well as the completeness and accuracy of the information included. Because the outline should reveal the author's assumptions about the topic, reviewers may also be able to correct any underlying logical errors.

1.83 There are two types of formal outlines: topic and sentence. Although both outline types use the same parts and groupings of ideas, a topic outline uses phrases to express the ideas while a sentence outline uses complete sentences.

1.84 The following example shows the beginning of a topic outline:

1. Data network benefits
    a. Shared software

b. Efficient data and file exchange
c. Shared equipment resources
2. Problems of evolving complex networks
a. Early networks
(1) Mainframe computers
(2) Terminals
(3) Single vendor
b. Today's networks
(1) Mainframe computers and terminals
(2) Microcomputers
(3) Minicomputers
(4) Local area networks (LANs)
(5) Multiple vendors
c. Compatibility issues
(1) Equipment/vendors
(2) Media
(3) Communication protocols

1.85 The following example shows how a sentence outline might present the same information:

1. Data networks provide modern computer environments with many benefits, including shared resources, efficient file transfers, and electronic mail.
a. Networks simplify support through shared software.
b. Exchanging data through networks provides accuracy and reduces costs, because all employees can access the same information.
c. Network users can also share equipment such as printers, plotters, and storage devices.
2. As complex networks grow, managers face problems.
a. Early networks consisted of mainframe computers and terminals from one manufacturer.
b. Now large organizations have also bought microcomputers, minicomputers, and local area networks (LANs), often from many different manufacturers.
c. But most network hardware and software work only with limited networks: one manufacturer's equipment, one or two programs, or one or two different media.

1.86 Use the sentence outline when reviewers will evaluate the final document's style and tone. Note how the sentence outline includes more detail and even implies a writing style and tone for the document.

Preparing an Informal Outline

1.87 Use an informal outline to list the controlling idea of each paragraph, but not to indicate the hierarchy of ideas.

1. Networks support modern computing facilities (include examples of network benefits).
2. Complex networks create problems for network managers (describe history and how it led to problems).

This sample outline uses complete sentences; it could have used phrases or simply key words.

Using the Nonlinear Outline

1.88 Some authors may feel uncomfortable with topic, sentence, and paragraph outlines, because they are linear approaches to document organization. The nonlinear outline approach, in contrast, is based on systems analysis, programming, and structured design/analysis.

1.89 Begin this outlining technique by writing the primary idea in the middle of a page. As related ideas occur, add them around the primary idea as relational nodes and draw connections among them. These connections extend to new ideas related to the secondary ideas, and so on. Work at random within this outline, to combine, add, and delete nodes in any order or direction.

1.90 Use the nonlinear outline to write the document by manually arranging index cards based on the nodes on a surface, some separate, some overlapping, and some piled on top of one another; then write the document by adding information to cards and adding cards to accommodate new relationships.

TWO

Solving Paragraph and Sentence Problems

GRAMMAR AND SCIENTIFIC AND TECHNICAL COMMUNICATION

2.1 To communicate effectively, scientific and technical writing must, more than anything else, be clear, consistent, and concise. To achieve these ends, a writer needs a thorough knowledge of grammar. Grammatical rules give structure to a piece of writing so that the writer's information and intentions will be understood by the reader.

It is important to recognize the relationships among words, phrases, clauses, sentences, paragraphs, sections, chapters, and even manuals. To do so, an author must be familiar with the structures available. This chapter offers guidance on these basic forms with emphasis on those particularly common or useful in scientific and technical writing.

PARAGRAPHS: CREATING FUNCTIONAL UNITS

The Nontechnical Paragraph

2.2 Typically, the nontechnical paragraph begins with a topic sentence that puts forth a central premise. The paragraph then develops that premise by using examples, illustrations, analogies, definitions, and classifications, or by analyzing and solving a problem.

The Technical Paragraph

2.3 Technical paragraphs tend to be short, often using a section heading or a summary in place of a topic sentence. As in journalistic paragraphs, they divide information into small conceptual units.

2.4 A reader performing a task or solving a problem will not be inclined to read a long and involved text. Thus, paragraphs for technical manuscripts should consider the

- Order in which the reader will need the information,
- Type of document necessary to support the tasks, and
- Purpose of the specific paragraphs (to illustrate, compare, and so on).

Paragraph Organization and Development

2.5 Paragraph structure depends on the purposes of the information contained in the paragraph. For instance, documents that support tasks for readers with varying levels of expertise should use shorter paragraphs than a scientific report written for a professional journal that has a well-defined style.

Headings, Topic Sentences, and Summary Paragraphs

2.6 Readers use headings to locate the sections they need. A heading should provide an efficient description of the content it introduces. Highlighting a heading with specialized typography such as underlining or boldface type will make it stand out on the page.

2.7 Use a topic sentence to elaborate on the heading and guide the reader into the paragraph.

2.8 When a document's complexity requires many subsections, a topic sentence should organize these small levels of information, as in this example

> There are five categories of processing operations: input, arithmetic, logical, output, and storage. Input operations place data in computer memory. Arithmetic operations include addition, subtraction, multiplication, and division. Logical operations compare data sorted in computer memory. Output operations transfer data to a screen or printout. Storage operations save data in electronic storage.

2.9 A summary paragraph can also function as a topic sentence by forecasting the content and organization of subsequent paragraphs. The preceding example listing five categories of operations, for example, could serve as an introductory paragraph to five more paragraphs, each dealing with one type of operation in detail.

Major and Minor Paragraph Supports

2.10 If a major point has several minor elements, each should be a separate paragraph. In the following example, for instance, the first paragraph states the three major functions without offering specific details about them. The next three paragraphs each discuss one of the major functions in detail:

> Disk drives perform three functions. They receive data from a variety of sources. They store that data for future use. And they send that data to other devices for storage, display, or manipulation.
>
> Disk drives can receive data from a keyboard, random access memory, another disk drive, scanning devices, modems, touch devices, pointing devices, and the like. These data can be either stored for later use or used immediately.
>
> Disk drives store data in blocks on a storage media—usually a floppy or hard disk—in a manner controlled either by the disk operating system or by a program. Data can be recalled from storage to use.
>
> Disk drives can send data to any device from which they receive data and to other devices that can display or act on data. Thus, a . . .

The three functions each offer minor support and have their own paragraphs.

General-to-Specific Paragraph Structure

2.11 Use a general-to-specific paragraph structure to focus on specific reader needs:

> Adobe building materials have three characteristics that make them susceptible to radon containment: lack of vacuoles for air exchange, high concentrations of vapor for gas entrapment, and reduced surface area for transpiration effects.
>
> Air exchange occurs wherever the structural integrity of a building . . .

The author can then give a general definition of air exchange characteristics in the first paragraph; the second paragraph can discuss problems created by a lack of adequate air exchange; and the third paragraph can relate this background information to the unique problems encountered with adobe materials. A final paragraph can provide engineering or scientific calculations for assessing or resolving these problems.

With this general-to-specific paragraph order, a reader must consider the entire text to discover the major point being made. For instance, the summary introduction would provide an overview, individual paragraphs would focus in detail on single aspects of the problem, and the final paragraph could help guide specific approaches for solving one or all of the problems.

Specific-to-General Paragraph Structure

2.12 Use the specific-to-general paragraph structure to present a number of specific points leading to a general statement or conclusion. This structure is particularly useful when the goal is to convince the reader of the concluding point by leading up to it with supporting detail. In the following example, the writer lists a series of problems to support the concluding point:

> Cognitive scientists have shown that the human mind experiences great difficulty in maintaining more than five to seven color elements simultaneously. Thus, the benefits of color in attracting attention, grouping information, and assigning value disappear if too many colors are used. To keep these benefits, limit screen displays to six unique colors. The general rule, then, is not to overuse color.

Specific-to-general paragraph structure is often used in executive summaries, marketing reports, proposals, and oral presentations.

Time and Order Paragraph Structure

2.13 Technical reports often report on processes that depend on a specific chronological or spatial order. In recording a laboratory test, for instance, the writer must follow the time order in which events occurred:

> To prepare a specimen for analysis, you must complete two preliminary steps:
>
> First, isolate two samples in individual test chambers with 45-percent humidity and a temperature of 20 degrees centigrade. Maintain these conditions for six hours before proceeding.
>
> Second, combine these samples with 5 centiliters of hydrogen sulfide in a 55-milliliter petri dish for 25 seconds.

Use the same paragraph structure for reporting any sequence of events that requires accuracy or repeatable results. It could, for instance, also be used by a police officer reporting the sequence of events that describes a crime or accident.

2.14 Use spatial organization to describe physical parts in a consistent direction (for example: top to bottom, left side to right, inside to outside). This organization helps the reader visualize the object or the progress of a process. The following paragraph describes common aircraft controls from the nose toward the tail:

> The control system has five functional parts starting from the aircraft's nose: manual instrument panel controls, cockpit yoke controls, overhead panel manual controls, flight engineer guidance panel, and a backup manual system in the rear bulkhead.

Process Paragraph Structure

2.15 Examples of specific operations might be illustrated by a simulation that shows how a mechanism changes as an operator works through a process. In this instance, develop the paragraphs by illustration and include both timing and location information:

> Users can place all or some modules in the overlay structure by using the overlay description file. The LINK command automatically places in the overlay structure modules not placed by the user. LINK places each module as low in the overlay structure as possible.
>
> When the placement process begins, there is only one module (module A) in the overlay structure. This module contains directory information and cannot be acted on by the user.
>
> If the user places four modules in the overlay structure, they will be named modules B, C, D, and E. E will be the last module the user added.
>
> In addition, if the user creates two additional modules and does not place them in the overlay structure, the LINK command will do so. In this instance, the user will find that there are seven modules in the overlay structure: module A (system defined), four modules (B through E) placed by the user, and two modules (F and G) placed by the LINK command.
>
> If the user wants to access her last module (E in this example), she must realize that it is not the last module and must remember its name.
>
> If the user does not place a module, the LINK command automatically places the module in the overlay structure.

These paragraphs illustrate the consequence of a person's actions as she proceeds through a process. They also depict the changes in the state of a process and tell the user their relationship to that process.

Logic Paragraph Structure

2.16 Develop paragraphs by reason or logic when arguing for a position or hypothesis. Such a development method is appropriate to journal articles, argumentative research reports, and proposals. In the following example, for instance, the author begins with a problem statement (containment site conditions and high-water flow rates) in the first paragraph. The next two paragraphs present data to support his hypothesis, and the final paragraph offers a series of counterproposals:

> While several proposals have indicated that a containment pond above the proposed bridge would eliminate the potential for structural damage during high-water conditions, both the soil characteristics around the containment site and the average high-water flow rate still present problems.
>
> The soil conditions around the proposed site have failed percolator tests at over 50 percent of the test wells. Given the porous nature of the containment walls, it is unlikely that they can successfully retain high-water levels such as those encountered in the spring of 1984. It is also unlikely that adequate bonding can be provided in this earthen fill because of the incompatible granular sizing of the proposed backfill sands.
>
> Similarly, the high-water flow rate has not been taken into consideration in the proposed design. While the present design does consider the high-water levels, it does not consider the flow rate. Since the containment pond is over 800 feet higher in elevation than the bridge pilings, the flow rate could be considerable during high-water conditions. Tentatively, a flow rate of 1,200 ft/lbs at the piling level has been calculated. This leaves a safety margin of less than 22 percent (the state's road and bridge commissions both require 35 percent).
>
> Thus, three steps must be considered before continuing work on this project:
>
> 1. Test sites both higher and lower than the current one to determine the percolation characteristics of a larger section of the streambed.
> 2. Select a backfill material more consistent with the in situ materials to improve the bonding characteristics of the containment wall.

3. Consider other measures of reducing the outlet flow rate during high-water conditions—that is, run-off spillways, laddered streambed upstream from the pilings, or additional pier work on the bridge.

Analysis Paragraph Structure

2.17 Use analytical paragraphs to divide a subject into parts and show the relationship among those parts. For example, in the following paragraph the author defines the kinds of operations a computer performs and separates those operations into distinct classes:

> Computers can perform only two types of operations: arithmetic and logic. Arithmetic operations include addition, subtraction, multiplication, and division. Logic operations compare data sorted in the computer's memory. While many computer programs use both of these capabilities, they operate independently. First one operation is performed and then the other, although in no special order.

Cause-and-Effect Paragraph Structure

2.18 Use the cause-and-effect paragraph to explain a given effect by the objects or events that caused it or by noting an effect and tracing its causes:

> If you try to print a lowercase letter by pressing the LOWR key, the computer will print it capitalized instead and will make it light green. If you try to print a dark-on-light capital letter by pressing the ^ and LOWR keys, the computer will instead make it dark blue–turquoise on a black background.

Comparison-and-Contrast Paragraph Structure

2.19 Use the comparison-and-contrast paragraph to compare or contrast items point by point. The following example compares two methods for creating computer networks. First the cabling method is described, and then the manner in which each network transfers information is illustrated. The intent is to offer a comparison/contrast of the relative strengths of the two methods:

> In a bus topology, a single cable connects all the devices in the network. In a ring topology, devices in the network are also connected to and share a single cable, but it forms a circle. Messages are sent from one device to another around the ring.
>
> In a bus topology, messages travel in either direction from one computer to another. An advantage of the bus topology is

> that devices can be attached or unattached from the network at any point without disturbing the rest of the network.
>
> The ring topology, on the other hand, cannot function if any one of the computers is disabled because as the message moves around the ring, each terminal electronically detects whether the message is for it, sending the signal to the next station if it is not.

Items should be matched consistently (point by point), and the purpose of the comparison or contrast should be apparent. Here the paragraph compares the two topologies in order to show the advantage of a bus topology over a ring topology.

Paragraph Coherence

2.20 A coherent paragraph is made up of a series of points that lead logically from one to the next and that all contribute to the central premise. The following paragraph lacks coherence because of poor organization and irrelevant details:

> Based on access to new technologies, the Centers for Disease Control identified the organism responsible for eurythromicotic sarcoma. Forty microbiologists worked on the project. Each researcher had access to an electron microscope, which offered a means of quick and accurate detection. While it was very difficult to isolate this organism, it was susceptible to a new staining agent that allowed more accurate visual assessment. The agent also allowed researchers to isolate several viable colonies of the organism. These colonies reacted well in new growth mixtures composed of agar and blood serum from selected carriers. Over five months of study was necessary to identify this organism.

The central premise in this paragraph is that new technologies have facilitated the identification of an organism. The new technologies described are electron microscopy, a new staining agent, and improved culturing media.

The unrelated details include the number of personnel involved in the study, their working conditions, and the amount of time it took to identify the organism. In addition, the order in which the relevant information is given is misleading. It is more logical that the improved culturing media and staining agent made electron-microscopic identification possible rather than the electron microscope being the central element in the study. Thus, the paragraph should be revised as

> Using new technologies, the Centers for Disease Control identified the organism responsible for eurythromicotic sarcoma. After

developing a new growth medium composed of agar and blood serum from selected carriers, the microbiologists working on the project cultivated several suspect organism colonies. These colonies proved susceptible to a new staining agent that allowed more accurate visual assessment by electron microscopy.

Transitions Between Structures

2.21 Transitions between sentences tend to be abrupt in technical writing. Typically, readers read only specific passages in technical information to solve problems. Even scientists often read only parts of research articles, perhaps the methodology and conclusion sections, to assess the validity of research reports for their interest. Nonetheless, the use of transitional words and phrases is important for creating continuity among document sections.

2.22 Common transitional words and phrases include conjunctive adverbs, coordinating conjunctions, and phrases:

Purpose	Word or Phrase
addition	again, also, and, and then, besides, equally important, finally, first, further, furthermore, in addition, in the first place, last, likewise, moreover, next, nor, second, third
comparison	in like manner, likewise, similarly
contrast	and yet, at the same time, but, even so, for all that, however, in contrast to this, nevertheless, nonetheless, notwithstanding, on the contrary, on the other hand, otherwise, still, yet
place	adjacent to, beyond, here, nearby, opposite to, on the opposite side
purpose	for this purpose, to this end, with this object
result	accordingly, consequently, hence, then, therefore, thereupon, thus
summary	for instance, indeed, in fact, in short, that is
time	after, at length, later, meanwhile, now, soon

2.23 Sentences can also be linked together with pronouns and key terms. Pronoun reference must be accurate in both number and person. Key terms should be identical throughout the document and clearly defined. The writer may want to attach a glossary.

Parallel Structure

2.24 Parallel structure presents items of equal importance in the same grammatical form and remains consistent in form throughout a series. The following sentence, for instance, contains a series of nouns: He replaced a valve, a hose, and a cap.

Parallel structure creates certain expectations, giving the reader a familiar structure from which to extract information and the writer a boilerplate (standard) form in which to insert similar pieces of information.

Using Parallel Structure to Describe a Process

2.25 When writing installation instructions, begin each step with an imperative form of the verb. This technique emphasizes the step-by-step nature of the process:

1. *Place* a copy of the system disk in the disk drive,
2. *Observe* the flashing prompt line on the screen,
3. *Type* FINSTAL, and
4. *Press* any key.

A parallel structure, such as this one, helps the reader locate and use information. Similarly, consistency in punctuation, sentence structure, and item names will make it easier for the reader to follow the instructions.

Using Parallel Structure Between Paragraphs

2.26 Although parallelism is most often used at the sentence level, in technical writing it can also be used in a larger context.

Use standard, or boilerplate, paragraphs for situations in which information remains similar in many contexts. For instance, in a diagnostic manual describing similar testing procedures, a boilerplate form might be used:

Test 34BX performs the following sequence of steps:

1. Resets the BIRP,
2. Clears the MDP pointer,
3. Increments the AP pointer, and
4. Calculates the pointer differences.

In this example the author has used active, imperative verbs and parallel sentence structure. Since this example represents a common user action, it can be repeated throughout the text in exactly the same form. In addition, similar listings can be used to describe other activities that have the same kinds of steps. Readers will soon learn to recognize the pattern and will use it to complete tasks more efficiently.

2.27 Including boilerplate information and parallelism depends on how readers will use the document. If they will use it to fix an engine, write a program, or troubleshoot a circuit board, they will appreciate easy-to-locate information. This ease in locating depends on predictability, knowing where to find what is needed for a specific task.

Using Parallel Structure to Organize a Document

2.28 Parallelism can also help locate information in sections, chapters, manuals, or even manual sets. When this information structure remains consistent, readers will be able to use the same retrieval methods each time they need to find additional information.

2.29 Begin planning good parallel structure in the outlining stage. Usually, in a larger document, the tables of contents—both manual and chapter level—and section headings develop from the outline. Good parallel structure in the outline puts related ideas together or effectively contrasts ideas. The outline hierarchy reflects these relationships.

Paragraph Consistency

2.30 Consistency is one of the hallmarks of good technical and scientific writing. Not only should organization and use of key terms be consistent, but paragraphs and sentences themselves must avoid basic shifts in

- Verb tense,
- Mood,
- Voice,
- Person and number, and
- Tone and point of view.

Shifts in Verb Tense

2.31 Use the present tense, though occasionally other tenses may be necessary. In either case, a writer should avoid unnecessarily changing verb tense.

> When the technicians *removed* the air intake valves from the combustion chamber, two testing methods *are repeatedly attempted.*

The author has shifted into the present *are repeatedly attempted* from the past *removed*. The correct version of the sentence should read

> After removing the air intake valves from the combustion chamber, the technicians used two methods to test them.

Shifts in Mood

2.32 The text determines how the verb is used. The following example illustrates the imperative and indicative forms:

> *Use* the Load Table to find LINKER settings. The settings *are used* to set up local telephone equipment.

Notice that this mood shift also creates a shift in voice. The imperative often helps you avoid passive voice constructions:

> Use the Load Table to find LINKER settings to set up local telephone equipment.

Shifts in Voice

2.33 Avoid using the passive voice unless it is simply more graceful than the active voice. In the following example, the writer uses unnecessary passive constructions:

> If the "automatic save" option is specified, the operating system will create file names. The "automatic save" description must include a limit on the file size. This number must be stated in bytes.

Using the active voice the paragraph becomes

> If you specify the "automatic save" option, the operating system creates file names. The "automatic save" description must include a file size limit stated in bytes.

Shifts in Person and Number

2.34 Use first, second, or third person in the singular or plural form consistently throughout a document. A change in person and number creates problems in subject and verb agreement and in pronoun reference. Attempts to avoid sexist language sometimes results in awkward shifts that are grammatically incorrect. Shifts in person and number often occur, as in this awkward example

> *You* can also move the cursor in other ways. *People* often hold the Ctrl key down and press the right arrow key, allowing *you*

> to move the cursor one word to the right. If a *person* wants to move quickly to the top of the screen, *he or she* should press the Home key and then the up arrow.

In the previous paragraph, person shifts from *you* to *people* to *you* to *a person* (*he* or *she*); that is, from second person, to third person plural, back to second person, to third person singular. A useful revision of this paragraph would be

> You can also move the cursor in other ways. You can hold the Ctrl key down and press the right arrow key to move the cursor one word to the right. If you want to move quickly to the top of the screen, you can press the Home key and then the up arrow.

Shifts in Tone and Point of View

2.35 Be consistent in the tone and point of view. The following example begins with a third-person narrator and a formal tone and then shifts abruptly to a second-person, informal style:

> During the detailed system investigation, the *analyst* should have gained a thorough understanding of the proposed new system. So *you* might want to *get going* on a prototype of the new system as you see the system design phase *starting up*.

Although either style is valid depending on the document's purpose, switching from one to the other within a document disorients the reader.

Structuring a Paragraph Based on Document Type

2.36 Paragraphing should also be considered a function of document type. For instance, organize reference texts for easy information retrieval with more numerous, short paragraphs. Paragraph tutorial texts to focus attention on small steps for completing a task. Journal articles should present information in the style appropriate for a particular discipline. Most journals have well-defined styles for presenting logical arguments.

Paragraphing Reference Books

2.37 Because reference books are written for a general audience, avoid specialized jargon and undefined terminology. Paragraphs should be written for nonspecialized readers to provide an overview of the subject. For example, a paragraph in *Collier's Encyclopedia* defining cholesterol reads

> . . . a fatty substance that normally occurs in all animal tissues. Cholesterol serves various functions in body physiology. It is a constituent of the membrane surrounding the cell, of soluble fat-protein complexes that circulate in the blood and other body fluids, and of the secretions of the oil glands of the skin. [Macmillan, NY, 1987, vol. 6, pp. 377–78]

Note that a description that uses generalized, familiar terms follows the definition. Reference information often begins with a technical definition and then offers other information that would be useful to the reader. In the preceding example, the article goes on to explain the importance of cholesterol, its chemical structure, human problems with high cholesterol levels, and recommended low-cholesterol diets.

2.38 A handbook provides a concise reference for specialists. Since handbooks are used daily for specific purposes, paragraphs must be well-organized, brief, and cross-referenced. For example, the following paragraph from a company style guide offers specific instructions concerning creating lists:

> Some lists consist of terms and their explanations.
>
> To format a list:
>
> 1. Enter the term at column 2,
> 2. Begin the explanation at column 10,
> 3. Insert a blank line between the text and the first entry,
> 4. Insert a blank line between entries, and
> 5. Insert a blank line between the last entry and the following text.

2.39 Reference manuals serve various kinds of users; thus paragraphs have to be cross-referenced carefully. Information often appears in tables and lists for easy retrieval. In the typical programmer's manual, for example, commands, functions, and utilities might be easier to find in alphabetized lists. Such books help users who need a great deal of related information in one place to answer specific questions.

Organizing Instructional Information

2.40 Instructional information should be short and divided into specific tasks, as demonstrated in these instructions for storing computer data:

> To use your computer efficiently, you must know how to store and retrieve information or data stored on a disk in a disk drive.

If you do not know how to place a disk into a disk drive, read Section 2 now and return here when you are done.

Now that you know how to put a disk in a drive, you can use that disk to store and retrieve information or data.

To see what a disk contains, type CAT after the > sign on your screen and press the ENTER key.

Organizing Scientific Journals

2.41 Scientific journals often address a broad readership. They have well-defined guidelines that require longer, more developed paragraphs giving enough background information to put the material in context. People reading the material for the first time will be disadvantaged if they know little about the topic. Thus, in the following example, the paragraph explains the implications of the problems it describes:

> Although Jackson tested common aphid infections, he neglected to include variables that could account for infection patterns under field conditions. This is a significant oversight because it means that his work has been performed under atypical conditions (Josephson et al., 1982a; Lavone and Jamerson, 1984; Tillis, 1983). To test these conditions, conditions typical of the Orinoco basin during aphid mating season were created. Those conditions are detailed in the methodology section of this study.

Technical Paragraph Format

2.42 Plan a paragraph's visual organization based on

- Length,
- Indentation, and
- Forms other than standard prose.

Length of Paragraphs

2.43 Use existing paragraphing conventions or standards for scientific and technical reports, for journal articles, and for the scholarly press whenever possible.

2.44 Plan paragraph length to present conceptual units and to organize information for easy retrieval and specific audience needs.

2.45 Most paragraphing should contain only one major or one minor supporting idea. If either of these ideas has several subdivisions, treat them independently in their own paragraphs.

Paragraph Indentation

2.46 Use indents to produce fast searching and reading rates. Use extra leading (additional line spacing) plus an indent:

- Indent all paragraphs or, if set flush left, provide an extra line of leading above the paragraph in scientific or technical manuscripts.
- Set technical journals flush left with additional leading before paragraphs.
- Provide paragraph indents in commercial publications. Use a minimum indent of two picas or four characters in printed material. No additional leading above paragraphs is needed in commercial publications.
- Provide one extra line between paragraphs in word-processed or typewritten documents.

2.47 Consider how a text will be used when creating a paragraphing convention. If readers will use the text to find specific information—a search target—then indents and extra leading will help them locate major breaks.

2.48 Paragraph indents should be consistent throughout a text. For instance, if text is indented, then notes, warnings, cautions, and the like should also be indented.

Other Paragraph Forms

2.49 Technical and scientific writing often uses such techniques as lists, matrices, charts, tables, or figures as paragraphs. These techniques act like sentences or paragraphs and support the text. Their connection with the text must be obvious and complete. If it is not, include an explanatory paragraph.

Paragraph Numbering

2.50 Number sections and subsections. Double numeration, the most common method, uses the first number to represent a major section and the number after the first period to indicate a subsection.

2.51 Variations on numbering include a three-part method that calls out first-degree heads followed by second- and third-degree heads with no specific reference to the major section or chapter. For example, 13.3.11 refers to the thirteenth first-degree head, its third second-degree head, and that second-degree head's eleventh third-degree head.

Using Paragraph Numbering

2.52 If paragraph numbering will help the reader retrieve information quickly, use them as retrieval aids in all but marketing material and job aids. Like paragraphing, paragraph numbering signals a change in topic:

1. Audience and organization.
1.1 Audience—A systematic approach to finding out what readers need to do their jobs and what a user does with a manual.
1.2 Jargon—Legitimate uses.
1.3 Simple language and concrete terms.

Avoiding Paragraph Numbering

2.53 Numbering need not be used in all technical and scientific documents. When a paragraphing method and heading system provide good retrieval cues, avoid numbering texts

- Intended for casual or business users, who generally find paragraph numbering too forbidding, and
- In short documents—less than 20 pages—and in commercial and journal articles.

Internal Numbering

2.54 Military specifications (MIL-STD) require internal numbering. While this method permits many numbers (typically five or more), most documentation does not have that many divisions; only three levels (analogous to three degree heads) are recommended.

Figure and Table Numbering

2.55 If a document uses a numbering scheme, number figures and tables. However, be certain that the reader can discriminate these numbers from those used for major text divisions. For instance, if the numbering system for text divisions uses a decimal system (1.1, 1.2), use a dash system for figures and tables (Table 1–1, Figure 1–1).

2.56 A further modification to this numbering practice is to select one term—such as *exhibit*—for anything that is not text. This method reduces the numbering pattern to two: section numbers and non-text numbers. Thus, Section 1.1 would have Exhibit 1–1, 1–2; and Section 1–1 would have Exhibit 1.1, 1.2, and so on.

SENTENCES: STANDARD AND TRUNCATED FORMS

2.57 The sentence offers a complete thought through a group of related words containing a subject and a verb. Technical writing often uses abbreviated sentences to communicate more efficiently. Use these functional sentence fragments carefully and only after you have tried standard forms.

Types of Sentences

2.58 English has four sentence types: simple, compound, complex, and compound/complex. The number and type of clauses in a sentence determines its type:

- Simple—one independent clause,
- Compound—two or more independent clauses,
- Complex—one independent clause and one or more dependent clauses, and
- Compound/complex—two or more independent clauses and one or more dependent clauses.

Simple Sentences

2.59 The simple sentence contains only one independent clause and, if not overused, can convey complex information because readers find its uncomplicated structure easy to read. An example of the simplest form of the simple sentence is: Jane wrote.

Compound structures will not necessarily change the sentence type: Jane wrote and edited.

The sentence now has a compound verb. Even if compound subjects and direct objects are added, the sentence remains simple:

> Jane and Sue wrote and edited the report and summary.

Modifiers can also be added to the sentence:

> Jane and Sue *carefully* wrote and edited the report and summary *for their supervisor.*

The adverb *carefully* and the adverbial prepositional phrase *for their supervisor* have been added. The sentence is still a simple sentence containing one independent clause with a subject and finite verb. Even the addition of a verbal phrase does not change the simple sentence structure:

> *Struggling to meet their many deadlines,* Jane and Sue carefully wrote and edited the report and summary for their supervisor.

Phrases never alter a sentence's basic structure.

Compound Sentences

2.60 The compound sentence brings together two or more independent clauses of equal importance. These clauses can be joined in three ways:

- Coordinating conjunctions (and, but, for, nor, or, so, yet),
- Semicolon (;), and
- Colon (:).

(See also Chapter 3.)

2.61 The semicolon is often used with conjunctive adverbs, which join independent clauses but do not function as coordinating conjunctions:

> This library provides maximum performance; *however,* it can also be used with the APC Math Library.

However is a conjunctive adverb joining the two independent clauses. By using the coordinating conjunction *but* instead, the sentence would look like this

> This library provides maximum performance, *but* it can also be used with the math library.

The colon can also join two independent clauses on occasion:

> The master program runs simultaneously with your program: this feature guarantees high-speed results.

2.62 It could be argued that the compound sentence is not a useful sentence type in technical writing. In technical writing, with its information-packed sentences and complicated interrelationships, the simple or complex types are more useful:

> The parameters for the READ option are [P, S, T, V], and the parameters for the WRITE option are [C, H, R, U], and both options use at least one parameter.

The preceding compound sentence should be broken up into three separate simple sentences (though a more useful presentation might be a table):

> The parameters for the READ option are [P, S, T, V]. The parameters for the WRITE option are [C, H, R, U]. Both options require use of at least one parameter.

2.63 Another example of a misused compound sentence shows a need for subordination (or a complex sentence type):

> The LED lights, and the user then presses the ENTER key.

A revision using a dependent clause shows the relationship much better:

> When the LED lights, the user should press the ENTER key.

2.64 However, there are times when the compound sentence works very well:

> The computer can read all of the register, *but* bits 0 through 15 are read only for diagnostic purposes.

The coordinating conjunction *but* sets up an obvious contrast highlighting an exception to the general statement of the independent clause. Generally speaking, compound sentences using the coordinating conjunction *and* will be the least effective in many technical writing situations, but writers must analyze each situation as it arises.

Complex Sentences

2.65 The complex sentence has at least one dependent clause and only one independent clause that should contain the main thought or point of the sentence. One of the most common writing errors is putting the sentence's main idea in the dependent rather than the independent clause:

> When a transposition error occurs, two numbers are switched.

The sentence's intent is more apparent and actually altered when revised as follows

> A transposition error occurs when two numbers are switched.

The complex sentence allows the writer to place several ideas in close relationship to one another by subordinating them to the main idea in the independent clause.

Compound/Complex Sentences

2.66 The compound/complex sentence contains two or more independent clauses and at least one dependent clause. As the most complicated sentence type, it demands more attention from the reader. Overuse of this sentence type can weary readers of complicated technical information. However, the compound/complex sentence can efficiently bring several ideas into a close and elegant relationship:

> When the consumer purchases an item, the sales receipt is made, and it becomes the source document.

2.67 Unfortunately, compound/complex sentences in scientific and technical prose too often look like this

> When you are evaluating a compiler, you should consider how fast the compiled code runs, and you should consider whether it compiles to assembly language or whether it goes directly to microcode, and you should look at the complexity of statements and length of code.

2.68 Check for overuse of the compound/complex sentence. To solve this problem, break up the sentences into simple and complex types:

> When evaluating a compiler, you should look for several capabilities. *First,* consider how fast the compiled code runs. *Second,* see if the compiler compiles to assembly language or whether it goes directly to microcode. *Third,* look at the complexity of statements and length of code.

Adding the linking words—*first, second, third*—to the preceding text organizes the text to help readers absorb the structure as discrete bits of information.

Intention of the Sentence

2.69 English sentences can be classified by intention as belonging to one of the four categories:

- Declarative,
- Interrogative (questions),
- Imperative (commands), or
- Exclamatory.

Declarative Sentences

2.70 The declarative sentence is the most common sentence in scientific and technical writing. Declarative sentences convey information and state facts, which accords well with technical writing's detached, factual tone:

> The J-15 port is located on the A-11 Board.

While declarative sentences are used in most documents, the other kinds of sentences also have their uses.

Interrogative Sentences

2.71 Interrogative sentences are questions, and they can be useful in a number of scientific and technical documents. They often create a less formal, more personal tone suitable for instructions and tutorials. A question can focus the subject and works very well as a topic sentence:

> Q: How do I reboot the PC?
> A: Turn the PC off. Insert the operating system disk . . .

Imperative Sentences

2.72 Imperative sentences are those in which the subject seems to be missing but is, in fact, an implied *you*. Use the imperative in procedures, instructions, and tutorials:

> Mark the end of a block with ^C and press the ENTER key.

2.73 Be careful with the imperative, however; used inappropriately, it can make readers feel uncomfortable, as if they are being ordered to perform continually. In addition, imperatives may be culturally unacceptable for some international markets. (See also Chapter 8.)

Exclamatory Sentences

2.74 Since they express strong emotion, exclamatory sentences have little place in technical and scientific writing:

> John is a horrible person!

However, many companies have found that adding these kinds of sentences in tutorials for beginning users prevents users from becoming intimidated by the technology:

> Congratulations! You have successfully completed MOD2.

Loose or Periodic Sentence Order

2.75 Sentences containing two or more ideas can be organized with the main idea, which is usually found in the independent clause, at the beginning or end of the sentence. In a loose sentence, the main idea comes at the beginning of the sentence; in a periodic sentence, the main idea comes at the very end of the sentence. The loose sentence is the most common:

> We checked the code before we started the program because we knew there were at least two mistakes.

In this sentence, the main idea of checking the code comes first in the main clause. However, the periodic sentence can be effective for emphasis if it is not overused:

Because we knew there were at least two mistakes before we started the program, we checked the code.

This example emphasizes the conditions, leading to a more emphatic statement, with the main clause at the end of the sentence.

Clause and Phrase Structure

2.76 All standard English sentences contain at least one independent clause. A clause is a group of related words containing a subject and a verb; that is, a performer and an action.

2.77 There are two kinds of clauses in English: dependent and independent. Dependent clauses usually begin with a subordinating conjunction that makes the clause depend on some other element in the sentence. Independent clauses, on the other hand, can be punctuated as sentences: The circuit failed.

This sentence is correct and consists of a simple independent clause. However, adding a subordinating conjunction makes

Because the *circuit failed.*

Now the sentence is incorrect. The word *because* is a subordinating conjunction that puts the subject, *circuit,* and verb, *failed,* in an adverbial relationship with a missing independent clause. Add an independent clause to create

The machine malfunctioned *because* the circuit failed.

Together the two clauses clarify why the machine malfunctioned.

2.78 Sentences may have any number of phrases, but these same phrases, when used alone, cannot provide meaning in the same way they can as part of a sentence. A phrase is a group of related words without a subject or verb. The following are all examples of phrases. They are not usually punctuated as sentences:

. . . in the die casing . . .
. . . running the program . . .
. . . to splice the Z cable . . .

None of these phrases contains both a subject and a finite verb (a verb that the subject is performing or being).

Faulty Sentence Parallelism

2.79 A sentence with faulty parallelism has grammatical elements that lack common structure. This faulty structure interferes with the presentation of the information, as in the following example:

The engine is overheated, corroded, and it is damaged.

This sentence can be rewritten in parallel as

The engine is overheated, corroded, and damaged.

Use parallelism to create balance between a set of words, phrases, or clauses within a given sentence:

The engine is overheated, corroded, and damaged. (words)

To reopen the file or to retrieve it, press "D." (phrases)

Press RETURN to leave a blank line after a paragraph; press CNTRL to indent a paragraph five spaces. (clauses)

2.80 In a long sentence, repeat a key word that introduces parallel elements to help the reader understand the relationship between the elements:

The processor *cannot* work automatically and *cannot* achieve good performance without some programmer involvement.

Sentence Faults

2.81 Basic errors in sentence construction are called sentence faults. These consist of fragments (incomplete sentences) and run-on sentences (sentences in which clauses are improperly joined).

Dependent Clauses as Fragments

2.82 Dependent clauses cannot stand alone as complete sentences. Instead, dependent clauses show the close relationship of two or more events, the dependence of one event upon another. A dependent clause punctuated as a sentence breaks the logical flow of the writing, separating yet joining elements and giving the reader a mixed message:

The I/O bus is an assembly of 150 signal lines. *That* connect the I/O Device Adapters to the I/O port.

That, a subordinating conjunction, acts in double duty as subject of the dependent clause. The dependent clause is acting as an adjective describing *signal lines.* The faulty period and capitalization break the close relationship between the two units. In this case, the closeness of the relationship shows in the doubling of *that* as a pronoun subject standing for *lines* and as a subordinating conjunction. The sentence is correct as

The I/O bus is an assembly of 150 signal lines that connect the I/O Device Adapters to the I/O port.

Phrases as Fragments

2.83 Unless consciously using the fragment in a list or other appropriate situation, do not punctuate a phrase as if it were a sentence. Phrase fragments should be used only with good reason.

Run-on Sentences

2.84 A run-on (or fused) sentence consists of two complete sentences run together without proper connectives. People often misunderstand the term *run-on sentence* and think it refers simply to any long, rambling sentence, but run-ons are grammatically incorrect and unacceptable. An example of a run-on sentence would be

> Five slides were prepared with solution *REB six* slides were prepared with solution TRP.

Use a connective between the words *REB* and *six*. These connectives include a

- Subordinating conjunction:

    > Five slides were prepared with solution REB, *while* six slides were prepared with solution TRP.

- Coordinating conjunction:

    > Five slides were prepared with solution REB, *and* six slides were prepared with solution TRP.

- Conjunctive adverb:

    > Five slides were prepared with solution REB; *however,* six slides were prepared with solution TRP.

The sentence could also be broken into two separate sentences or joined with a semicolon:

> Five slides were prepared with solution REB. Six slides were prepared with solution TRP.
>
> Five slides were prepared with solution REB; six slides were prepared with solution TRP.

Choices would depend on exactly what relationship the writer wishes to express between the thoughts in the two clauses.

Fragments as Used in Lists

2.85 A list is an indented series of words or phrases that present a group of similar items. The list can make complex material easier to read by breaking up information so that the reader can select important points at a glance. The sentence fragment leading into a list often can act as a subject with the list providing a series of predicates:

This test

1. Sets the counter,
2. Increments the pointer to 1, and
3. Checks the register contents.

A list can also be a series of noun phrases preceded by a complete sentence:

Shared processor systems have three features:

- A number of keying systems,
- A controlling processor, and
- Disk-stored data.

Sometimes a list includes a mixture of fragments and complete sentences. In the following two-column list, the fragment on the right describes the text on the left. Since the words "is the" are implied by the space between the columns, the fragment is punctuated as a complete sentence. For a full discussion of list punctuation, see Chapter 3.

LSTMSG	Address of the last message received. The supervisor uses this address when answering a message from an unspecified address.
RCLOCK	Right pointer for this task in the clock queue. If this task is waiting at the clock queue, RCLOCK points to the next task in the queue. If this task is not waiting at the clock queue, RCLOCK points to itself.

Fragments as Used in Data Displays and Illustrations

2.86 Sentence fragments are often necessary in nontextual elements to make efficient use of space. Use key words that appear in the text and highlight them in the data display or illustration. These key words help the reader make connections that may be grammatically missing. Detailed explanation can be found in the accompanying text.

Fragments in Procedures and Processes

2.87 Use procedural fragments when the subject remains the same, while the actions change. In the following example, *test* is the subject of four listed verbs:

This test

1. Resets the BP,
2. Clears the AP pointer and the MDP pointer, and
3. Increments the AP pointer.

Other Uses for Fragments

2.88 Use sentence fragments in headings, in headers and footers, and in any situation in which a few key words can summarize a body of text.

Phrases

2.89 Phrases are groups of related words without a subject or finite verb or both. Phrases cannot make a complete statement as a clause can, but instead act as nouns, verbs, adjectives, or adverbs in a sentence. There are six kinds of phrases

- Verbal,
- Noun,
- Appositive,
- Verb,
- Prepositional, and
- Absolute.

Verbal Phrases

2.90 Verbal phrases include participial, gerund, and infinitive phrases. Verbals are created from nonfinite verb forms that look like verbs but do not act like them.

2.91 A participial phrase is a verb formed from the verb stem and given an *ing* for the present form; the past form for regular verbs ends in *ed* or *d*. The participle can function as part of a verb phrase: The machine is running.

2.92 If a participle functions as a finite verb, it will always have a helping verb that is a form of the verb *to be*. Without the helping verb, the participle always acts as an adjective:

The *corroded* wire must be removed.

The past participle *corroded* modifies the noun *wire* as an adjective.

2.93 Participial phrases can function as adjectives. The participle is the headword (first word) in the phrase, followed by its object and modifiers:

Providing directional indicators for data exchange, the signal DMAACX can also request data.

In this example, *providing* is the participle, *indicators* is the participle's object, and the other modifiers include the adjective *directional* and the prepositional phrase. The entire phrase modifies the noun *signal* (and its appositive, DMAACX).

2.94 A dangling participial phrase has no noun or pronoun to modify. A common problem in technical writing is an introductory participial phrase followed by a clause in passive voice, often one in which the subject has been dropped:

Being high true, the bus is granted access by the modem signal.

This sentence says that the bus is high true when, in fact, it is the signal that is high and true. After rewriting the passive voice, the participle modifies the correct noun:

Being high true, the modem signal grants the bus access.

2.95 A gerund phrase, a verb form (usually ending in *ing*), acts as a noun. Note that this form is identical to the present participle, but the gerund phrase acts differently in the sentence. A gerund phrase contains a gerund and its objects or modifiers or both. Gerund phrases are often used incorrectly when there are many modifiers:

The interrupt enable mask bit reading proceeds as follows.

Like other stacked nouns, these modifiers must be unstacked to make the gerund phrase obvious:

The reading of the interrupt enable mask bit proceeds as follows.

2.96 An infinitive phrase is a nonfinite verb form without person and number and usually preceded by the word *to*. An infinitive phrase contains the infinitive, its object if it has one, and any modification:

To service the machine *accurately*, use this code.

The infinitive is *to service*, the object of the infinitive is *machine*, and the modifier is the adverb *accurately*.

2.97 The infinitive or the infinitive phrase can act as a noun, adjective, or adverb. Usually, infinitive phrases present problems when they are used as adjectives. As with participial phrases, they can dangle or modify the incorrect term(s).

2.98 Avoid using the split infinitive (inserting an adverb in the middle of an infinitive):

> To *correctly* build the solar home, you need this manual.

The sentence is clearer as

> To build the solar home *correctly*, you need this manual.

In recent years rules have relaxed on this issue.

Noun Phrases

2.99 Noun phrases include nouns and their adjectival modifiers:

> The *system program manual* contains *program code*.

This sentence has two noun phrases: the *system program manual* and *program code*.

Appositive Phrases

2.100 An appositive or appositive phrase, another noun or pronoun that renames or defines its referent, directly follows the noun or pronoun to which it refers:

> The SAVE command writes files in ASCII format to a disk file, *DSK in Version 3.1*, in the system directory.

DSK in Version 3.1, an appositive phrase, renames or elaborates on *disk file*.

Verb Phrases

2.101 Verb phrases consist of any finite verb that is not in a simple tense (in which the verb is only one word).

> The corporation *may have had* trouble at that point.
>
> The line *is housed* by the adapter.
>
> Laboratory personnel *will monitor* the test.
>
> This office *is consuming* too much coffee.

In the first sentence, *may have had* is the verb phrase. The second sentence verb phrase, *is housed*, is a passive form. The verb phrase, *will monitor*, in the third sentence is simply in future tense.

Finally, the fourth sentence uses a progressive tense verb phrase, *is consuming.*

Prepositional Phrases

2.102 Prepositional phrases act as adjectives and adverbs. They consist of a preposition as a headword (first word of the phrase) and a noun or pronoun acting as the object of the preposition:

> The preliminary report should go *to the Marketing Department.*

The prepositional phrase *to the Marketing Department* acts as an adverb modifying the verb phrase *should go.*

Absolute Phrases

2.103 Absolute phrases modify the whole sentence and therefore act as adverbs. They can be a confusing structure in the midst of complex information because they come close to being elliptical constructions:

> *The error messages being correct,* I can finish executing.

Turn the absolute phrase *the error messages being correct* into a dependent clause to create a closer connection:

> Because the error messages are correct, I can finish executing.

Subject and Verb Agreement

2.104 Within the sentence, the subject and its verb must agree in number. Several situations can cause problems with subject and verb agreement.

Collective Nouns

2.105 Collective nouns are plural in form but take singular verbs when thought of as one unit or entity:

> The *number* of calculations is extraordinary.

However, if the noun is thought of as a plural entity, the verb is plural:

> A *number* of calculations were made.

Sometimes writers get confused when a collective noun follows a singular subject, but the verb is singular because the subject is:

All of the testing equipment was broken.

The testing equipment is broken.

Subject Complements

2.106 In sentences with a subject complement, the verb agrees with the subject, not the subject complement:

The *object* of his studies was molecules.

Compound Subjects

2.107 Compound subjects (nouns or pronouns joined by coordinating or correlative conjunctions) can be problematic. If *and* is the coordinating conjunction, then usually the verb is plural:

Ray and Charles are presenting a paper.

However, sometimes the subject is thought of as a single entity:

This ship and product release date is ridiculous.

2.108 If the words *each* or *every* modify the compound subject, the verb is singular:

Every circuit and board *has* been tested.

Each circuit and board *has* been tested.

2.109 A compound subject joined by *or* or *nor* may contain a singular and plural noun or pronoun. In that case the verb agrees with the noun or pronoun closest to it:

Jerry or the other managers *are* able to sign for it.

Nouns

2.110 Nouns name persons, places, things, ideas, qualities, or even actions. The following sections describe certain problems with nouns that occur in technical or scientific writing.

Abstract Versus Concrete Nouns

2.111 When possible, limit the use of abstract nouns; when this is impossible, further define the abstract noun with concrete nouns. Abstract nouns refer to things that cannot be discerned by the senses, while concrete nouns refer to things that can be discerned by the five senses.

Plural Versus Possessive Nouns
(See also Chapter 4.)

2.112 Add *s* (*es*) to nouns to create the plural form. Add *s* with an apostrophe (*'s*) to create the possessive form. Sometimes the writer does not make a distinction between the plural and the possessive. This occurs frequently in instances in which the apostrophe is used with acronyms in plural forms (DMA's).

2.113 Use an apostrophe after plural nouns ending in *s* (girls') to show possession. For compound nouns, use the apostrophe only with the final noun:

John, Terry, and Joe's board . . .

Possessive nouns always function as adjectives within the sentence.

Appositives and Parenthetical Definitions

2.114 An appositive is a noun, pronoun, or noun phrase that renames a noun, pronoun, or noun phrase it directly follows. Technical writing uses appositives to define parenthetically an acronym, mnemonic, abbreviation, or new term:

The host control status register (HCSR) . . .

Gerunds and Infinitives

2.115 Gerunds are nouns with the form of a verb plus *ing* and function as nouns even though they derive from verb forms:

Loading a program is tiresome.

Loading a program is a gerund phrase acting as the noun subject of the sentence.

2.116 Infinitive constructions can function as nouns, adjectives, or adverbs. Usage in the sentence determines the function of any part of speech, particularly infinitives:

To write a program in OCCAM can be difficult.

The infinitive phrase *to write a program in OCCAM* is the noun subject of the sentence.

Stacked Nouns

2.117 In technical writing many nouns can be stacked next to each other to act as adjectives for one noun at the end. If there are too many nouns, it is difficult to know which noun is modifying what word:

> request acknowledgment exchange data bit

Unstack these nouns and use some of them in prepositional phrases:

> data bit for request acknowledgment exchanges

Pronouns

2.118 A pronoun stands for a noun. There are many kinds of pronouns: personal, demonstrative, relative, reflexive, indefinite, interrogative, intensive, and reciprocal. Possessive pronouns, like possessive nouns, always function as adjectives in a sentence. Pronouns present three major problems: reference, agreement, and sexist language.

Accurate Reference for Pronouns

2.119 Use accurate referents for pronouns. Technical writing often discusses many topics in a single text. A paragraph discussing hardware, for instance, might mention registers, buses, LEDs, and clocks. Since constantly repeating an item's name might irritate or bore the reader, the pronoun is often substituted for a noun to prevent repetition.

2.120 The pronoun's antecedent is the noun it stands for. Every pronoun should have an antecedent, and there must be no doubt which antecedent goes with which pronoun. Two troublesome cases are *this* and *it*. Avoid using the word *this* by itself as a pronoun. *This* nearly always stands for a situation or event and never does so obviously enough. Instead, follow the word *this* with a clarifying noun (turning *this* into an adjective):

> This sequence is repeated until all errors are detected and a message is sent announcing that all subroutines have been completed. *This* allows the user to run the computer most efficiently.

In the second sentence of the example, *this* can refer to the sequencing, to the message, or to both. Adding a noun to refer to the original subject (with an additional adjective for extra clarity) solves the problem:

> *This automatic sequencing* allows the user to run the computer most efficiently.

In like manner, keep the pronoun *it* specific, unambiguous, and close to its antecedent.

Agreement of Pronouns

2.121 Pronouns must agree with their antecedents in number and gender. Confusion can result when number is not clear:

> A *writer* is always sure *they* are right.

Compound Antecedent Pronouns

2.122 When the pronoun has a compound antecedent in which the nouns are joined with *and,* use a plural pronoun:

> Ms. Robbins and her secretary are at a meeting *they* had to attend.

2.123 If a compound antecedent is joined by *or* or *nor,* use a singular pronoun if both nouns are singular and a plural pronoun if both nouns are plural. When one noun is singular and one plural, the pronoun agrees with the noun closest to it:

> A slide or a printout will be available; it should be sufficient.
>
> The slide or the printouts will be available; they should be sufficient.
>
> Slides or printouts will be available; they should be sufficient.

Collective Nouns as Pronoun Antecedents

2.124 Collective nouns can be either singular or plural, depending on the sense of the noun:

> The project group decided to delay *its* plans for ship date.
>
> The project group met and returned to *their* cubicles.

Sexist Language and Pronouns

2.125 Avoid sexist language through the judicious use of pronouns. Usually antecedents such as *each, either, neither, one, anyone, anybody, everyone, everybody, a person* take singular pronouns, but be careful of sexist language. Traditionally, the singular male pronoun has been used:

> Anyone should be able to find *his* way out of the building.

Some writers have tried to remedy this in several ways:

> Anyone should be able to find *his or her* way out of the building.
>
> Anyone should be able to find *their* way out of the building.
>
> Anyone should be able to find *her* way out of the building.

The first solution is acceptable if the pronouns are not repeated. The second solution, although common, is unacceptable because the pronoun does not agree in number with the antecedent. The third solution is acceptable. However, some writers have taken to varying the pronoun, using *his* in one sentence, *her* in the next, *his* in the next, and so on. An easier solution to the problem is to make the antecedent plural whenever possible. Thus, in the preceding example, the solution could be

> People should be able to find *their* way out of the building.

Verbs

2.126 Verbs have five aspects: tense, voice, mood, number, and person. Problems with number involve subject and verb agreement, and problems with person are not common. However, tense, voice, and mood are often involved in awkward writing situations. Avoid basic shifts in verb tense, voice, mood, number and person, and tone and point of view. (See also Sections 2.30–2.35.)

Modifiers

2.127 Modifiers are adjectives, which modify nouns and pronouns only; and adverbs, which modify verbs, adjectives, and other adverbs. Modifiers can be individual words, phrases, or clauses:

> The *complex* plan was *carefully* developed.
>
> The switch *under the panel* should not be touched *without permission.*
>
> He requested the report *that described the project's status because he needed to study the facts.*

In each of these sentences, the first example is an adjective and the second an adverb. In the third sentence, the adverb clause begins with the word *because.*

2.128 Occasionally, writers confuse adjective and adverb forms. In the case of complements, this can be particularly confusing:

> She smells good.

This sentence says that her smell is good. But the following sentence, which uses the adverb form, produces

She smells well.

The sentence now says that her smelling ability is highly developed. *Well* is an adverb modifying the verb *smell.*

2.129 Again, in an object complement position, the same problem can occur:

We considered the project hurried.

The sentence says that the project was done in haste. But the adverb form produces a different meaning:

We considered the project hurriedly.

Now the sentence says that we were in a hurry when we considered the project. *Hurriedly* is an adverb modifying the verb *consider.*

Misplaced Modifiers

2.130 Avoid misplaced modifiers. Modifiers must be placed in relation to what they modify. In technical or scientific writing of any complexity, unclear modifiers can seriously undermine the clarity of the writing:

The technician reported that the sample was ready with alacrity.

Place the modifier as close as possible to the item it modifies. The preceding sentence should read

The technician reported with alacrity that the sample was ready.

Dangling Modifiers

2.131 A common problem is a dangling modifier that has nothing to modify. This problem often occurs when the writer uses passive voice and drops the subject:

Executing a program, code must first be checked.

The participial phrase *executing a program* modifies the noun *code,* but actually a missing subject is executing the program. The sentence can be rewritten as

Executing a program, the user must first check code.

Squinting Modifiers

2.132 Modifiers "squint" when they can modify either the word preceding or following them:

> Performing *often* tires him.

Often can function as an adverb modifying *tires* or as an adjective modifying *performing* (a gerund). The solution is to move the modifier:

> Often, performing tires him.

Or:

> Performing tires him often.

Telegraph Style

2.133 In telegraph style writers leave out function words—conjunctions and prepositions—and articles in favor of brevity. In a figure or table, this practice can be acceptable if the meaning is obvious, but such a style should be avoided in most writing:

> Move dial and enter switch to 1 position.

It is unclear whether

- There is a dial-and-enter switch,
- Both the dial and the enter switch need to be moved to the 1 position, or
- The dial should be moved and the enter switch turned to the 1 position.

A rewrite might be

> Move the dial and turn the enter switch to the 1 position.

Conjunctions

2.134 Conjunctions connect words, phrases, clauses, or sentences. There are three kinds of conjunctions: coordinating, correlative, and subordinating.

2.135 Use coordinating conjunctions to connect words, phrases, and clauses with the same grammatical form: nouns with nouns, prepositional phrases with prepositional phrases, or any other structures. There are only seven coordinating conjunctions: and, but, for, so, yet, or, nor.

2.136 Use correlative conjunctions to connect parallel structures within the sentence. They consist of the following pairs of words used within a sentence: either . . . or, not only . . . but also, neither . . . nor, both . . . and, whether . . . or.

2.137 Use subordinating conjunctions to signal a dependent clause and connect it to an independent clause within a sentence. Some subordinating conjunctions are: because, if, although, when, as, while.

Prepositions

2.138 Use prepositions to link a noun or pronoun to some other word in a sentence. All prepositions must have a noun or pronoun, which functions as the object of the preposition within a prepositional phrase. Prepositional phrases almost always act as modifiers.

2.139 The old myth about a preposition never ending a sentence is untrue. Occasionally, prepositions follow instead of precede their objects, and they can be placed at the end of a sentence:

> What is this program for?

THREE

Punctuating Scientific and Technical Prose

Colon
  Elements in a Series and Colon
  Amplification and Colon
  Long Quotation and Colon
  Ratios and Colon
  Time of Day and Colon
Dash
  Amplification or Illustration and Dash
  Em Dash and Other Punctuation
  Range of Numbers or Time and En Dash
  Two-Noun Combinations and En Dash
Period
  Vertical Lists and the Period
  Abbreviations and Periods
  Raised Periods
  Ellipsis Points
Question Mark
Exclamation Point
Hyphen
  Unit Modifier Containing Adjectives and the Hyphen
  Unit Modifier Containing Number and Unit of Measure
  Unit Modifier Containing Adverb and Hyphen
  Unit Modifier Containing Comparative or Superlative
  Unit Modifier Containing a Proper Name
  Unit Modifier Longer Than Two Words
  Foreign Phrase Used as Adjective
  Predicate Adjective
  Self- *Compound and Hyphen*
  Suspended Hyphen
  Numbers and Hyphens
  Technical Terms and Hyphens
  Prefixes and Suffixes and Hyphens
Slash
Apostrophe
  Possessives and Apostrophes
  Contractions and Apostrophes
  Plurals and Apostrophes
Quotation Marks
  Single Versus Double Quotes
  Placement of Quotes With Other Punctuation
  Specialized Usage for Quotation Marks
  So-Called *Expressions*
Parentheses
  Abbreviations and Parentheses

3.1 Writers have considerable flexibility in choosing the appropriate punctuation mark (or lack thereof). Most of the punctuation rules have one or two qualifications, and few can be applied inflexibly in all situations. Several might seem arbitrary and even illogical. However, most are sensible, and all of them foster ease in reading and help clarify the text.

COMMA

3.2 The comma, semicolon, colon, dash, and parenthesis prevent parts of a sentence from running together in such a way as to obscure the author's meaning. The comma—the most frequently used of these marks—provides the smallest interruption in sentence structure. It also has, by far, the most rules.

Compound Sentence and Comma

3.3 Insert a comma before the coordinate conjunction (*and, but, nor, whereas,* and similar words) in a compound sentence containing two independent clauses:

> The rules control the use of nonflammable liquids, but provisions exist for handling flammable liquids as well.

This comma is optional for a sentence with short clauses:

> The samples have arrived [no comma needed here] and testing will begin shortly.

Compound Predicate and Comma

3.4 Do not separate compound predicates (two verbs for a single subject) by punctuation:

> The samples *were suspended* by a wire [no comma here] and *held* in the center of the study chamber.

3.5 Compound predicates also appear within dependent clauses. These do not require a comma:

> We began our work with methane, which *is also known* as marsh gas [no comma here] and *is* a primary part of the natural gases.

3.6 Occasionally, inserting a comma before the conjunction separating a compound predicate is needed for clarity:

> This version of the programming language provides extensive tracing and debugging capabilities, [comma helpful here for clarity] and has windows that can be used simultaneously for editing code.

Introductory Phrase or Clause Punctuated With a Comma

3.7 Use a comma after a phrase or clause that begins a sentence:

> For sorting more than one database, use the sort program in the batch mode.

3.8 Omit the comma after very short phrases if no ambiguity results:

> In 1992 [no comma needed here] we will investigate the effect of water flow on bridge pilings.

Introductory Expression After Conjunction and Comma

3.9 Separate by commas an introductory word, phrase, or clause that immediately follows the coordinate conjunction of a compound predicate:

> In the late 1960s, the National Radio Astronomy Observatory implemented FORTH and, *even today,* uses this popular programming language.

3.10 An introductory word, phrase, or clause following a coordinate conjunction in a compound sentence leaves the writer with several options:

> The curves in Figure 2 are less familiar geometrical objects than those in Figure 1, but, *other than the top curve,* they have a reasonable shape.

> . . . in Figure 1, but *other than the top curve,* they have . . .

> . . . in Figure 1, but *other than the top curve* they have . . .

The first option includes all the punctuation that the sentence structure suggests. However, at least one comma could be eliminated (second option) without sacrificing clarity. Use the third option when the introductory expression is short and its function in the sentence is apparent without the final comma.

Restrictive Phrase or Clause and Comma

3.11 Set off a nonrestrictive phrase or clause (one that is not essential to the sentence's meaning) with a comma:

> Sodium sulfate changes to sodium phosphate by addition of potassium oxide, *a typical reagent chemical.*

3.12 Do not separate a restrictive phrase or clause (that is, one that is essential to the meaning of the sentence) by commas unless it precedes the main clause:

> You will be more productive *if you invest time in learning how to use these tools.*

But:

> *If you invest time in learning how to use these tools,* you will be more productive.

Noun Clause as Subject and Comma

3.13 A noun clause does the work normally reserved for a noun alone. Do not set off a noun clause used as the subject when it immediately precedes the verb:

> *Storing batch jobs on card decks* [no comma here] has gone the way of the dinosaur.

Items in Series and Comma

3.14 In a series of three or more items with a single conjunction, use a comma after each item except the last one:

> The Computer Services Group provides assistance in graphics applications, equipment procurement, data analysis, and personal computer networking.

3.15 Some authorities recommend omitting the comma before the conjunction except when misreading might result. They argue that the conjunction signals the final break in the series and, therefore, the final comma is unnecessary. However, including the final comma leaves no doubt about the author's meaning.

3.16 Do not use a comma if conjunctions (". . . one and two and three") join all the items in a series.

Phrases and Clauses in Series and Comma

3.17 With three or more parallel phrases or dependent clauses in a series, insert a comma after each item in the series except the last:

> This program will automatically read the file, apply the user's instructions, and create an updated file.

Transitional Word or Phrase and Comma

3.18 Use a comma after

- Transitional adverbs (*accordingly, but, consequently, furthermore, hence, however, indeed, instead, moreover, nevertheless, so, therefore, then, thus,* and similar transitional terms):

  > Thus, the machine did not operate.

- Transitional phrases (*for example, in other words, that is*):

  > In other words, we could not complete the project.

- Organizational transitions (*namely, first, secondly, next, lastly, finally*):

  > Finally, the repairs were completed.

Place commas around transitional adverbs and phrases or organizational transitions when they are embedded in the sentence and record a distinct break in thought:

> This reaction, *in turn,* leads to the development of other opinions.
>
> This reaction, *as well as* other social pressures, leads to the development of other opinions.
>
> This reaction, *too,* leads to the development of other opinions.

Transposed Terms and Comma

3.19 Set off any words, phrases, or clauses placed out of their normal position for emphasis or clarity:

> If, *after a period of time,* metal crystallization appears, then metal failure will occur.

Apposition and Comma

3.20 An apposition is a grammatical construction in which two adjacent nouns (or noun equivalents) refer to the same person, thing, or concept. Separate the second part of the apposition by commas (or parentheses) if it is not restrictive:

> A peritectic reaction, *the reverse of a eutectic reaction,* occurs in many metals.

3.21 Follow this same rule with mathematical symbols and expressions:

> Substitute the initial temperature, T_i, by . . .

3.22 Do not separate the second part of an appositive if it is restrictive:

> Darwin's treatise *On the Origin of Species* was debated at an 1860 meeting of the British Association for the Advancement of Science.

Contrasting Phrase and Comma

3.23 Separate by commas any phrase inserted to contrast with a preceding word or phrase:

> This unlikely, *though at the same time logical,* conclusion started the debate.

Deliberate Omissions and Comma

3.24 A comma can indicate the deliberate omission of a word or words for the sake of brevity:

> The first two digits indicate flying time in hours; the next four, the fuel usage in pounds.

3.25 A comma indicating omission is unnecessary when the sentence reads smoothly without it:

> The first sample came from France, the second from Japan, and the third from Canada.

Similar or Identical Words and Comma

3.26 To avoid confusion, separate two similar or identical words with a comma, even when the grammatical construction does not dictate such a break:

> In 1988, 377 coal samples were analyzed for sulfur content.

Two or More Adjectives in Series and Comma

3.27 Separate two or more adjectives modifying the same noun by a comma or commas:

> The material consisted of dense, coarse particles.

This comma can be deleted when the expression can be understood readily without it:

> For these tests, we designed and built a high-temperature [no comma needed here] high-pressure apparatus.

3.28 Do not use a comma if the first adjective modifies the idea expressed by the second adjective and noun combined:

> The material consisted of coarse dust particles.

In this example, "coarse" refers to "dust particles," not just "particles."

Transitional Terms (*That is* and *Namely*) and Comma

3.29 Use a comma after transitional terms inserted to introduce material that provides more information about a point already made, unless it is expressed in an independent clause (see section 3.46). Such terms include *that is* (*i.e.*), *namely* (*viz.*), *for example* (*e.g.*), and *in other words*:

> Sulfides can be found in many valuable metals, *for example,* chrome, magnesium, and nickel.

Direct Quotations and Comma

3.30 Use a comma to introduce direct quotations that are a complete sentence or two in length:

> In his opening address, the physicist Henri Poincaré asked, "Does the ether really exist?"

However, in several similar situations, do not use a comma:

1. If the quotation is the subject of the sentence or a predicate nominative (that is, it completes the meaning of a linking verb such as *is* or *was*):

   > Marshall McLuhan's message was "The medium is the message."

2. If the quotation is a restrictive appositive (that is, it follows and further describes a noun or noun substitute and is essential to the sentence's meaning):

Louis Sullivan followed the dictum "Form follows function."

3. Or if the conjunction *that* introduces the quotation:

Although I am not sure what it means, the rule states *that* "Nonstandard replacements violate company standards."

Direct Questions and Comma

3.31 Direct questions are usually set off by a comma (see Section 3.79):

The question is, how do we get there from here?

Dates and Comma

3.32 Punctuation for dates depends on the date style. If the date is given as month, day, year, then place commas after the day and year:

A paper printed in *Nature* on April 22, 1990, asserts that sea turtles produce fewer than two hundred eggs per gestation.

3.33 Do not use commas when the date style is given as day, month, and year ("... on 22 April 1990 asserts ...") or when the day does not appear in the date:

Plutonium was discovered in March 1940 at the Lawrence Radiation Laboratory.

3.34 Except in tables of data, the preferred style is to spell out the month rather than reporting it as a number:

Basic Forms	Acceptable	Preferred
MM/DD/YY	09/25/90	September 25, 1990
DD/MM/YY	25/09/90	25 September 1990
YY-MM-DD	90-09-25	1990 September 25

If a document will be distributed in a country other than the U.S., follow the appropriate date conventions for that country.

Specifying Names with Commas

3.35 Set off a term that makes the preceding reference more specific:

The thermal data are shown in Table 4, column 3, and in Figure 2, top curve.

3.36 Use a comma to separate the elements in a geographical location:

The meeting in Chicago, Illinois, was postponed because of a blizzard.

Numbers and Commas

3.37 Counting from right to left, insert commas between groups of three digits in numbers greater than or equal to one thousand:

1,731 345,657 3,972,999,029,388

However, be sure to follow the appropriate decimal conventions if the document will be distributed in other countries:

4,586,782.51 (U.S.) 4.586.782,51 4 586 782,51

3.38 Note that the U.S. convention applies only to digits to the left of a decimal point. Digits to the right of a decimal should not be separated by commas:

11,496.01 0.0040

3.39 The style for some publishers of technical documents is not to insert commas for numbers with four digits or less:

1000 9999

Reference Numbers and Commas

3.40 Set off reference numbers by commas unless they signify a range of numbers, which requires an en dash (see Section 3.64; see also Chapter 10):

Recent research[6, 9, 11–15] has revealed that . . .

SEMICOLON

3.41 The semicolon divides parallel elements in a sentence when a separation stronger than that indicated by a comma is needed.

Clauses Joined Without Conjunction and Semicolon

3.42 Insert a semicolon between the clauses of a compound sentence when they are not connected by a conjunction:

With this software, you need not write a program to create a graph; after a little training, nearly anyone can produce graphs.

Clauses Joined by Conjunction and Semicolon

3.43 Use a semicolon to separate the clauses of a compound sentence joined by a coordinate conjunction when they are long or are themselves internally punctuated:

A vapor analyzer measured atmospheric moisture, humidity, and dewpoint; and a sniffer instrument sampled heavy metals, pollen, and gases.

3.44 Use a comma in place of the semicolon if the sentence will still be easily understandable:

These hardness tests are accurate and reproducible, and if good judgment is used, they give a good estimate of the strength for most metals.

Clauses Joined by Transitional Adverb and Semicolon

3.45 Use a semicolon to precede transitional adverbs *(accordingly, consequently, furthermore, hence, however, indeed, instead, moreover, nevertheless, so, then, therefore, thus,* and similar words) and follow these words with a comma when they link clauses of a compound sentence:

The values from the two studies disagree; *therefore,* definitive results cannot be given at this time.

Before *so, then,* and *yet,* a comma can be used instead of a semicolon. No comma is required after these terms:

The experiment confirmed that the particles were electrons, *yet* too much was at stake to publish immediately.

Transitional Terms (*That is* and *Namely*) and Semicolon

3.46 Use a semicolon before expressions such as *for example, in other words, namely,* or *that is* and follow these words with a comma when they introduce an independent clause:

No refinement is necessary; *that is,* one merely has to derive a single set of rules to describe the wing's final shape.

Elements in a Series and Semicolon

3.47 Use a semicolon to separate parallel elements in a series if they are long or internally punctuated:

The zinc ores include sphalerite, ZnS; smithsonite, $ZnCO_3$; willemite, Zn_2SiO_4; calamite, $Zn_2(OH)SiO_3$; and zincite, ZnO.

COLON

3.48 The colon usually introduces an illustration or elaboration of something already stated.

Elements in a Series and Colon

3.49 Use a colon to introduce words, phrases, or clauses in a series if the clause preceding the series is grammatically complete:

The metals tested were treated with three chemicals: Ac, Ar, and Cl.

3.50 Do not use a colon when the elements in series are needed to complete the clause introducing them:

The metals tested were [no colon here] Mo, Hg, and Mn.

3.51 Use a colon to introduce words, phrases, or clauses presented as a vertical listing based on these guidelines (see also Section 2.85):

- If the introductory clause preceding the list is grammatically complete, use a colon and punctuate the list as running text.

  These lectures on technical communication offer instruction in three areas:

  1. Constructing readable illustrations and data displays,
  2. Reducing technical information to its simplest terms without compromising the author's meaning, and
  3. Eliminating technical jargon and ambiguous expressions that may confound translation.

- If the introductory clause preceding the list is grammatically complete, use a colon but eliminate all other punctuation at the end of each item except the final period.

  These lectures on technical communication offer instruction in three areas:

1. Constructing readable illustrations and data displays
2. Reducing technical information to its simplest terms without compromising the author's meaning
3. Eliminating technical jargon and ambiguous expressions that may confound translation.

- If the introductory clause preceding the list is not grammatically complete, do not use a colon. Terminal punctuation practice can vary, as in the preceding examples.

This text provides rules on

1. Constructing readable illustrations and data displays,
2. Reducing technical information to its simplest terms without compromising the author's meaning, and
3. Eliminating technical jargon and ambiguous expressions that may confound translation.

Amplification and Colon

3.52 Use a colon to separate two statements, the second of which is introduced by the first and amplifies or illustrates it:

> Two simple observations underlie relativity theory: a body's mass depends on its total energy, and a body's energy is proportional to its mass.

3.53 An introductory phrase such as *the following* or *as follows* signals that an amplifying or illustrating statement is about to be made and a colon is thus needed:

> One of the simple modifications to the ideal gas law is *the following* equation: $pV = RT + \alpha P$.

If *the following equation* is omitted from the example, then the colon is also dropped:

> . . . modifications to the ideal gas law is $pV = RT + \alpha P$.

3.54 If the material introduced by a colon contains more than one sentence, capitalize the first letter in the first complete sentence after the colon:

> An outline of our experimental procedure follows: First, the samples were heated at high temperature for one hour. Second, a new reactant was used. Third . . .

3.55 Use the colon to introduce a vertical list or example when a short introductory word or phrase offers a contrasting or emphatic alternative. Such terms could include *but, not, use, for example, on the other hand, note,* and the like:

But this test:

1. Sets the counter,
2. Increments pointers, and
3. Checks register content.

Long Quotation and Colon

3.56 Use a colon to introduce long quotations:

In the spring of 1915 Einstein wrote to a friend: "I begin to feel comfortable amid the present insane tumult, in conscious detachment from all things which preoccupy this crazy community." [trans. by Abraham Pais.]

Ratios and Colon

3.57 Use a colon between the parts of a ratio:

The procedure calls for an atmosphere with an H_2O:H_2 ratio of 1:115.

Time of Day and Colon

3.58 Use a colon between the hours and minutes in expressing time on a 12-hour scale:

Our experiment ran from 7:05 A.M. to 9:43 P.M.

3.59 By convention, no colon separates the hours and minutes on a 24-hour scale:

Our experiment ran from 0705 hours to 2143 hours.

DASH

3.60 The two kinds of dashes are the em dash and en dash. The characters "—" and "–" signify the em and en dash, respectively. If these characters are not available on your typewriter or word processing software, then use a double hyphen, - -, for the em dash and a single hyphen, -, for the en dash.

Amplification or Illustration and Dash

3.61 Em dashes can introduce and conclude a short enumeration or amplification. They can also enclose brief interruptions in thought and asides. Unlike commas (or parentheses), em dashes direct the reader's attention to what has been set off:

> Users have available to them several interactive systems—batch, Wylbur, and TSO—24 hours a day.

Em Dash and Other Punctuation

3.62 Normally an em dash both precedes and follows an enumeration or amplification. However, the second dash should not be followed immediately by any other punctuation mark. If the grammatical structure indicates a comma, use the dash alone:

> Because this alloy bears a misleading name—beryllium copper—[no comma here] the uninformed reader might assume that it is a beryllium-rich alloy.

3.63 If the grammatical structure calls for a semicolon, colon, or period where the second dash belongs, use that punctuation mark and leave out the dash.

Range of Numbers or Time and En Dash

3.64 Use the en dash for a range of numbers or a time span unless it is preceded by the prepositions "from" or "between":

> in the years 1987–1988
> for the period December–June 1989

But:

> from 1987 to 1988
> between December and June 1989

Two-Noun Combinations and En Dash

3.65 Use the en dash to link a compound term serving as an adjective, where the elements of the compound are parallel in form (that is, the first does not modify the second):

> heat capacity–temperature curve
> metal–nonmetal transition

A hyphen or slash can also be used for this purpose.

PERIOD

3.66 The period marks the end of a sentence or sentence fragment. For example, the following footnotes to a table should all end with periods:

[a]From studies by Wheeler.[9]
[b]Calculated results.

The period also has several specialized uses.

Vertical Lists and the Period

3.67 Use a period after numbers or letters that enumerate the items in a vertical list:

These lectures on technical communication offer instruction in three areas:

1. Constructing readable and attractive tables and figures,
2. Reducing technical material to its simplest terms without compromising the author's meaning, and
3. Eliminating technical jargon and wordy expressions.

Instead of periods, you can also use parentheses around the numbers or replace the numbers by bullets.

In identifying items within a displayed list, consistently use either bullets or numbers. Normally use numbers with lists that indicate order of importance or a sequence and bullets with lists in which order has no particular significance.

3.68 There are several ways to punctuate items in a vertical list. If the items are not complete sentences, the options are to

1. Punctuate as if the items were running text (as in the example in Section 3.67 above),
2. Eliminate the commas or semicolons separating the items in series and retain the final period, and
3. End all items in series with periods.

With all three options, the first letter of each item is usually uppercased.

3.69 If one or more items contain a complete sentence, then all sentences and sentence fragments should end with periods:

Users have access to three types of data storage:

- HAL 3350–equivalent disk drives. There are 35 such drives.
- HAL 3330–equivalent disk drives. There are 80 such drives.
- HAL 2314 disk drives. There are two such drives, which must be set up as needed for a job.

Abbreviations and Periods

(See also Chapters 5 and 6.)

3.70 Many abbreviated words end with periods. These include

Academic degrees (Ph.D., A.B., B.S.),
Parts of firm names (Corp., Co., Inc., Ltd.),
Names of countries (U.S., U.K., W. Ger.),
Scholarly Latin abbreviations (e.g., i.e., ibid.), and
Journal names used in references (J. Electrochem. Soc., Phys. Rev., J. Chem. Phys.).

3.71 Abbreviated Latin names for microorganisms, plants, and animals end with periods: *E. coli* for *Escherichia coli.*

3.72 Abbreviations for nearly all units of measure and many technical terms do not include periods unless they spell a word:

ft for feet
s or sec for second
kg for kilogram
mp for melting point
wt% for weight percent
H for hydrogen

But:

in. for inches
at.% for atomic percent
anal. for analysis
b.i.d. for twice a day

3.73 In general, periods are not included with abbreviations formed by combining the first letters of each word (initialisms) in a technical expression:

NMR for nuclear magnetic resonance
DNA for deoxyribonucleic acid

3.74 If an abbreviating period ends a sentence, do not add another period to mark the end of the sentence:

> The battery was manufactured by Gould Inc.

Raised Periods

3.75 The raised period (also called "center dot") serves several purposes in scientific prose. It indicates multiplication ($Z = X{\cdot}Y$) in mathematics, an addition compound in a chemical formula ($Ni_2O_3 \cdot 2H_2O$), the unshared electron in the formula of a free radical ($HO\cdot$), and associating base pairs of nucleotides ($T{\cdot}A$). You should become familiar with any specialized usage for this unusual punctuation in your own specialty.

Ellipsis Points

3.76 Inside a quotation, use three ellipsis dots plus any other needed punctuation to indicate omission of a word, fragment of a sentence, a whole sentence or more. This rule is mainly followed in scholarly writing in which the author extensively quotes the work of others.

3.77 A simpler rule to indicate ellipsis is to use three dots without any other punctuation for any omission. This rule is used in writing in which there are few quotes, as is usually the case for science and technical writing.

3.78 The ellipsis is also used in mathematical expressions:

$$x_1, x_2, \ldots, x_n$$
$$x_1 + x_2 + \ldots + x_n$$

Note how the commas and mathematical symbols enclose the ellipsis points.

QUESTION MARK

3.79 The question mark indicates a direct question. Research problems are sometimes expressed in the form of a question:

> The question we examined is, what is the effect of these market conditions on corporate profits?

3.80 An indirect question within a statement does not end with a question mark:

> We had to ask ourselves why their calculated results coincided so closely with the measured ones.

EXCLAMATION POINT

3.81 The exclamation point signifies that a vehement or ironical statement has been made:

> Mixing these chemicals together at higher temperatures could result in an explosion!

3.82 Science and technical writing—especially that in specialized technical periodicals—assumes a restrained tone. Emotional statements followed by exclamation points are usually out of place. However, the exclamation point can be used effectively when the author assumes a folksy tone, as is often done in science and technical writing aimed at a wide audience:

> TeX's new approach to this problem (based on "sophisticated computer science techniques"—whew!) requires only a little more computation than the traditional methods, and leads to significantly fewer cases in which words need to be hyphenated.
>
> —Donald E. Knuth

HYPHEN

3.83 The hyphen links words in compound terms and unit modifiers and separates certain prefixes and suffixes. A *compound term* expresses a single idea with two or more words ("cross-contamination," "face-lift," "time-sharing"). Two or more words modifying a noun or noun equivalent constitute a *unit modifier* ("one-dimensional computer model"). In general, science and technical writing relies more heavily on unit modifiers than do other kinds of writing.

3.84 The hyphen helps readers identify the relationship between any words modifying a noun.

Unit Modifier Containing Adjectives and the Hyphen

3.85 Use a hyphen in a unit modifier consisting of an adjective and a noun or noun equivalent:

> dual-wavelength radar
> random-access disk file

3.86 Where the meaning is obvious, use of the hyphen is unnecessary:

light water reactor
molecular orbital calculation

3.87 Achieving consistent hyphenation in technical documents can be difficult. If an author decides to hyphenate "long-term leaching" in a report, then she should hyphenate not only all instances of this expression but also any similar expressions, such as "long-term durability."

Unit Modifier Containing Number and Unit of Measure

3.88 Insert a hyphen between a number and its unit of measure when they are used as a unit modifier:

15-g sample
5-m thickness

3.89 Leave out this hyphen if confusion might result:

450°C electrolyte bath
10 cm/s velocity

3.90 If a number precedes a number with its unit of measure, write out the first number:

twenty 10-g samples
two 20-mL aliquots

Unit Modifier Containing Adverb and Hyphen

3.91 Do not use a hyphen in unit modifiers containing an adverb ending in *ly:*

previously known conditions
recently given procedure

3.92 Hyphenate unit modifiers when the adverb does not end in *ly:*

still-new battery
well-kept laboratory

3.93 If yet another adverb is added to this type of construction, then no hyphens at all should appear: *very well defined* structure.

Unit Modifier Containing Comparative or Superlative

3.94 Do not insert a hyphen in a unit modifier if the first element is a comparative or superlative:

higher temperature experiments
lower conductivity ceramic

Unit Modifier Containing a Proper Name

3.95 Do not insert a hyphen in a unit modifier if one of the elements is a proper name:

Fourier transform infrared analysis
Raman scattering experiment

Unit Modifier Longer Than Two Words

3.96 Use a hyphen to join multiple-word unit modifiers:

up-to-date files
easy-to-learn software

Foreign Phrase Used as Adjective

3.97 Do not hyphenate foreign phrases used as adjectives unless they are hyphenated in the original language:

a priori argument
in situ spectroscopy

Predicate Adjective

3.98 Do not hyphenate unit modifiers that serve as predicate adjectives:

This step in the procedure is *time consuming*.
The package must be *vacuum sealed* before shipment.

3.99 Do not hyphenate compound verbs formed with a preposition acting as an adverb:

We *burned off* the binder after the tape-casting procedure.

Self- Compound and Hyphen

3.100 Hyphenate all *self-* compounds:

self-consistent
self-discharge

Suspended Hyphen

3.101 A suspended compound is formed when a noun has two or more unit modifiers, all linked by the same basic element. Use a hyphen after each such modifier:

10-, 20-, and 30-mm thick
long- and short-term tests

Numbers and Hyphens

3.102 Use a hyphen to connect numerator and denominator in fractions unless either is already hyphenated. Hyphenate whole numbers between twenty-one and ninety-nine:

Fractions:

three-eighths
five and six-sixteenths

Already hyphenated:

fifty-five thousandths
five thirty-seconds

Whole numbers:

thirty-three hundred
three thousand thirty-three

Technical Terms and Hyphens

3.103 If a compound technical term (names of chemicals, diseases, plants, and the like) is not hyphenated when used as a noun, it should remain unhyphenated when used as a modifier unless misreading might result:

potassium chloride vapor
lithium oxide solid

But:

current-collector material

Prefixes and Suffixes and Hyphens

3.104 In general, do not use a hyphen if a prefix or suffix is added to a word:

stepwise
nonuniform
multifold

3.105 Consult a dictionary to determine whether or not to use a hyphen with a word containing a prefix or suffix. Use hyphens

- After prefixes added to proper names or numbers (non-Newtonian, un-American, post–World War II, mid-1990),
- After prefixes added to unit modifiers (non-radiation-induced effect, non-temperature-dependent curve, non-tumor-bearing tissue, non-load-bearing material),
- When the suffix "fold" is added to a number (40-fold, 25-fold (*but:* tenfold),
- With prefixes or suffixes to avoid a confusing repetition of letters or similar terms (non-negative, bell-like, sub-micromolar, anti-inflation),
- If their omission would alter the meaning (multi-ply, re-solve, re-pose, re-cover, re-fuse),
- With prefixes that stand alone (pre- and postoperative examination), and
- With prepositions combined with a word to form a noun or adjective: built-in, trade-off, on-line (though some words may not be hyphenated under these same conditions: buildup, cleanup, setup).

In all other instances, spell the word out without a hyphen.

SLASH

3.106 Use the slash (solidus or slant) to signify division in fractions, mathematical expressions, and units of measure:

2/3	x/y	100 m/s
4/5	$x^{3/4}$	200 Wh/kg

3.107 Use the slash to link two terms where an en dash or hyphen would be inadequate or confusing:

IBM 370/195
VAX 11/785
electrode/electrolyte interface

APOSTROPHE

3.108 The apostrophe indicates the possessive case or a contraction. It is also used in forming certain plurals.

Possessives and Apostrophes

3.109 Form the possessive of a singular noun by adding apostrophe *s*:

What could be lovelier than a calm sea on a *summer's* day?

3.110 Form the possessive of a plural noun that ends in *s* by adding the apostrophe alone:

The government failed to heed the *scientists'* warnings.

3.111 If the plural ends with a letter other than *s*, add apostrophe *s* to form the possessive: *spectra's* variance.

3.112 Form the possessive of closely linked proper names by changing the last name only:

Einstein and Planck's law
Lewis and Randall's classic textbook

3.113 Time expressions are formed in the same way as possessives: three weeks' time.

Contractions and Apostrophes

3.114 An apostrophe can also indicate the omission of a letter or letters:

cont'd = continued
it's = it is

Plurals and Apostrophes

3.115 Use apostrophe *s* to form the plural of abbreviations with periods, lowercase letters, numerals, and abbreviations where confusion would result without the apostrophe:

Btu's
Ph.D.'s
x's and y's

3.116 The *s* alone is usually sufficient to form the plural of all-capital abbreviations and years, although some style guides recommend apostrophe *s* here too:

CPUs
1960s

QUOTATION MARKS

3.117 Quotation marks enclose material spoken or written by someone other than the author, or they give a word or expression special emphasis. They are also used around the titles of certain types of publications, such as journal articles and book chapters. (See also Chapters 9 and 10.)

Single Versus Double Quotes

3.118 Use double quotation marks to enclose primary quotations and single quotation marks to enclose quoted or emphasized matter within a quotation:

> The senator's question was simple and direct: "What do you think about when you hear the words 'nuclear energy'?"

Placement of Quotes With Other Punctuation

3.119 Common practice is to place closing quotation marks after commas and periods, before colons and semicolons:

> Science involves much more than the application of "organized common sense," according to Steven Jay Gould.

3.120 Quotation marks or exclamation points always appear within a closing quote when they are part of the quoted matter; otherwise, they appear outside the quotes:

> Who said that "Editing is like bathing in someone else's water"?

Specialized Usage for Quotation Marks

3.121 Put quotation marks around words or phrases that constitute new or special usage. These quotes need not be repeated for a given word or phrase after its first appearance in the text:

We determined the effect of the income loss on the "corporate profile" for various sales campaigns. The required corporate profile for acceptable losses was determined to be the Free-zone campaign.

3.122 Quotes are also used in naming characters the reader should enter to execute some computer commands:

Type "HELP ME" for on-line assistance in changing your password.

If a specialized character string ends a sentence, place the period outside the quotation marks to indicate that the period is not part of the character string the reader must enter:

After an error message, you have several options: (1) type "help!"; (2) type "x" or "X"; or (3) type "e" or "E".

3.123 Make certain that the meaning of all new or special terms is obvious from the context in which the terms appear; if not, define them:

X-rays passing through crystals undergo "diffraction"; that is, the rays bend from a straight-line motion.

3.124 Do not use quotation marks around terminology that the intended audience should be familiar with. An article appearing in a specialized physics or chemistry journal should read

The near-neighbor coordination number of A^{3+} ions was determined.

Instead of:

The "near-neighbor" coordination number . . .

So-Called Expressions

3.125 The meaning of words or phrases following *so-called* is usually apparent without quotation marks:

The data illustrate the *so-called* principle of insufficient reason.

PARENTHESES

3.126 Parentheses—like commas, colons, and dashes—separate material that clarifies, elaborates, or comments on what preceded them. Parentheses usually enclose material that is of secondary importance:

We recovered the remaining gases (SO_2, CO_2, and O_2) by changing our procedures.

Abbreviations and Parentheses

(See also Chapters 5 and 6.)

3.127 Place acronyms and other abbreviations in parentheses when they are preceded by the spelled out term:

> The Source Term Experiments Program (STEP) consists of four accident simulations run in the Transient Reactor Emulation Analysis Test (TREAT) facility.

References and Parentheses (or Brackets)

(See also Chapter 10.)

3.128 One method for citing reference numbers in text is to enclose them in parentheses (or brackets):

> Further details on their experimental technique can be found in the literature (2, 5–8).
>
> . . . in the literature [Tillis 1982a].

Numbered Lists and Parentheses

3.129 Use parentheses in pairs, ( ), when numbering items in a list within running text:

> After an error message, you have several options: (1) type "help!"; (2) type "x" or "X"; or (3) type "e" or "E".

3.130 If the house style is to cite references by numbers enclosed in parentheses, use of a single parenthesis, instead of a pair, helps avoid confusion in numbering items:

> Burris (3) attributes this discrepancy to one of three factors: 1) the absorption influence identified by Johnson (4), 2) the wettability factor, or 3) the effect of high temperature on the system.

Punctuating Parenthetical Comments

3.131 Do not place a period inside the closing parenthesis of a sentence in parentheses inserted into another sentence:

The reaction $2SO_2 + O_2 = 2SO_3$ (sulfuric acid is obtained by dissolving SO_3 in water) occurs in the desired direction only at low temperatures.

3.132 A sentence enclosed by parentheses but not placed in another sentence should end with a period inside the closing parenthesis; it should also begin with a capital letter:

Hadrons are subatomic particles called quarks. (The word *quark* was adapted from a line in a novel written by James Joyce, *Finnegans Wake*.) Murray Gell-Mann was . . .

BRACKETS

3.133 Brackets set off an editorial interjection or parenthetical material within parentheses.

Editorial Interjections and Brackets

3.134 Use brackets to enclose editorial interjections within quoted matter:

Robert Boyle told young scientists that they would likely "meet with several observations and experiments which, though communicated for true by candid authors or undistrusted eyewitnesses, or perhaps *recommended by your own experience* [italics added], may, upon further trial, disappoint your expectation . . ."

Parentheses Within Parentheses

3.135 Use brackets to enclose a parenthetical remark within parentheses:

(We must still examine the mixture's effects on several materials [particularly Ba, Cs, and Sr].)

3.136 Use brackets to enclose a parenthetical remark containing a term requiring parentheses:

Initial experiments were performed with nonradioactive solutions [that is, solutions containing Ho(III)].

PUNCTUATING EQUATIONS

(See also Chapter 7.)

3.137 Equations pose special problems in punctuation. When equations occur in running text, punctuate them according to their grammatical function in the sentence. The whole equation can be treated as a noun

If $x = 0$ is taken to be a boundary condition . . .

or as a clause with the equals sign acting as a verb

If $x = 1$, the boundary-layer model is taken to be a . . .

3.138 When equations are displayed separately from running text, many technical publications prefer that no punctuation marks appear immediately after the equations:

One of the simple modifications of the ideal gas law

$$pV = nRT \tag{1}$$

is the equation

$$pV = RT + \alpha P \tag{2}$$

In Eq. 2, α is a function of temperature.

3.139 Other technical publications, however, require the same punctuation marks as would be employed in running text (in that case, a period would be placed after (2), the logical end of the first mathematically placed complete sentence in the preceding example). Note that the number given to a displayed equation or reaction is put in parentheses flush with the right-hand margin.

FOUR

Spelling

SPECIALIZED SPELLING

4.1 Everyone who writes should own a suitable dictionary. The latest version of a college dictionary, preferably an unabridged version, is probably the most easily referred to source. For professions that use specialized words, a technical dictionary should also be available, perhaps one directed to a specific technical or scientific discipline.

DICTIONARY PREFERENCE

4.2 If a dictionary provides more than one spelling, the first spelling is preferable. American rather than British spelling should be used for material published in the United States. Periodically updated dictionaries can be especially useful in tracking words that have entered the language recently.

Electronic Dictionaries

4.3 Many computerized word processors have spelling checkers, but be cautious of them. Although an automated spelling checker may question words it does not recognize, it will not question a word that is incorrect in context. For example, *casual* is a perfectly good word, but if you intended to write *causal,* the spelling check will not catch this error.

PLURALS

Endings and Plurals

4.4 Form plurals by adding *s* or *es* to the singular. Use *es* for words ending in *j, x, z, sh:* appendixes, arcs, masks, rushes.

4.5 Some words that end in *f, ff,* or *fe* may change the plural to a *ve* form; some even have two forms: halves, leaves, lives, staffs (staves) (different meaning, however), wolves.

4.6 Add *s* to words ending in *o*, if a vowel precedes the *o* (as in ratios), but if a consonant precedes the *o*, add *es* (mosquitoes). Exceptions do occur, however (zeros).

4.7 Words ending in *y* generally change the *y* to an *i* and add *es:* counties, harpies, puppies.

4.8 Some words change form altogether:

child	children
foot	feet
louse	lice
tooth	teeth
woman	women

Foreign-Language Words and Plurals

4.9 Many technical words come from foreign languages and form the plural irregularly. Sometimes, though, meanings can be changed if a "plural" form is assumed.

4.10 Retain the original-language plural for these foreign terms:

Singular	**Plural**
crisis	crises
datum	data
erratum	errata
ovum	ova

4.11 Use the original-language plural or an English version for such words as

Singular	**Foreign Plural**	**English Plural**
appendix	appendices	appendixes
criterion	criteria	criteria
formula	formulae	formulas
matrix	matrices	matrixes
stratum	strata	strata
vortex	vortices	vortexes

4.12 Some words change their meaning based on their ending:

indexes (tabular content)
indices (algebraic signs)

SINGULAR/PLURAL DILEMMA

4.13 Some spelling rules must be memorized. For example, some singular words can also be used as plurals: deer, fish, moose.

4.14 Some words that can be made plural in the usual way also can be plural without any change:

couples	couple
dozens	dozen
heathens	heathen
peoples	people

4.15 Some words can be either singular or plural with no change: aircraft, chassis, series, species.

PROPER NAMES AND ABBREVIATIONS

4.16 Add *s* or *es* to proper names (capitalized forms) to make them plural: Johns, Joneses, Marches, Marys, Thompsons.

4.17 Add a simple *s* to letters or numbers used as words:

ICBMs
IRAs
1920s
'40s

4.18 If a simple *s* will be confusing, then use an *'s:*

MOS's
Ph.D.'s
S's
x's

POSSESSIVES

4.19 Add an *apostrophe* and an *s* to form the possessive case of most singular nouns (see Section 3.109 and following):

cow's	mother's
dog's	tree's

4.20 Use an *apostrophe* and an *s* for plural nouns not ending in *s;* those ending in *s* take an *apostrophe only.* (See Section 3.109 and following.)

aunts'
children's
mothers'
women's

4.21 Treat almost all proper names in the same way:

Davises'
Douglas's
John's
Stan's

4.22 Form compound possessives (e.g., fool's gold) with the first portion of the compound as a singular possessive. Often terms such as *spider web,* even though they imply a possession, do not use the apostrophe. In other compounds, such as *someone else's,* add the apostrophe to the second term.

4.23 Phrases can pose a problem and often can be rewritten. For example, "The President of the United States' statement . . ." might much more logically be "The statement made by the President of the United States. . . ."

4.24 The idea that an inanimate object cannot show possession is outdated. But, some editors and readers might object to

aircraft's propellers
destroyer's wake
operations' results

COMPOUND WORDS

4.25 Compound words, in general, are treated as one word, a hyphenated word, or two words. Compound words tend to become single words. This trend can be seen in computer terminology (data base and database, for instance, are in transition right now).

Some General Hyphenation Guidelines

1. Hyphenate *self-* and *quasi-*. (See Section 3.100.)
2. Do not hyphenate *non, de, extra, re, over,* and *pseudo.*
3. If a word does not appear in any reference source as one word, use two words and do not hyphenate unless these words are a unit modifier, a temporary compound formed by adjectives preceding a noun.

4.26 Many dictionaries may not list separate definitions of all the words that can be created by adding prefixes; however, many do include word lists at the bottom of the appropriate page to provide guidance about the use of hyphens.

Trend Toward Nonhyphenation

4.27 The nonhyphenation trend extends to such words as reentry, reexamine, reexpose. However, use hyphens when not using one changes a term's meaning (for example, recreation in contrast to re-creation).

4.28 When a temporary compound is formed by adjectives preceding a noun, the construction is called a unit modifier. For example, a rising market might be of interest to an economist; a fast-rising market could cause panic on the stockmarket floor. In the adverbial form, a rapidly rising market, no hyphen is used.

WORD DIVISION

4.29 Divide most words according to the American system of pronunciation.

General Word Division Guidelines

1. Do not divide at single letters (such as i-dea), and never divide the *ed* from the parent word. Neither one gains a space advantage on the line, and both look ridiculous on the page.
2. Avoid breaking any word in such a way that only two letters carry to the following line.
3. Words should not be divided between one page and the next.
4. Although it sometimes cannot be avoided, try not to break proper names, and if possible, do not even break between initials and the last name, although this is permissible.
5. Already-hyphenated compounds should not be broken except at the hyphen, and once-compounded, but now solid, words ordinarily should be broken at the natural compound:

   over-all
   high-energy
   low-temperature
   data-base

6. If prefixes exist, break at that point rather than elsewhere in the word:

   quasi-technical
   anti-matter
   super-conducting
   pre-exponential

7. If a text uses words that are not English, recognize that word division rules are different for foreign languages: weg-werf-en, not weg-wer-fen.
8. If it is absolutely necessary, large numbers expressed in figures can be broken after a comma, but not after a decimal point:

   500,-000 but not 500.-45
   100,000,-000 but not 458.-5

9. Abbreviations used with numbers that record measurements (such as a line 74 mm long) should not be separated from the figure with a hyphen. However, some number combinations have become generic terms (such as 35-mm camera) and these terms use a hyphen. (See Section 3.88 and following.)
10. For numbered lists in text, carry the number to the following line to appear with the material directly following the listing:

    The elements included (1) hydrogen, (2) oxygen, and (3) various unidentified solids.

    Not:

    The elements included (1) hydrogen, (2) oxygen, and (3) various unidentified solids.

SUFFIXES

4.30 Words ending with a silent *e* usually drop the *e* before a suffix beginning with a vowel, with some exceptions based on preferred American style:

create, creating
force, forcible

But:

mile, mileage
enforce, enforceable

And, to avoid misreading:

dyeing, not dying
singeing, not singing

4.31 If the silent *e* precedes a consonant, the *e* is usually retained:

movement
wholesome

Again, of course, there are exceptions:

wholly
truly

4.32 Most confusing, perhaps, are words for which dropping or retaining the *e* are both correct:

acknowledgment, acknowledgement
judgment, judgement

The first usage is preferred in each case.

Soft Endings (*ce, ge*)

4.33 Words with soft endings—*ce* or *ge*—retain the *e* before suffixes beginning with *a, o,* or *u,* to retain the soft sound:

advantageous
changeable

The *y* Ending

4.34 Most words ending in a vowel followed by *y* keep the *y:*

buy, buyer
enjoy, enjoyable

4.35 When a consonant precedes *y,* an *i* usually substitutes for the *y:*

creepy, creepier
fatty, fattiness
fly, flier

Consonant Endings

4.36 Many words that end in a single consonant preceded by a single vowel double the final consonant:

control, controlled
occur, occurrence
prefer, preferred

But:

transfer, transferable
travel, traveled

4.37 For words ending in a hard *c*, a *k* is often added to ensure proper pronunciation:

mimicking
picnicking
shellacked

4.38 The *ise* ending is the commonly accepted American ending for many words (and the preferred British ending for words that often end in *ize* in America—such as characterize or criticize):

advertise
comprise
exercise
improvise

4.39 It is increasingly common to create a verb from a noun by adding *ize*, as in

finalize
prioritize

But these words tend to be jargon for which meaningful words already exist:

complete
rank

Ceed, Sede, Cede Rule

4.40 The rule for the *ceed, sede, cede* endings is relatively easy to remember: Three words in English end in *ceed* (exceed, proceed, succeed). One ends in *sede* (supersede). Others with the same pronunciation end in *cede* (concede, for example).

CAPITALIZATION

Beginning Sentences

4.41 The first word in a sentence is normally capitalized (capped). Capitalize all proper nouns. Caps are also sometimes used for emphasis (as in a Good and Loyal subject).

4.42 When a complete sentence follows a colon, the first word is normally capped:

> The mainspring was rusted: Excessive forces created the possibility of an equipment breakdown.

Listings

4.43 Depending on the format, listings can be capitalized or lowercase. Ordinarily, listings within a sentence are not capped:

> . . . consist of (1) methods and materials, (2) procedures, and (3) test results.

Listings that are displayed, however, whether they consist of complete sentences or not, are usually capped. Incomplete sentence listings do not need end punctuation, though end punctuation is preferred (see Sections 3.67–3.69); sentence listings do:

consist of

1. Methods and materials
2. Procedures
3. Test results

But:

consist of three parts

1. Methods and materials will be listed.
2. Procedures will be followed according to specification.
3. Test results will be completely reported.

Parentheses

4.44 A sentence enclosed in parentheses within another sentence need not be capped. However, if the parenthetical sentence is long or complex, perhaps it would be more understandable if the entire construction could be rewritten as two complete sentences. Both of these would be capped, even though one remains in parentheses. (See Sections 3.131–3.132.)

Quotations

4.45 When quoting a passage, use caps where the quotation uses caps. Begin a complete quotation with a cap. If the quotation begins in the middle of a sentence, it should be so indicated by using ellipses. (See Section 3.117 and following.)

> ". . . using results found in earlier tests. However . . ."

Salutations and Closings

4.46 In the salutation and closing of letters, usually the first word only is capped:

My dear colleague,
Yours sincerely,

4.47 However, cap the noun used in place of a name, or a formal title:

Dear Professor Adams:
Dear Sir:

Book or Article Titles

4.48 Capitalize titles based on individual publishing style. For instance, either of the following could be correct (see also Chapter 10):

Flight Aerodynamics in the Gunnell Wind Tunnel
Flight aerodynamics in the Gunnell wind tunnel

4.49 Most American usage requires the capitalization of all important words in a title, including short words. Prepositions of four or more letters are generally capitalized, although, again depending on the publisher, *From* or *from* may be equally correct.

4.50 Although articles (a, the, an, for example) are usually not capped, if they follow a definite break, as in a subtitle, they are:

Flow Fields: A Study of Werner Methods
Workshop on Thermodynamics—The Feumer Factor

4.51 If hyphenated compounds occur in a title, cap those terms that would not ordinarily be hyphenated (such as unit modifiers), but leave lowercase those that would be hyphenated. (See Section 3.83.)

Analysis of Wind-Driven Fans

But:

Statistical Analysis of Twenty-five Air Blowers

Proper Names

4.52 One of the more confusing areas of capitalization lies in the use of proper nouns and adjectives. Undoubtedly, one would capitalize American, Farouk, Fourier, Texas.

4.53 When proper nouns are used to produce a common classification, the tendency is toward lowercase, and sometimes the distinction is not easy to make. Use a dictionary to resolve these confusions:

afghan robe
dutch oven
italic (or roman) type
turkish bath

4.54 Measurements derived from proper names are usually lowercase, but note that the abbreviated version retains the cap. (See also Chapter 7.)

ampere, A
farad, F
gauss, G
hertz, Hz
kelvin, K
newton, N
pascal, Pa
tesla, T
watt, W

4.55 Capitalize formal titles or names, particularly if they refer to specific offices or people:

Court of Claims
Kemal Pasha
King James
the Senate
the President . . . Army . . . Navy (of the United States)

4.56 Honorary titles and academic degrees are capped if they follow a proper name:

Frank Smith, Fellow of STC
Paul Nash, Ph.D.

4.57 Names of organizations are usually capped:

Boston Pops
Giants
Girl Scouts
Odd Fellows

4.58 Names of college colors and national flags are usually capped:

the Black and Orange
Old Glory
Union Jack

4.59 Names of games are not capped unless they are trademarked:

checkers
chess
golf
Monopoly™
Scrabble™

4.60 The names of planets and other heavenly bodies are usually capped (see also Chapter 6):

Comet Halley
Great Nebula
Polaris
Venus

4.61 The names of holidays or other special days are generally capped, as are the names of the months and days of the week:

Christmas
July Fourth
Passover
Thanksgiving

But the seasons (spring, summer, fall, winter) are lowercase.

4.62 The names of races and groups are generally capped:

African (or Black to refer to native African)
African-American (or Black to refer to African ancestry)
Amerind (Native American Indian)
Asian (Asian ancestry)
Caucasian (or White to refer to European ancestry)
Indian (ancestry in the country of India)
Kiowa (ancestry in a specific Amerind tribe)
Masai (ancestry in a specific African tribe)

Trade Names

4.63 Trade names are protected by law, but many fall into common use. A good example is Xerox; no matter what kind of a copy a person wants, a Xerox copy is commonly requested. Use generic terms if they are sufficiently descriptive. For example, *photocopy* or *xerographic copy* can be used in place of *Xerox* or *tissue* can be used instead of *Kleenex*.

4.64 However, in reporting test results based on specific equipment, accurate research reporting requires the use of the equipment's trade name to help duplicate a procedure:

Benzedrine (amphetamine sulfate)
Demerol (meperidine)
Fiberglas (fiber glass)
Pyrex (heat-resistant glass)
Xerox (copying machine, photocopy)

Some companies require a disclaimer—often as a footnote—to indicate that mention of a product does not imply endorsement.

Foreign Terms

4.65 Some foreign languages have capitalization practices that are very different from American style. German, for instance, capitalizes all nouns regardless of their position in the sentence. In using specialized terms from German (or any other language), retain their capitalization practice:

> The wing strut can be supported through the addition of a Werzeugkasten at station 77.55.

TROUBLESOME WORDS

Sound- and Look-Alike Words

4.66 English has a number of words that sound alike but are not spelled alike, or are spelled alike but do not sound alike:

beet, beat
feet, feat
might, mite
night, knight
rough, ruff

4.67 English has a number of endings that are the same but have different sounds:

cough
enough
plough
through

4.68 Some words look alike and sound alike but have different meanings:

duck (a bird)
duck (dodge)

4.69 And some words, although they look alike, neither sound alike nor have the same meaning: A *lead* pipe is considerably different from a *lead* singer. The similar sound of some terms can be equally confusing: To *alight* from a train is different from needing *a light*.

Spelling Variations

4.70 The trend toward simpler spellings can sometimes be confusing (as ax or axe). When either is correct, use a consistent form throughout a document.

4.71 Avoid some simplified spellings:

light, not lite
night, not nite
through, not thru

4.72 Because of an interest in health, everyone knows what *lite* means—whether applied to beer or mayonnaise. And people who commute understand what *thruway* means on a road sign. However, these spellings should not appear in formal writing.

American Versus British Spelling

4.73 The differences between American and British spelling often lies in the endings. (See also Chapter 8.)

American	**British**
center	centre
check	cheque
connection	connexion
defense	defence
labor	labour
program	programme

Easy Misspellings

4.74 Some words, because of letter doubling, pronunciation variations, or assumptions about vowel use, are almost predictable misspellings. These words include

absorbent	chromatography	lightning	privilege
accommodate	commercial	liquefy	prominent
adsorption	desiccate	maintenance	recur
adviser	embarrass	nevertheless	seize

all right	fluoride	nickel	separate
analogous	forty	occurred	siege
arctic	fulfill	parallel	specimen
benefited	grammar	personnel	supersede
breathe	hangar	preceding	
canister	irrelevant	preventive	

Foreign Words

4.75 Many foreign words and phrases are in common use. In the past, the style has been to italicize uncommon terms and to set in roman those terms that have been accepted into English. The following list contains the latter words or phrases:

ad hoc	caveat	modus operandi	prima facie
ad libitum	de facto	par excellence	pro forma
alma mater	dossier	passé	quasi
a priori	en route	per annum	résumé
bona fide	ex officio	per capita	versus
briquette	in situ	per diem	via
carte blanche	in toto	per se	vice versa

FIVE

Abbreviations

(See also Chapters 6 and 7.)

5.1 Abbreviated terms—whether abbreviations, initialisms, acronyms, or symbols—save time, avoid repetition, emphasize information, and require less space in tables and illustrations. Abbreviated terms, which are derived by shortening longer terms or combinations of terms (UHF for ultrahigh frequency), are easily understood in context. Initialisms are formed by grouping the first letter of each term (DTP for desktop publishing). Acronyms derive from pronounceable initialisms (scuba for self-contained underwater breathing apparatus). Symbols are a special class of abbreviations defined by international or professional agreements (for example, metric measure and *Système Internationale* [SI]). All of these abbreviated terms are easily understood in context.

ALPHABETS AND SYMBOLS FOR ABBREVIATED TERMS

5.2 Base abbreviations on

- English alphabet (written in combinations of upper- and lowercase),
- Specialized typography (boldface, italic, or roman),
- Punctuation conventions (periods, commas, sub- and superscripting).

5.3 Science, engineering, and mathematics add

- Greek alphabet,
- Some graphic symbols, such as arrows, mathematical operation signs, biological indicators, and the like.

When to Use Abbreviated Terms

5.4 Abbreviations save space and act as a technical shorthand for specialized terms. Using abbreviated terms shortens texts only slightly: they eliminate about 2.3 words per acronym while making the text somewhat more difficult to understand.

5.5 Use abbreviated terms for technical material that will be used by a technical audience.

Frequency of Use for Abbreviated Terms

5.6 Spell out words to avoid too many abbreviated terms. Since readers regard any unusual typographic or spelling convention as an emphasizing technique, establish an optimal and maximum ratio for these terms. Any emphasis technique has a threshold (about 1 in 300, or one abbreviated term per page of text) beyond which the reader will begin to ignore the technique.

Beginning a Sentence with Abbreviated Terms

5.7 Although a sentence does not usually begin with an abbreviated term, it can start with an initialism:

AAA members can get service for their cars on the road.

Abbreviated Terms and Hazards

5.8 Write out terms that indicate hazards. Don't create an abbreviation that contradicts an accepted standard. For instance, use micrograms (μg)—based on SI units (see Chapter 7)—rather than mcg, an abbreviation that has been suggested to avoid misreading μg for milligram (mg).

STYLE GUIDES FOR ABBREVIATED TERMS

5.9 Organizations that create or use abbreviations often in their work should develop an abbreviation database or glossary to insure consistency. This database will also serve as the basis for decisions about abbreviations among different disciplines and organizations. Typical disagreements such as the following can be resolved with a large database:

Exhibit 5–1:
Typical Abbreviation Disagreements

Term	American Institute of Physics	National Aeronautics and Space Administration
National	Natl.	Nat.
Ultrahigh frequency	uhf	UHF (IEEE)
Ultraviolet	UV (ACS, IEEE)	uv
Electromagnetic radiation	x ray	x-ray (ACS)
ampere hour	A • h or A h	Ah (IEEE)

ACS—American Chemical Society. IEEE—Institute of Electronics and Electrical Engineers.

In writing for a client, find out if that organization follows the abbreviation style of a particular professional organization or if there are local preferences.

GLOSSARY FOR ABBREVIATED TERMS

5.10 Any abbreviation that the reader is unlikely to recognize must be explained in the text. This explanation should occur in three places:

- When the term first appears in the text,
- If it occurs again away from the first explanation, and
- In a glossary.

5.11 Identify all specialized abbreviated forms a reader is unlikely to recognize. Accepted units of measure need not be included, but unusual signs or symbols should be defined.

5.12 Place the glossary in the text's front matter if it is short. Longer glossaries belong in the appendixes.

CREATING SUITABLE ABBREVIATIONS

5.13 Consider the audience's professional background when selecting abbreviations. Also, find out if this audience usually reads documents that conform to an existing style guide. (See also Chapter 10.)

Abbreviating Single Words

5.14 Abbreviations consist of one or more letters of the abbreviated word, sometimes followed by a period. Use of the terminating period depends on the preferences of the person or organization that sets the style.

Single-Letter Abbreviations

5.15 Single-letter abbreviations consist of the first letter of the word:

t. temperature
T. absolute temperature

5.16 Use these abbreviations in tables and illustrations. To use the same abbreviation in a mathematical or physical sense, present them in unpunctuated italics:

t	temperature
T	absolute temperature

(See also Chapter 7.)

Two-Letter Abbreviations

5.17 Two-letter abbreviations may consist of the first two letters of the word:

Fr.	French
sl.	slight

Or the first and last letter: Ct. Court.

5.18 Less common combinations include

cp.	compare
ff.	following
pp.	plural of page

5.19 Among two-letter combinations usually printed in all capital letters (or, if available, small caps) are

A.D.	(AD)	anno Domini (in the year of the [our] Lord)
B.C.	(BC)	before Christ
C.E.	(CE)	common era

Three-Letter Abbreviations

5.20 Three-letter abbreviations might consist of the first three letters of a word or a combination of letters, usually a consonant from the beginning, middle, and end of the word:

abs.	absolute	pos.	positive
col.	column	res.	research, resolution
Hon.	Honorable	Rev.	Reverend
int.	international	sec	secant (never with a period)
neg.	negative	vol.	volume

Other, more novel, combinations include

ctr.	center
mgr.	manager

Mme.	madame
rpt. (rep.)	report
Twp.	Township

Abbreviations of Four Letters or More

5.21 In abbreviations of four letters or more, any easily recognizable combination of letters may be used:

acct.	account	mfre.	manufacture
Blvd.	Boulevard	mfrg.	manufacturing
chap. (ch.)	chapter	Msgr.	Monsignor
ency.	encyclopedia	soln.	solution
engr.	engineer	strd.	standard

5.22 There is no agreement on this abbreviation method. For instance, the American Institute of Physics uses Assoc. for Association, while a contemporary, commercial publisher uses Assn. Either choice can be defended.

Abbreviating Two Consecutive Words

5.23 When abbreviating two words that form one concept, leave a space between the abbreviated terms:

at. wt.	atomic weight
sp. gr.	specific gravity
sq. in.	square inch(es)
sq. ft.	square foot (feet)

5.24 A space also occurs between a number and the unit of measure or symbol that follows it:

3 g
5 m

Some organizations, such as the American Institute of Physics, remove the space when reporting temperature: 67°C.

The American Chemical Society retains the space: 67° C.

The symbols for degree (of angles) and percent share similar confusions.

5.25 There are many exceptions to both the spacing convention and the use of periods. In the following abbreviations, no space occurs between terms and they always use periods:

e.g.	for example (exempli gratia)
etc.	and others (et cetera)
i.e.	that is (id est)
q.v.	which see (quod vide)
v.i.	verb intransitive
v.t.	verb transitive

PUNCTUATING ABBREVIATED FORMS

5.26 Two punctuation marks deserve special attention when creating abbreviated forms: the period (.) and the slash (/).

Periods in Abbreviated Forms

5.27 Many specialized measurement systems, such as metric and SI, omit periods after unit symbols:

ft	foot (feet)
g	gram (grams)
m	meter (meters)
oz	ounce (ounces)
sq ft	square foot (feet)

5.28 Some abbreviations, including symbols, are punctuated with periods to avoid confusing them with similar words:

at.	atomic
fig.	figure
in.	inch(es)
no. (NO.)	number (Number)

This practice creates its own confusions. Weight, for instance, is usually abbreviated as wt, without a period. Thus, atomic weight would be abbreviated as at. wt—without a terminating period. To avoid confusion, both the National Aeronautics and Space Administration (NASA) and the American Society of Mechanical Engineers use at. wt. for atomic weight; however, NASA still abbreviates weight, when used alone, without the terminating period.

5.29 Professional organizations add their own, nonpunctuated, abbreviations to these lists. Thus, the American Chemical Society uses

sp ht	specific heat
sp vol	specific volume

But also uses:

e.g.	for example (in tables and figures)
(e.g.)	for example (in text, or written out if not placed in parentheses)
i.d.	inside diameter
i.e.	that is (in tables and figures)
(i.e.)	that is (in text, or written out if not placed in parentheses)
o.d.	outside diameter

The American Medical Association uses both

fig (Fig)	figure (Figure)

Slash in Abbreviated Forms

5.30 In recording abbreviated forms associated with sound, use the slash (/) to replace the letter *p* (per):

c/m	cycles per minute
c/s	cycles per second
r/m	revolutions per minute

ABBREVIATED TERMS IN TECHNICAL AND SCIENTIFIC TEXTS

5.31 Abbreviated terms common to all technical and scientific disciplines are initialisms, acronyms, and symbols. The principles underlying these terms and the stylistic rationale of various organizations are illustrated in the following sections. More detailed information on specific applications can also be found in Chapters 6 and 7.

Initialisms

5.32 Initialisms are abbreviations formed by grouping together the first letters (initials) of two or (usually) more consecutive words. Almost all initialisms are written without periods, as in DTP, desktop publishing.

Exhibit 5–2:
Sample Initialisms

Initialism	Meaning
AAO	American Academy of Ophthalmology American Academy of Optometry American Academy of Osteopathy
CERN	European Center of Nuclear Research
CFC	Chlorofluorocarbon
CIC	Combat Information Center
DJIA	Dow Jones Industrial Average
DoD (DOD)	Department of Defense
ECT	Encyclopedia of Chemical Technology
fob (f.o.b.)	Free on board
GPO	Government Printing Office
NMFS	National Marine Fisheries Service
PNG	Papua New Guinea
PVC	Polyvinyl chloride

Plurals of Initialisms

5.33 A plural is often indicated by adding a lowercase *s* to an initialism:

CFCs chlorofluorocarbons
TAs teaching assistants

Some organizations, however, still prefer an apostrophe before the lowercase *s:*

CFC's chlorofluorocarbons
TA's teaching assistants

5.34 Unless required to follow a specific style, reserve the apostrophe for the possessive case:

The AAA's rules require a driver to obey the law.

Acronyms

5.35 Acronyms are initialisms that can, and are, pronounced as one word, as in CAM, computer-aided manufacturing.

5.36 Nested acronyms contain one or more other acronyms:

flir	forward-looking infrared radar
radar	radio-assisted detection and ranging

Unless the intended audience is highly trained, nested acronyms may be confusing and should be avoided.

5.37 Some acronyms have become words in their own right and may be used routinely in technical and scientific text. Some more common acronyms include

laser	light amplification by stimulated emission of radiation
maser	microwave amplification by stimulated emission of radiation
quasar	quasi-stellar radio source

Indirect Articles and Acronyms

5.38 To decide whether to use *a* or *an* before an acronym, the author must decide on a pronunciation rationale for initialisms:

Exhibit 5–3:
Indirect Articles and Acronyms

Initialism	Pronunciation	Indirect Article
CAM	CAM cee-a-em	a an
FET	FET eff-ee-tee	a an

Symbols (See also Chapter 7.)

5.39 Symbols are a special type of abbreviation. A symbol has often been adopted or created by international organizations, such as the *Système Internationale* (SI), or the modern metric system. Some organizations still have not resolved the difference between symbols and other abbreviated terms.

5.40 Usually a symbol is composed of a single letter or specialized sign selected from the symbol set available for all abbreviated terms, and is not punctuated. Symbols also do not form the plural by adding *s;* for example, the symbol for gram—g—also stands for grams.

While symbols composed of letters seem to be classified as abbreviated terms, mathematic symbols do not share the same confusion. They are always classified as symbols.

Typographic Treatment of Symbols

5.41 Few organizations agree on casing practices—upper- or lowercase characters—for symbols:

Exhibit 5–4:
Typographic Treatment of Symbols

Term	Symbol	Organization
continuous wave	cw	American Institute of Physics
	CW	Institute of Electronic and Electrical Engineers
intermediate frequency	i.f.	NASA
	IF	American Institute of Physics; Institute of Electronic and Electrical Engineers

Arbitrary Changes in Symbols

5.42 Some organizations change symbols depending on where the symbols occur. For instance, the American Society of Mechanical Engineers uses lowercase symbols in text and uppercase in illustrations. This practice creates notations such as ML on drawings, which means megaliter, rather than ml—milliliter. The Institute of Electronic and Electrical Engineers has similar guidelines.

These guidelines seem to be aimed at authors who do not have access to special typographic symbol sets. However, these shorthand practices should

- Be directed only to internal specialized audiences,
- Be fully explained in a glossary, and
- Not appear in public documents.

SIX

Specialized Terminology

- *Personal Names*
  - *Given Names*
  - *Natural Appellations*
  - *Abbreviated Forms*
  - *Nicknames*
    - *Familiar Names*
    - *Epithets*
  - *Family Names*
    - *Multiple Names*
    - *Names Assumed Through Marriage*
    - *Née*
- *Titles*
  - *Common Title Forms*
  - *Academic Title Forms*
- *Honors and Awards*
  - *Civil Honors*
- *Foreign Names*
  - *Civil and Social Factors*
  - *Prefixes in Foreign Names*
- *Foreign Names and Forms of Address*
    - *Arabic Names*
    - *Asian Names*
    - *French Names*
    - *Germanic Names*
    - *Italian Names*
    - *Portuguese Names*
    - *Russian Names*
    - *Spanish Names*
- *Geographic Names*
  - *States*
  - *Canadian Provinces*
  - *Cities and Towns*

Phenomena
Atomic and Nuclear Particles
Chemistry Terminology
Isotopes
Elements
Chemical Compounds
Formulas
Biological Terms
Geological Terms
Mathematical Terms
Astronomical Terms
Planets
Moons
Comets
Meteors
Constellations
Stars
Classes of Stars
Nova
Galaxies
Technology Terms
Computer Terms
Trends in Terminology
Hardware Terminology
Software Terminology
Commercial Programs and Application Packages
Personal Computer Software
Mainframe Computer Software
Releases and Versions of Products
Computer Commands
Computer Messages
Software Development Documentation
Programming Languages
Backus-Naur Form
Procedures
Electronics
Circuit Technologies
Passive Electronic Components
Active Electronic Components
Logic Circuits
Integrated Circuits
Telecommunication
Abbreviations and Acronyms
Transmission and Control Characters
Device States

6.1 Scientific and technical communication relies on writers and readers sharing a common language. Use terminology familiar to readers in an accepted and consistent manner. This chapter discusses the terms used for specific subjects, the meaning conveyed by those terms, and standardized conventions used to print those terms.

PERSONAL NAMES

Given Names

6.2 Use the term *given name* rather than *Christian name* which, although common, has an obvious ethnic bias. This section discusses given names in a North American context. (See Section 6.26 for treatment of foreign names.)

Natural Appellations

6.3 The natural appellation is the full official form of a person's first name, as distinct from forms such as the abbreviated, familiar, epithet, and nickname forms.

Regard given names as proper nouns and capitalize them:

Ann, Robert, Phillip, William, Zoe

Variations may commonly occur in spelling:

Ann or *Anne, Phillip* or *Philip*

Abbreviated Forms

6.4 Certain given names have abbreviated forms. Although these forms are no longer in common use, they still appear in such areas as commercial signage and business cards. Use a period after the abbreviated form:

Chas.	Charles
Wm.	William
Jas.	James
Thos.	Thomas

Nicknames

6.5 A nickname replaces a proper name. Nicknames may take the form of familiar names or epithets.

Familiar Names

6.6 A familiar name is usually a shortened and modified form of a proper name. Familiar names commonly occur in spoken form; they are less usual in formal writing. The list of familiar names in English is very extensive. Almost every language has a set of familiar names:

Art	Arthur
Becky	Rebecca
Jill	Gillian
Joe	Joseph

Although they are abbreviated forms, familiar names do not end with a period.

Epithets

6.7 An epithet, a characteristic word (or phrase), replaces or is used together with a person's proper name. Sports and entertainment are among the many rich sources of colorful epithets.

6.8 Capitalize epithets, but do not use quotation marks:

> Buffalo Bill, Billy the Kid, Leadbelly, Jelly Roll Morton, Babe Ruth, the Sultan of Swat, Goose Gossage, the King of Swing, Doctor J., and Murray the K.

Note the terminal period in the last two names to indicate an abbreviation.

6.9 Epithets may often be placed between a person's first and last proper names. In such cases, do not use punctuation marks (such as quotes or parentheses) around the epithet:

> Nat King Cole, Dennis Oil Can Boyd, Earl Fatha Hines

Family Names

6.10 A family name (also known as surname) is the name borne in common by members of one family. The individual proper names of the family members precede the family name:

> John Fisher, Maryann Fisher, Karen Fisher

6.11 Where a man bears the identical full name to his father, the older man is identified as *Senior* and the younger man as *Junior:*

> Joseph Smith, Sr., and Joseph Smith, Jr.

Note the use of comma following the name.

6.12 Where a succession of men in a family line bear the same name, identify each person's name (except the first) by roman numerals:

George Hamilton IV.

Multiple Names

6.13 Some family names may be complex in that they comprise multiple names. Family names may also be hyphenated.

Debrah Ormsby-Gore, George St. John-Quimby, Helena Maria Devalier Ortega Querrero (See Section 6.26.)

Names Assumed Through Marriage

6.14 North American civil practice encourages women to adopt their husband's surname on marriage. Therefore, Miss Jane Smith may become Mrs. Jane Bailey. In formal address, a married woman is referred to by her husband's full name: Mrs. David Bailey.

6.15 Where a married woman wishes to retain her father's family name, this precedes her married family name:

Christine Evert Lloyd, Olga Hughes-Roberts

Family names for married women in civil systems other than the English can be quite involved. (See Section 6.26.)

Née

6.16 Nee or née (French for *born*) identifies a woman by her maiden (original) family name in addition to her name by marriage:

Olga Roberts née Hughes

TITLES

Common Title Forms

6.17 The basic forms of common title include Mr., Mrs., Miss, and (the modern) Ms. Note that Miss does not take a period:

Mr. David Webster	Mrs. Catherine Parsons
Miss Alice Marley	Ms. Charmaine Geer

Some organizations do not use periods after titles. The American Medical Association, for instance, uses Dr, Mr, Mrs, and Ms.

In joint address, use the man's first name:

Mr. and Mrs. David Webster
Mr. and Mrs. D. J. Webster

Academic Title Forms

6.18 Unless linked with a proper name, do not capitalize academic titles:

the dean of Engineering
the president of the University of Nebraska
associate professor of English
chair of the Department of Language and Literature
a professor emeritus

6.19 Capitalize academic titles when they directly precede a proper name:

Professor Smith
Dean Jarvis
President Maclaren
Prof. Smith

Note that the title *chairman* is never used in this way. If a gender-neutral term is required, use *chair* or *head.*

Do not include the rank of a professor (that is, assistant, associate, full, or adjunct) in this direct form of title.

In the case of professor emeritus, the title may precede or follow the professor's name:

Professor Emeritus George Smith lectured the audience . . . The lecture was given by George Smith, Professor Emeritus of History at Queen's College.

6.20 Address academics who have a doctorate as *doctor:*

Dr. Smith
Dr. John Smith

6.21 Some organizations use periods in professional titles:

M.D., Ph.D., D.Sc.

6.22 In citing someone as a doctor, specify the kind of doctorate:

Medical Doctor	M.D.
Doctor of Dental Surgery	D.D.S.
Doctor of Veterinary Medicine	D.V.M.

Doctor of Philosophy	Ph.D.
Doctor of Science	D.Sc.

Include this specification the first time the doctor citation appears:

Enrico Ferraro, M.D.
Margaret Wooldridge, M.D., Ph.D.

Not: Dr. Harold Arosnyth, Ph.D.

6.23 Collective address is handled as

Doctors Smith and Jones performed the initial study.
The seminar was arranged by Professors Langley and Weston.
Doctors Brenda and Philip Snebur lectured on hypertext.

HONORS AND AWARDS

Civil Honors

6.24 Outstanding service to the country and bravery are recognized by such awards and grades as

Companion of the Order of Canada	CC
Cross of Valour	CV
Medal of Bravery	MB
Member of the Order of Canada	MC
Officer of the Order of Canada	OC
Star of Courage	SC

6.25 Append the abbreviation, without additional punctuation, for the award after the holder's name:

Jeanette Blais CC
Raymond Paul Dupré MB
Suzanne Leblanc SC
J. Kenneth Owen OC

FOREIGN NAMES

6.26 The main concern with foreign names is to achieve an accurate spelling in English. This section illustrates how to write the more common foreign names in English.

6.27 Most European languages use the roman alphabet, usually with accents or modification on certain letters. The letters most commonly accented are vowels. The following exhibit summarizes the major European languages that use the roman alphabet and indicates the accented letters:

Exhibit 6–1:
Common European Languages Using the Roman Alphabet

Language	Accented Letters	Special Letters
Danish	å, ø	(diphthong)
Dutch	none	ij (diphthong)
Finnish	ä, ö	
Flemish	none	
French	é, è, ô, ç	
German	ä, ö, ü	ß
Italian	none	
Norwegian	å, ø	
Portuguese	ã, ñ, ê, á, é, í, ó, ç	
Spanish	á, é, í, ó, ú, ü, ñ	
Swedish	å, ä, ö	

Virtually all of these accents can be created with English-language electronic publishing systems, word processors, or even typewriters. For specialized work in foreign languages, use a foreign-language keyboard.

6.28 Accents are usually omitted from all capital-letter renditions: CAFE not CAFÉ; ECOLE not ÉCOLE.

6.29 Languages such as Arabic, Russian, and Greek use alphabets other than roman. For example, Russian uses the Cyrillic alphabet and Greek uses the Greek alphabet.

6.30 Asian languages—the various forms of Chinese, Japanese, and Korean—are character-oriented. In such languages, one character usually represents a complete word or concept.

However, some of these languages use a bewildering array of character sets. Japanese, for instance, has three methods for representing language: Kanji, Katana, and Roman. All three of these character sets sometimes appear simultaneously in the same text. Chinese also uses multiple character sets. Thus, it is very difficult to establish a uniform method for representing these languages.

Civil and Social Factors

6.31 Civil regulations and social factors influence conventions for foreign names. Regulations include adopting a husband's family name on marriage, retaining a maiden name after marriage, and children carrying both father's and mother's family names. Most European languages other than English have explicit gender forms. In some languages, such as Russian, diminutive name forms are very common.

Prefixes in Foreign Names

6.32 Prefixes are a very common feature among western European family names. These prefixes have the general meaning of *from* or *of* and are used to indicate genealogy:

Ian son of Donald, hence, Ian MacDonald

6.33 The following exhibit summarizes the more common prefixes. Capitalization of prefixes is variable. Although lowercase is common, prefixes may be capitalized:

Exhibit 6–2:
Common West European Family Name Prefixes

Country	Common Family Name Prefixes
Dutch, Flemish	de, den, der, in't, ten, ter, van, van't
French	de, de la, de l', des, du
German	van, vom, von, vor
Irish	fitz (derived from Norman names), o'
Italian	del, della, di
Portuguese	da, de
Scottish	mac, mc
Spanish	de, de la, de las, de los, y
Welsh	ab, ap

Note that Scandinavian languages (Danish, Norwegian, and Swedish) use the suffix *son* or *sen:*

Sven son of Johann, hence, Sven Johannson

Foreign Names and Forms of Address

6.34 Titles and forms of address vary from language to language.

Arabic Names

6.35 Arabic is the language of numerous Middle Eastern countries including Algeria, Morocco, Egypt, Iran, Iraq, Lebanon, Jordan, Syria, and the Yemen.

6.36 Except for works such as linguistics studies or syntactic analyses, modifiers such as ^ and ′ may be omitted when Arabic names are transliterated into English. Retain punctuation such as - and ’.

6.37 Many Arabic family names include the word for *the* as a prefix, which is *l-*. To make this word pronounceable, it must be preceded by a vowel, generally *a.* Dialectic variations (such as Egyptian Arabic) have introduced *e* and *i.* Thus the possible forms are: l-, al-, el-, il-. Other prefixes include ibn-, abu-, abd-.

All prefixes are combined with the following parts of the family name by a hyphen and are regarded as part of the name. Examples of full Arabic names are

ali ibn-al-Abbas al-Majusi, abu-Ali al-Husayn, Fahima al-Hadan, Hunayn ibn-Ishaq, Loftia Labib ach-Shahed, Yuhanna ibn-Masawayh, Aisha abdel-Rahman, Ali ibn-Sahl Rabban al-Tabari, Ali al-Tabari

Shortened forms of these names are

abu-Ali al-Husayn, ibn-Ishaq, ibn-al-Abbas al-Majusi, ibn-Masawayh, ibn-Sahl Rabban al-Tabari, al-Tabari

Shortened forms are often used for the many important figures in Arab art, literature, science, mathematics, and astronomy.

Variations in spelling are also common. For instance, Husayn may be seen as Hussein or Ghossein. All these forms are phonetically equivalent.

Asian Names

6.38 Romanization of Asian names is based on a phonetic equivalence of characters in the native language to a string of letters that form a name in the roman alphabet.

6.39 The major differences between English personal names and Asian personal names is the order of those names. In Chinese cultures, the family name precedes the given name or names and each part of the name is capitalized. Examples of Asian names are

> Chiang Ch'ing, Yang K'asi-Hui, Ho Chi Minh (originally Nguyen Tat Thanh), Mao Zedong

6.40 Many Asian immigrants to North America adapt their names to the English system. Thus names such as Wong, Fong, Chin, and Cheung, which were originally given names, are used as family names. Examples of such names are

> Tze Wah Cheung, Miau Yin Chin, Fook Sun Chow, My Le Dung

French Names

6.41 French names are based on the roman alphabet. The accented letters are é, è, ç, and ô. Retain these accents. Most electronic publishing systems do provide them. Examples of common French names are

> Leo Delibes, César Franck, Vincent d'Indy, Adrienne Lecouvreur, Pierre de Chamblain de Marivaux

The short form of these names, quoting the family name only, is

> Delibes, Franck, d'Indy, Lecouvreur, de Chamblain de Marivaux

French-Canadian family names in common use include

> Belisle, Blais, Dargenais, Delisle, Demers, Denis, Dennis, Desbiens, Deschamps, Deschenes, Desjardins, Desmarais, Desrosiers, Dubé, Dubois, Ladouceur, Larmour, Lavoie, Leblanc, Leclerc, Seguin, St. Laurent

Many of these names exist in several forms:

> Desrosiers, DesRosiers, Des Rosiers

French first names are also commonly double. Hyphenate such names and treat them as one name:

> Jean-Marie, Jean-Paul, Jean-Pierre, Marie-Claire, Marie-Parisse

Germanic Names

6.42 The Germanic languages include Dutch (the language of the Netherlands), Flemish (one of the languages of Belgium), and German (spoken in Austria, Germany, and Switzerland).

6.43 In the Netherlands and Belgium, many family names have prefixes. The prefix may be attached to part of the family name or may be separate. When the prefix is separate, invariably it is not capitalized:

> Jan de Boer, Wim DeBoer, Karla den Hartog, Willy in 't Venn, Andrea ter Elst, Lies ten Broek, Vincent van Gogh, Anje van der Voort

6.44 In the Netherlands, married women append their maiden name to their husband's family name. Hyphenation is used:

> Lies van Arnhem-ter Elst, Anje in 't Veld-van 't Loo

In Belgium, married women retain their maiden name. Thus, *Marijke van Damm* remains *Marijke van Damm* after marriage.

In both Belgium and the Netherlands, children take their father's family name.

6.45 Many German names include the prefix *von*. This prefix is not usually capitalized, though it should be at the beginning of sentences:

> The problem, von Essen said, was to provide adequate propulsion.
>
> Von Essen said that the problem was to provide adequate propulsion.

As required by particular gender forms, *von* changes to *vom* or *vor*, though the former is rare in names. Common German names with the prefix *von* include

> Rudi von Essen, Klaus von Finkelstein, Adolph von Gronau, Hermann von Karajan, Andreas Vorstandlechner

A notable exception is Ludwig van Beethoven.

The prefix *van* is generally associated with the Low Countries (the Netherlands and Belgium).

If the umlaut modifier (ä, ö, or ü) is not available, a widely accepted convention is to add the letter *e* after the modified letter:

Name	**May be written as**
Grönlander	Groenlander
Gässler	Gaessler
Müller	Mueller

Another common convention is that the Gothic ß may be written as *ss:* Meßner may be written as Messner.

Italian Names

6.46 Italian is spoken in Italy and also in the southern areas of Switzerland. Italian names, which have no accented letters, are based on the roman alphabet. Common Italian names include

> Arturo DelBosco, Udo del Fabro, Enzo Delfino, Alicia Della Bianca, Maria Della Penta, Gina Dellavalle, Victoria Di Franco, Joe DiMaggio, Karla Dimillio, Guiseppe Ferrante

Portuguese Names

6.47 Portuguese names are based on the orthography of the roman alphabet. The accented letters are ã, ñ, á, é, í, ó, ç, and ê. These accents may be omitted in an English context without serious loss of meaning. However, many electronic publishing systems do provide these accents. Examples of Portuguese names include

> Maria Helena Vierra da Silva, João Babtista de Almeida Garrett, Vasco da Gama, Francisca de Sá de Miranda, Antonia Salazar, Dominga de Sequiera, Manuel Caetano de Sousa, Diogo do Couto

Russian Names

6.48 The Cyrillic alphabet is the formal script of the Russian language. There is a well-established system of transliteration from the Cyrillic alphabet to the Roman.

6.49 In the Russian language, family names may have some gender dependency. Male members of a family may have a different ending on their family name than females. For example:

> Igor Pavlov's sister would be called Irena Pavlova.

6.50 Russian middle names are referred to as patronyms. Children take these patronyms from their father's ancestral name. Again, there are male and female forms of patronyms:

> Ivan Ilyich Romanov's sister would be called Olga Ilyinichna Romanov.

6.51 The diminutive form in Russian is equivalent to the familiar form in English. Common diminutives are

Diminutive	Full Name
Petya	Peter
Natasha	Nataly
Nikolushka	Nikolay

Examples of well-known Russian names, with patronyms in parentheses, are

Anton (Pavlovich) Chekov, Sofya (Vladimiroyna) Giatsintova, Galiya (Bayazetova) Izmaylova, Dmitri (Ivanovich) Mendeleyev, Ivan (Petrovich) Pavlov, Sofya (Petrovna) Preobrazhenskaya, Faina (Grigorevna) Ranevskaya, Valentina (Ksenofontovna) Rastvorova, Leo (Nikolayevitch) Tolstoy, Vladimir (Ilyich) Ulyanov

6.52 Technical and scientific publications do not include patronyms or diminutives in Russian names.

Spanish Names

6.53 Spanish names are based on the roman alphabet. The accented letters are á, é, í, ó, ú, ñ, and ü. Accents may be omitted in an English context with little loss of meaning. However, many electronic publishing systems do provide these accents. Examples of common Spanish names include

Juanita del Enciña, Juan de Fuca, Bartolomé de las Casas

Short forms for these names (family name only) are

del Enciña, de Fuca, de las Casas

6.54 Some Spanish family names are formed from the father's family name and the mother's family name. In such cases the father's family name precedes the mother's family name and the two names are joined by the word *y* (meaning *and*):

Jordi Lopez y Mendoza, Juanita Sanchez y de Falla

GEOGRAPHIC NAMES

States

6.55 Except when standing alone, when they should be spelled out, all names of states, territories, and possessions of the United States should be abbreviated with a recognized U.S. Postal Service standard abbreviation. Note there is no period at the end of the abbreviation:

Alabama	AL	Montana	MT
Alaska	AK	Nebraska	NE
American Samoa	AS	Nevada	NV
Arizona	AZ	New Hampshire	NH
Arkansas	AR	New Jersey	NJ
California	CA	New Mexico	NM
Canal Zone	CZ	New York	NY
Colorado	CO	N. Carolina	NC
Connecticut	CT	N. Dakota	ND
Delaware	DW	Ohio	OH
Florida	FL	Oklahoma	OK
Georgia	GA	Oregon	OR
Guam	GU	Pennsylvania	PA
Hawaii	HI	Puerto Rico	PR
Idaho	IH	Rhode Island	RI
Illinois	IL	S. Carolina	SC
Indiana	IN	S. Dakota	SD
Iowa	IA	Tennessee	TN
Kansas	KS	Texas	TX
Kentucky	KY	Utah	UT
Louisiana	LA	Vermont	VT
Maine	ME	Virginia	VA
Maryland	MD	Virgin Islands	VI
Massachusetts	MA	Washington	WA
Michigan	MI	W. Virginia	WV
Minnesota	MN	Wisconsin	WI
Mississippi	MS	Wyoming	WY
Missouri	MO		

District of Columbia (DC) is a federal district.

The following abbreviations, however, are the forms occasionally still found in some print media. Sometimes no abbreviated form exists. Note the period at the end of each abbreviation:

Alabama	Ala.	Montana	Mont.
Alaska		Nebraska	Neb.
American Samoa	Amer. Samoa	Nevada	Nev.
Arizona	Ariz.	New Hampshire	N.H.

Arkansas	Ark.	New Jersey	N.J.
California	Calif.	New Mexico	N.M.
Canal Zone	C.Z.	New York	N.Y.
Colorado	Colo.	North Carolina	N.C.
Connecticut	Conn.	North Dakota	N.D.
Delaware	Del.	Ohio	
Florida	Fla.	Oklahoma	Okla.
Georgia	Ga.	Oregon	Ore.
Guam		Pennsylvania	Pa.
Hawaii		Puerto Rico	P.R.
Idaho		Rhode Island	R.I.
Illinois	Ill.	South Carolina	S.C.
Indiana	Ind.	South Dakota	S.D.
Iowa		Tennessee	Tenn.
Kansas	Kan.	Texas	Tex.
Kentucky	Ky.	Utah	
Louisiana	La.	Vermont	Vt.
Maine		Virginia	Va.
Maryland	Md.	Virgin Islands	V.I.
Massachusetts	Mass.	Washington	Wash.
Michigan	Mich.	West Virginia	W.Va.
Minnesota	Minn.	Wisconsin	Wis.
Mississippi	Miss.	Wyoming	Wyo.
Missouri	Mo.		

District of Columbia (D.C.) is a federal district.

6.56 Do not use quotation marks with the popular names for states or groups of states:

> the Carolinas, the Hoosier state, the Green Mountain state, the New England states, New England

Canadian Provinces

6.57 Use all caps for the recognized standard abbreviations for all Canadian provinces:

Alberta	AB	Nova Scotia	NS
British Columbia	BC	Ontario	ON
Manitoba	MB	Prince Edward Island	PEI
New Brunswick	NB	Quebec	QC
Newfoundland	NF	Yukon Territory	YK
North West Territories	NWT		

6.58 The unofficial system is compatible with the US system of abbreviations for state names. The traditional Canadian system is

Alberta	Alta.	Nova Scotia	N.S.
British Columbia	B.C.	Ontario	Ont.
Manitoba	Man.	Prince Edward Island	P.E.I.
New Brunswick	N.B.	Quebec	P.Q.
Newfoundland	Nfld.	Yukon Territory	Yuk.
North West Territories	N.W.T.		

Cities and Towns

6.59 Capitalize the names of all cities and towns:

Denver, Moose Jaw, Palo Alto, San Francisco, Toronto

Except those names containing a foreign-language genitive:

Marina del Rey, Val d'Or

Some cities may have the word *city* as part of their name:

Jersey City, Kansas City

Most cities having the word *Saint* in their name use the abbreviated form: St. Johns, St. Louis. But: Saint John (New Brunswick).

Hyphenate the names of twin cities:

Dallas–Fort Worth, Minneapolis–St. Paul

Capitalize popular names for towns and cities:

the Big Apple, the City of Brotherly Love, the Windy City

Abbreviations such as L.A. for Los Angeles are strictly colloquial and have no place in formal writing.

6.60 Append the name of the state or province to the name of a town or city to avoid any confusion:

Paris, Texas (as distinct from Paris, France)
London, Ontario (as distinct from London, England)

Countries

6.61 Capitalize the names of countries and political unions and spell them out in the text:

Canada, the United States, the United States of America, Russia, the Commonwealth of Independent States

Abbreviate the names of countries and political unions based on dictionary preference:

France	Fr.
Germany	Ger.
Sweden	Swed.
United Kingdom	U.K.

Among the European countries note

United Kingdom	Formal name of the British Isles (United Kingdom of Great Britain and Northern Ireland). Great Britain includes England, Wales, and Scotland.
the Netherlands	Not Holland; Holland is a province within the Netherlands.
Scandinavia	Collective term for the countries of Denmark, Norway, Sweden, and Finland.

6.62 It is not uncommon for countries in Africa and Asia to have changed from colonial form to meaningful ethnic names:

Rhodesia became Zimbabwe, Ceylon became Sri Lanka

6.63 The term *European Economic Community* (*EEC*), commonly known as the *Common Market,* refers to a number of countries in Europe that have an economic and trading union. The term *Benelux,* or *Benelux countries,* refers to the union of Belgium, the Netherlands, and Luxembourg.

Islands

6.64 Capitalize the name of islands:

Ellis Island, Prince Edward Island

Inverted form is sometimes used in singular and plural form:

Isle of Man, Isles of Sicily

The abbreviated form of *island* is I. or Is.:

Manitou I., Ellesmere Is.

Mountains

6.65 Names for individual mountains take the form

Mount Everest, Mount Washington

Mountain can also be abbreviated Mt.:

Mt. Everest, Mt. Washington

Inverted form is also possible:

Table Mountain

Some mountains do not contain the word *Mount* or *Mountain* in their name:

the Eiger, the Matterhorn

Still other mountains do not have the definite article or the word *Mount* in their name:

Annapurna, Ben Nevis, Caedr Idris, Mam Tor, Mont Blanc, Snowdon, White Nancy

Names of mountain ranges are always plural in form:

the Andes, the Himalayas, the Rockies

Generic terms for mountains in the plural form take a lowercase mountains:

Berkshire and Catskill mountains

Bodies of Water

6.66 Capitalize the names of oceans, seas, lakes, bays, gulfs, channels, canals, and rivers:

the Atlantic Ocean, the Black Sea, Botany Bay, the Caspian Sea, Chesapeake Bay, the English Channel, the Gulf of Bothnia, the Gulf of Mexico, the Indian Ocean, Lake Huron, the Ottawa River, the Panama Canal, Pickerel Lake, the River Rhine

Certain bodies of water are well known enough for the common noun such as *ocean* or *sea* to be omitted. This is acceptable in formal writing: the Atlantic, the Mediterranean, the Mississippi.

In other cases, the proper noun may be omitted. *The Channel* is known from context to refer to the *English Channel.* If the context is not well established, use the full term.

Note that rivers that join have a hyphenated name:

Mississippi-Missouri

Note the inclusive form: Lakes Huron and Michigan.

Generic terms for bodies of water in the plural form take a lowercase:

along the Potomac and Patuxent rivers
along the shores of the Atlantic and Pacific oceans

ORGANIZATION NAMES

6.67 Since they are proper names, capitalize all organization names. When present, prepositions take lowercase.

Government and Legislative Names

6.68 Capitalize common names for departments and offices when they form part of the name of a government or legislative agency. Where acronyms or abbreviations exist, periods are omitted. Examples of government and legislative names include

> Congress, Department of Defense (DOD), Department of External Affairs, Department of National Defence (DND [Canadian]), the Foreign Office (British), the House of Commons, the Commons (colloquial), the House, this House (in direct speech), House of Representatives, Parliament, the Senate.

> Canadian Security and Intelligence Service (CSIS), Central Intelligence Agency (CIA), Federal Bureau of Investigation (FBI), MI5 (Military Intelligence 5, British). Note for the last term the full form is seldom used.

> United Nations (UN), United Nations Educational and Scientific Council (UNESCO), World Health Organization (WHO).

6.69 Normally, include the full name of a department, office, or corporation within the text and add the abbreviation or acronym immediately after that citation in parentheses:

> The National Aeronautic and Space Administration (NASA) is the agency primarily responsible for space exploration.

After this reference, the abbreviated term can be used routinely unless a considerable amount of text occurs between references—typically, more than a chapter. In that case, repeat the full citation of the abbreviated term as a reminder for the reader.

Judicial Bodies

6.70 The following are common names for different types of court. None of these terms is capitalized when used in a generic sense:

family court, juvenile court, military court, night court, provincial court, traffic court

6.71 The following are specific unique courts, hence names are capitalized:

Alberta Provincial Court, the Court of Queen's Bench, Munroe County Court, the Supreme Court, the Supreme Court of Canada, United States Supreme Court

Administrative Organizations

6.72 Names of administrative organizations include

the Census Bureau, the National Labour Relations Board, Revenue Canada, the State Department, Statistics Canada (StatsCan), United States Department of the Interior

Commercial Organizations

6.73 All company names, from large corporations to small independent consultants, are registered with some level of government agency.

6.74 The full official form of company names may include the legal status of the company, such as *Incorporated* (*Inc.*) or *Limited* (*Ltd.*). Generally, company names are capitalized. However, note that many companies may use lowercase, hyphenation, parentheses, and other devices for legal or marketing reasons.

6.75 Many companies also have a registered abbreviated form of their company name:

AT&T™	American Telephone and Telegraph
DEC™	Digital Equipment Company
IBM™	International Business Machines Corporation
nt™	Northern Telecom

6.76 Beginning a sentence with an acronym is permissible:

IBM introduced its newest mainframe computer today in New York City.

However, some style guides, notably the American Medical Association, do not allow acronyms. The following sentence is unacceptable:

TSS has been identified in four commercial products.

6.77 Many companies have separate operating companies or divisions. These will be apparent in the name:

AT&T Technologies
Bell South Enterprises
Bell South Media Technologies

6.78 Indicate legal status after the company name:

Northern Telecom, Inc.
Northern Telecom Ltd.

Note particularly the placement of the comma with the designation *Inc.*

6.79 Common legal designations for companies in European countries include the following, most of which have the meaning of Limited or Incorporated:

Country	Incorporated Symbol
Belgium, Netherlands	NV and BV
Belgium, France and Switzerland	SA
Britain	plc (note lowercase)
Germany	GmbH (note mixed case)
Scandinavia	AB

For example:

Herman Gaessler GmbH
Plessey plc

The definitive reference source for company names is literature issued by the company. Caution may be necessary, as it is not uncommon for a company to have an inconsistent corporate image. Standard trade directories are also a good reference source. Stock exchange listings should not be used as reference since companies are identified by marketing codes that have no legal status.

Academic Organizations

6.80 Capitalize the names of universities, colleges, and schools:

City University, King's College, London University, Merchant Tailors' School, Northeastern University, Thomas Aleyn's Grammar School, Trinity College, University of London

Many universities are named after towns or cities. These names can often be inverted: Toronto University or the University of Toronto.

Where universities are named for individuals, these names cannot be inverted: Simon Fraser University, Yale University.

Certain universities are often well known by their abbreviated or mnemonic form:

MIT	Massachusetts Institute of Technology
SMU	Southern Methodist University

Such abbreviated forms should not be used without an initial definition no matter how well known the institution may be.

It is acceptable practice to refer to extremely well known universities in their simple form: Harvard, Yale, Princeton.

Where a university has campuses in different towns or cities, indicate this as

University of California at Los Angeles (UCLA)
University of California at San Diego (UCSD)
University of Texas at Arlington (UTA)

6.81 Typical terms for different departments or functions within a college or university are

the Faculty of Engineering, the Engineering Faculty, the School of Journalism, the Sloane School of Business Management

Societies and Associations

6.82 Names for societies and associations take the following forms. Many of these organizations also have abbreviated forms of their name:

Boy Scouts of America, the British Medical Association (BMA), the Institute of Electrical and Electronic Engineers (IEEE) (often shown incorrectly as IE^3), the National Geographic Society, the National Union of Broadcast Journalists Local 626, the Royal Academy, the Royal Society, the Society for Technical Communication (STC)

Note that a society may not use the term *Royal* unless it has received Royal charter.

Religious Organizations

6.83 The list of religious organizations, churches, sects, and persuasions is extensive:

the Anglican Church, the Catholic Church, the Church of Rome, the Church of Jesus Christ of Latter-day Saints, the Church of Scientology, the Society of Friends, the Unitarian Universalist Church, the United Church

Colloquial names such as *Mormons* and *Moonies* have no place in formal writing.

The Amish, the Mennonites, and *the Shakers* are the correct formal names for well-known religious communities in North America.

TITLES OF WORKS

Publication Titles

Publication Titles Cited in Texts

(See also Chapters 9 and 10.)

6.84 In text, for proper discrimination, capitalize publication titles and set them in italics; quotes are not used:

> Fowler's *Modern English Usage* is the standard reference . . .
>
> The standard reference work *Modern English Usage* by Sir Henry Fowler is . . .

6.85 In the absence of italics, the title may be set in roman (normal) text and underscored:

> Fowler's Modern English Usage is the standard reference . . .

6.86 Where the title of a publication contains an article (*a* or *the*), for fluency in text the article is omitted. In the following example, the full title of the publication is *A Digital Computer Primer:*

> In his *Digital Computer Primer,* John Martin explains the evolution of the . . .

In the titles of periodicals, the definite article (*The*) is omitted: *Journal of Applied Psychology.*

Publication Titles Cited in Bibliographies

6.87 References to publications in bibliographies are much more formalized. Sources must be identified accurately. Many publishing groups have a house-style for bibliographies. (See also Chapter 10.)

6.88 In the absence of a house style, set the author's name in roman text and the title of the work in italic:

Bradley, F. J. *Analytic Methods.* Cambridge, MA: MIT Press, 1987.

In the absence of italics, the title may be set in roman text and underscored:

Bradley, F. J. <u>Analytic Methods.</u> Cambridge, MA: MIT Press, 1987.

6.89 If the work involved is an article in a journal, set the article title in roman within quote marks and the journal title in italics:

Allen, C. E. "Aspects of polymorphism in mammals." *Journal of Genetics* 15 (4):150–158.

In the absence of italics, the journal title may be set in roman text and underscored:

Allen, C. E. "Aspects of polymorphism in mammals." <u>Journal of Genetics</u> 15 (4):150–158.

6.90 Most journals have a standard abbreviation. (See also 10.16–.19.)

Audio and Video Titles

6.91 Audio and video materials are published information recorded on magnetic media. Such materials can include promotional, training, tutorial (*Macintosh Guided Tour*), instructional, procedural, diagnostic, and creative.

Style rules for these materials follow those for printed publications. Examples of the naming of such materials are

Getting Started with the Meta II Personal Computer, Meta Computers, Palo Alto, CA. (audiovisual tape)

Matrix Algebra for Electronic Circuit Analysis, McMaster, F. E., Cottage Publishing, Flower Station, Ontario, Canada. (video instruction tape)

Yesterday's Spy, Brown, G., Centipede Books, London, England. (audio recreational tape)

Artistic Work Titles

Plays, Poems, and Films

6.92 The titling conventions for plays, poems, and films (motion pictures or movies) are identical. Plays may be performed on stage (in the theater) or broadcast on radio or television.

Set the title of the work in italic. Set the remainder of the citation, such as author or producer, in roman. Examples of such titles include

> Shakespeare's *Midsummer Night's Dream* is always the opening play at the . . .
>
> In Susan Seidelman's *Desperately Seeking Susan,* the role of . . .

6.93 Where the title of the work contains the definite or indefinite article, this is omitted when the citation occurs in line: Brendan Behan's *Quare Fellow* evokes. . . .

Musical Compositions

6.94 There are a variety of forms of musical composition both instrumental and vocal, all of which have formal names. Many of these terms have obvious Italian origins. Some of these names refer to a part of a composition. The names include

> opera, operetta, concerto, overture, allegro, anthem, Agnus Dei, mass, requiem, canon, cantata, capriccio, carol, canticle, chorale, prelude, aria, fugue, nocturne, largo, oratorio, sonata, symphony, suite

6.95 Certain other works may be named after dance forms. These include fandango, pavan, polonaise, tarantella, and waltz.

6.96 Titles for complete musical works are set in italics. Prepositions, articles, and conjunctions in English titles are not capitalized:

> *Pomp and Circumstance, Peter and the Wolf, Enigma Variations, Variations on a Theme by Paganini*

6.97 Many titles are cited in their original language:

> *Eine Kleine Nachtmusik, Die Fledermaus, Cappriccio stravagante, Cavalleria Rusticana, Les Sylphides*

6.98 Titles for masses are invariably in Latin.

6.99 The composer's family name always appears in roman text:

> Elgar's *Pomp and Circumstance* provides a rousing . . .

6.100 With such a vast range of compositions, many do not have unique descriptive titles. Such titles, set in roman, include the form of composition followed by the musical key:

Fantasia and Fugue in G minor for Two Violins, Sonata in E-flat, Minuet in G, Impromptu in C-sharp minor

Note that the special characters for *sharp* and *flat* are not used, since they would require a special musical typeface.

6.101 The term *opus,* a Latin word meaning *work,* refers to specific works of a composer. Works are identified as

op. 25—meaning opus 25
Dvorak's *Noonday Witch* op. 108

Often several similar short pieces may be grouped in one work. These are identified by

op. 25, no. 1
op. 25, no. 2
Haydn's op. 76, no. 3 is a sparkling . . .

6.102 Append catalog references (such as K for Köchel) directly to the title of the work. Köchel references are based on a cataloging system created by Ludwig von Köchel to classify Mozart's work:

Mozart's *Magic Flute* K. 286

Graphic Arts, Sculpture, and Other Art Forms

6.103 Set titles for works such as paintings, drawings, sketches, cartoons (in the traditional sense), sculptures, and any other art form in italic. Set the artist's name in roman:

Rembrandt's *Night Watch*
The Thinker by Rodin
Rain, Steam and Speed by Turner

MILITARY TERMINOLOGY

6.104 Clear communication is of the utmost importance in the military. Terminology is therefore carefully controlled and frequently revised.

Military Standards Related to Terminology

6.105 A number of United States military standards govern terminology. These include

MIL-STD-109B	Quality Assurance Terms and Definitions
MIL-STD-188-120	Military Communication Systems Standards—Terms and Definitions
MIL-STD-280A	Definitions of Item Levels, Item Exchangeability Models and Related Terms
MIL-STD-721C	Definition of Terms for Reliability and Maintainability
MIL-STD-429C	Printed Wiring and Printed Circuit Terms and Definitions
MIL-STD-1165	Glossary of Environmental Terms (Terrestrial)
MIL-STD-1241A	Optical Terms and Definitions

6.106 The U.S. Department of Defense issues military standards and indexes them in *The Department of Defense Index of Specifications and Standards, Part I and II.*

Organizational Units for the Military

6.107 Always capitalize the names of military organizational units. Most of the terms also have a common abbreviated form that is capitalized:

US Armed Forces	
US Air Force	USAF
US Army	US Army
US Navy	US Navy
US Marine Corps	US Marine
Canadian Armed Forces	
Royal Canadian Air Force	RCAF
Canadian Army	
Royal Canadian Navy	
Royal Air Force (British)	RAF
Royal Navy (British)	RN
Fleet Air Arm (British)	

Use the adjectival form for the generic form for the armed forces of other countries:

the French Navy, the Indian Air Force, the Israeli Army

Also note that military with a mandate from the United Nations are referred to as UN Forces or United Nations Forces. It is not meaningful to refer to the UN Air Force or a similar unit.

6.108 Within the three main services the following organizational units exist: army, battalion, corps, detachment, division, fleet, regiment, and squadron.

When used as part of a proper name of a military unit, capitalize these terms. Numeric identifiers are also commonly used:

Afrika Korps
Army of the Rhine
8th Army or Eighth Army
84th Foot Regiment, Royal Highland Emigrants
501st Parachute Battalion
7th Panzer Division
U.S. Third Fleet

6.109 Some units are also identified by mnemonic forms:

4CMBG 4th Canadian Mechanized Brigade Group

Names for such units may be further abbreviated by using the cardinal form of number instead of the ordinal:

4 Canadian Mechanized Brigade Group

Ranks for the Military

6.110 The full terms and abbreviations for various military ranks in international armies are

Army		Navy		Air Force	
Field Marshall		Admiral of the Fleet		Marshall of the Air Force	
General	Gen.	Admiral	Adm.	Air Chief Marshall	
Lieutenant General	Lt. Gen.	Vice Admiral	V. Adm.	Air Marshall	
Major General	Maj. Gen.	Rear Admiral	R. Adm.	Air Vice Marshall	
Colonel	Col.	Captain	Capt.	Group Captain	Gr. Cpt.
Lieutenant Colonel	Lt. Col.	Commander	Cmdr.	Wing Commander	
Major	Maj.	Lieutenant Commander	Lt. Cmdr.	Squadron Leader	Sq. Ldr.
Captain	Capt.	Lieutenant	Lt.	Flight Lieutenant	Fl. Lt.

Army		Navy		Air Force
Lieutenant	Lt.	Sub-Lieutenant	Sub. Lt.	Flying Officer
Sergeant	Sgt.	Chief Petty Officer	Chief P.O.	
Corporal	Cpl.	Petty Officer	P.O.	
Lance Corporal	L. Cpl.	Able Seaman		

Military Awards

6.111 Capitalize the names of particular medals and awards. Many awards also have an abbreviation:

British Empire Medal	BEM
Commander of the Order of Military Merit	CMM
Congressional Medal of Honor	CMH
Distinguished Conduct Medal	DCM
Distinguished Flying Cross	DFC
Distinguished Flying Medal	DFM
Distinguished Service Cross	DSC
Distinguished Service Medal	DSM
George Cross	GC
Member of the Order of Military Merit	MMM
Military Cross	MC
Military Medal	MM
Officer of the Order of Military Merit	OMM
Purple Heart	
Victoria Cross	VC

6.112 Recognize Canadian and British awards by appending the abbreviation, without periods, for the award after the holder's name. Where a bar is awarded, to indicate a second instance of the same award, this is also shown in the abbreviation:

Lt. John Davidson CMM
Capt. W.E. Johns DFC and Bar
Jack Cornwall VC

Wars and Campaigns

6.113 Use the general form for the names of wars and campaigns:

WW1	World War I	First World War
WW2	World War II	Second World War

And:

American Civil War, Arab-Israeli War, Battle of Arnheim, Battle of Britain, the Campaign of the Peninsular War, 1815 campaign, Franco-Prussian War, Gulf War, Italian campaign, Korean War, Russian campaign, Six-Day War, Vietnam War

And:

European Theater, Pacific Theater

Use single quotes for code names:

Operation 'Anvil,' Operation 'Market Garden'

Naval Vessels

6.114 Warships are the only military equipment to have individual formal names. Note that the names are italicized, but the designations such as USS are set in roman:

USS	United States Ship	USS *Cuttlefish*
HMS	His/Her Majesty's Ship/ Submarine	HMS *Ark Royal*
HMCS	His/Her Majesty's Canadian Ship	HMCS *Endeavour*

Use lowercase for the common types of vessels:

aircraft carrier, battleship, cruiser, destroyer, frigate

Indicate vessel classes as

Ticonderoga-class cruiser
Tribal-class destroyer
Yankee-class submarine

Armament

6.115 Types of armament include

Armament	Abbrev.
anti-ballistic missile	ABM
air-to-surface missile	ASM
air-to-air missile	AAM
intermediate-range ballistic missile	IRBM
medium-range ballistic missile	MRBM
multiple re-entry vehicle	MRV
multiple independent re-entry vehicle	MIRV

Armament	Abbrev.
submarine-launched ballistic missile	SLBM
surface-to-air missile	SAM

Armament type names in use include

Minuteman, Polaris, Snark, Titan 2

Type identifiers are alphanumeric and include

SSN6, SS13, UGM

Aircraft Terminology

6.116 Aircraft designators include

F-111, FB-11A, CF-18 (C indicates Canadian), B-52D, TU20 Vulcan, MiG21 (note lowercase *i*), Mirage F2, Wasp helicopter

Note plural form with no apostrophe and lowercase *s:*

F-111s, CF-18s

Allied Groups Terminology

6.117 Military terminology uses many mnemonics and acronyms. Mnemonics, by definition, should be memorable. They need not be pronounceable, but well-formulated mnemonics are pronounceable and hence more easily remembered. In the following list, only NATO, SEATO, SHAPE, and ASEAN are true acronyms (initialisms), where the abbreviated form comes from the initial letters of the term. (See also Chapter 5.)

Association of South-East Asian Nations	ASEAN
Canadian Air-Sea Transportable force	CAST
Commander-in-Chief Channel (English Channel)	CINCHAN
Maritime reconnaissance command	MARAIRMED
NATO Air Defence Ground Environment	NADGE
North American Aerospace Defence	NORAD
North Atlantic Treaty Organization	NATO
South East Asian Treaty Organization	SEATO
Standing Naval Force Atlantic	STANAVFORLANT
Supreme Allied Commander Atlantic	SACLANT
Supreme Allied Commander Europe	SACEUR

Terms such as *NATO* may be seen in the form *Nato;* this is incorrect.

Surveillance Terms

6.118 Some of the terminology for surveillance equipment is well established, even in lay language. For example, *radar* is a prime example of an acronym term that has become a common noun. Generic terms for surveillance equipment are

radar	radio detection and ranging system
sonar	sound navigation ranging system

The term radar may be qualified as

- Airborne radar,
- Land-based radar,
- Over-the-horizon back scatter radar,
- Shipborne radar, and
- Space-based radar.

Examples of complete radar systems and installations include

AWACS	Airborne Warning And Control System aircraft
DEW line	Distant Early Warning (line of radar stations)

HEALTH AND MEDICAL TERMS

6.119 The medical profession uses Latin and Latin-derived terms extensively. Because the terminology is standard for the discipline, the terms are generally set in lowercase roman.

Examples of typical health and medical terms are tibia, thorax, appendectomy, tracheotomy, pulmonary vein, and gluteus maximus.

Plurals are formed according to the rules of Latin grammar:

vertebra	vertebrae

6.120 Some medical vocabulary—such as heart, eye, ear, lung, kidney, liver, and brain—are everyday terminology.

6.121 Some Latin terms should be set in italics to avoid confusion with the word *in;* however, base any italic decision on the dictionary treatment of the term. *In articulo mortis* is italicized, but in vitro, in vivo, and in utero are not.

Anatomical Terms

6.122 Set anatomy terms in lowercase roman unless they include a proper name: thyroid, pancreas, islets of Langerhans, aorta, ventricle, Achilles tendon.

Diseases and Conditions

6.123 Set terms for diseases and medical conditions in lowercase roman unless they include a proper name: axanthopsia, anemia, apoplexy, dyslexia, rheumatic fever, Alzheimer's disease.

Drugs

6.124 Use generic names for medical drugs and set them in lowercase roman. Examples of generic names include aspirin, digitalis, tetracycline, diazepam.

6.125 If proprietary drugs must be referred to in print, capitalize the trade or brand name:

> In North America Valium is probably one of the most widely prescribed drugs in the treatment of anxiety disorders.

Infectious Organisms

6.126 Capitalize and italicize names for infectious organisms (such as bacteria): *Bacillus, Streptococci, Staphylococci.*

When using the name as an adjective, set it in lowercase roman:

> *Staphylococci* causes acute staphylococcal enterocolitis, an infection of the gastrointestinal tract.

6.127 Abbreviate members of a genus (see also Chapters 3 and 5):

> *Escherichia coli* becomes *E. coli* or *Esch. coli.*
> *Staphylococcus epidermidis* becomes *S. epidermidis.*

Use the abbreviated term in most text:

> Most *E. coli* strains ferment lactose readily.

Medical Equipment

6.128 Set terms for common medical equipment such as stethoscope and centrifuge in lowercase roman. As with most electronic equipment, terminology for advanced medical equipment involves many initialisms:

CT scan	computerized tomography (phonetically corrupted to CAT scan)
EEG	electroencephalogram

Medical Procedures and Treatments

6.129 Set terms for medical procedures and treatments in lowercase roman unless they include a proper name: chemotherapy, appendectomy, Coombs' test, Credé's method.

Vitamins and Minerals

6.130 Refer to vitamins by their name or letter convention. Names are lowercase; letter designations are uppercase. Distinguish the vitamin B complex by numerical suffixes. In typeset text, set the numerical suffix as a subscript: vitamin B_6. If subscripting is unavailable, the number may be set on line: vitamin B6.

Vitamin		Vitamin	
retinol	A	complex	B
thiamine	B_1	ascorbic acid	C
riboflavin	B_2	calciferol	D
niacin	B_3	riboflavin	G
pyridoxine	B_6	biotin	H
cobalamin	B_{12}	phylloquinone	K_1
folic acid	B_c		

6.131 Set the names of minerals and trace elements in lowercase roman: calcium, phosphorus, potassium, iron.

SCIENTIFIC TERMS AND SYMBOLS

6.132 Scientific terminology is well established because of a long history of scientific discovery and a tradition of published works. Some areas of scientific terminology are closely related to symbols and notations.

6.133 Science requires well-defined symbols to convey precise information. These symbols, often only a single character, can be grouped together in specific ways to describe a complex idea or operation. (See also Chapter 7.)

6.134 In chemical notation, depending on the amount of information needed about an element, the symbol for the element could conceivably carry an attached notation at all four corners:

1. Leading superscript: ^{1}H—mass number of an atom
2. Leading subscript: $_{6}C$—the element's atomic number
3. Trailing superscript with three distinctive indicators: ionic charge (Na^{+}), excited electronic state (He^{*}), and oxidation number (Pb^{II}), and/or
4. Trailing subscript: H_2—number of atoms in the molecule.

Chemical notation has a compact symbology and well-defined conventions. Since conventions change, follow the current ones. For instance, mass number once appeared at the upper right of the element symbol: U^{235} represented the fissile isotope of uranium-235. It is now written as ^{235}U. (See Chapter 7.)

6.135 Indicate the mass number of an atom by a leading superscript:

^{1}H stands for mass-1 hydrogen (one proton in its nucleus)

Mass-2 hydrogen, with one proton and one neutron in its nucleus, also exists in nature, and it would be indicated by

^{2}H (although it is preferably labeled as D, for deuterium)

A third isotope of hydrogen, a mass-3 hydrogen atom (one proton and two neutrons in its nucleus), does not exist in nature but can be made in the laboratory; its designation is

^{3}H (although normally it is labeled T, for tritium)

6.136 Reserve the leading subscript position for the element's atomic number. For hydrogen, this number is 1 ($_{1}H$), which is usually omitted. The three hydrogen isotopes have the same atomic number (1, for the proton each has in its nucleus) but different mass numbers (they have 0, 1, or 2 neutrons in their nucleus). A few other examples of chemical element symbols with their atomic numbers are

Carbon ($_{6}C$),
Sulfur ($_{16}S$), and
Uranium ($_{92}U$).

6.137 Use the trailing superscript position (upper right) for any of three different indicators:

1. Ionic charge, indicated by plus or minus signs (Na^{+}, Ca^{2+}, O^{3-}),

2. Excited electronic state, indicated by an asterisk (He^*, NO^*), or
3. Oxidation number, indicated by a roman numeral (Pb^{II}, Pb^{IV}).

6.138 Attach the trailing subscript 2 to the H—H_2O—to indicate the number of hydrogen atoms in the molecule; the lack of a subscript on the O indicates there is only one oxygen atom.

STANDARD PHYSICS SYMBOLS

A	Area	*K*	Bulk modulus of a material
a	Attenuation coefficient	l	Wavelength
c	Speed of light in vacuum	*m*	Mass
g	Shear strain	*P*	Power
d	Diameter	*r*	Radius
d	Damping coefficient	*R*	Resistance (electrical)
E	Energy	r	Density (mass/unit volume)
e	Linear strain,	*t*	Time
	relative elongation	*T*	Temperature
F	Force	*V*	Voltage (electrical)
g	Acceleration due to gravity	W	Work
h	Height	w	Angular velocity
I	Current (electrical)	(*x*,*y*,*z*)	Space coordinates

Physics Terminology

Laws, Principles, and Theories

6.139 Since many laws, principles, theories, equations, constants, and hypotheses are named for their originators, capitalization and apostrophes are used:

- Bernoulli's Principle
- Heisenberg's Relative Uncertainty Principle
- Newton's Laws of Motion, Newton's First Law

6.140 Other laws and principles are named for the topic or concept to which they relate. The names are lowercase:

- The laws of thermodynamics
- The principle of conservation of energy
- The second law of thermodynamics

Phenomena

6.141 Phenomena are named for their discoverer:

Hall effect or van der Waal's force

Atomic and Nuclear Particles

6.142 In atomic and nuclear physics, identify types of radiation and particles with letters of the Greek alphabet:

α-particle radiation (alpha-particle radiation)
ß-radiation (beta radiation)

6.143 Set the names of atomic and nuclear particles in lowercase roman. Use correct Latin grammatical forms: atom, nucleus, nuclei (plural), nucleons (collective term for protons and neutrons), proton, neutron, electron, positron, neutrino, anti-neutrino.

6.144 Identify quantum numbers as *j, l, m,* and *n* and set them in italic. If italic is not available, they can be set in roman without loss of meaning:

The *l* quantum number determines the shape of the orbit.
The l quantum number determines the shape of the orbit.

Identify electron shells by the letters *k, l, m, n,* and *p*. (The letter *o* is omitted to prevent confusion with the digit *0*.) Set shell identifiers in italic:

There are only two available electron orbits in the *k* shell.

Do not hyphenate quantum number and electron shell identifiers.

Chemistry Terminology

Isotopes

6.145 The universal notation for indicating isotopes of an element is a numeric superscript *preceding* the element's symbol. Thus show isotopes of uranium and lead as ^{235}U, ^{208}Pb.

Elements

6.146 Set the full names of chemical elements in lowercase. Represent elements by their chemical symbols, such as C for carbon and Au for gold. The symbols are uppercase, or uppercase and lowercase roman.

6.147 Most element names are Latin. The more common elements—gold, silver, copper, tin, iron—are based on the vernacular. Note, however, that the symbols for these elements come from the Latin form of their names.

6.148 Chemical symbols are universally used in the equations of chemical reactions. They are never used in line in text.

6.149 The following table includes the full name of the chemical elements and their chemical symbols. Note that even though some of the elements are named for famous scientists or famous research institutes, the full names are not capitalized:

Symbol	Name	Symbol	Name
Ac	actinium	F	fluorine
Ag	silver	Fe	iron
Al	aluminum	Fm	fornium
Am	americanium	Fr	francium
Ar	argon	Ga	gallium
As	arsenic	Gd	gadolinium
At	astatine	Ge	germanium
Au	gold	H	hydrogen
B	boron	He	helium
Ba	barium	Hf	hafnium
Be	beryllium	Hg	mercury
Bi	bismuth	Ho	holmium
Bk	berkleyium	I	iodine
Br	bromine	In	indium
C	carbon	Ir	iridium
Ca	calcium	K	potassium
Cd	cadmium	Kr	krypton
Ce	cerium	La	lanthium
Cf	californium	Li	lithium
Cl	chlorine	Lu	lutecium
Cm	curium	Lw	lawrencium
Co	cobalt	Md	mendelevium
Cr	chromium	Mg	magnesium
Cs	cesium	Mn	manganese
Cu	copper	Mo	molybdenum
Dy	dysprosium	N	nitrogen
Er	erbium	Na	sodium
Es	einsteinium		
Eu	europium		

Symbol	Name	Symbol	Name
Nb	nobium	Sb	antimony
Nd	neodymium	Sc	scandium
Ne	neon	Se	selenium
Ni	nickel	Si	silicon
No	nobelium	Sm	samarium
Np	neptunium	Sn	tin
O	oxygen	Sr	strontium
Os	osmium	Ta	tantalum
P	phosphorus	Tb	terbium
Pa	prothactinium	Te	tellerium
Pb	lead	Th	thorium
Pd	palladium	Ti	titanium
Pm	promethium	Tl	tallium
Po	polonium	Tm	thulium
Pr	praesodymium	U	uranium
Pt	platinum	V	vanadium
Pu	plutonium	W	tungsten
Ra	radium	Xe	xenon
Rb	rubidium	Y	yttrium
Re	rhenium	Yb	ytterbium
Rh	rhodium	Zn	zinc
Rn	radon	Zr	zirconium
Ru	ruthenium		
S	sulfur		

Chemical Compounds

6.150 Form the names of chemical compounds from their combining elements. Lowercase each word in the term for the compound. Do not hyphenate these terms:

copper sulfate
ferrous oxide (not iron oxide)
ferric oxide

6.151 Use the prefixes *bi, di, tri, tetra,* and *penta* to indicate chemical combinations in units of two, three, four, and five:

bi, di	2
tri	3
tetra	4
penta	5

Other prefixes include *meta,* meaning *above* or *beyond.* There is no numeric value associated with *meta.*

6.152 Lowercase the common names of compounds:

sulfuric acid, hydrochloric acid, prussic acid, caustic soda, potash, bleach

Formulas

6.153 Chemical reactions are described by formulas written in chemical symbols for the reacting elements and compounds:

$$Cu + H_2SO_4 = CuSO_4 + H_2$$

Set the numeric suffixes (such as H_2) as subscripts. Set numeric prefixes on line (such as $2H_2$).

(See also Chapter 7.)

Biological Terms

6.154 All animal life is subdivided according to the following hierarchy:

- Phylum,
- Class,
- Order,
- Family,
- Genus, and
- Species.

6.155 Capitalize terms for phylum, class, subclass, order, family, and genus. Set terms for species in lowercase. For example, the common European squirrel would be classified as

phyla	class	order	family	genus	species
Vertebra	Mammalia	Rodentia	Sciuridae	Sciurus	vulgaris

Geological Terms

6.156 Names for generic geological features are considered common nouns, hence set them in lowercase:

anticline, fault, graben, igneous rocks, incline, magma, monocline, pyroclastic rocks, rift, strata

6.157 Names of minerals are lowercase, except for some vernacular forms that may contain proper names. One such example is *Blue John* (a specific type of fluorspar). Typical mineral names are

calcite, felspar, magnetite, mica, pyrite, quartz, silica

6.158 Geological eras, periods, and revolutions are all proper nouns, hence they are capitalized:

ERAS
- Precambrian
- Paleozoic
- Mesozoic
- Cenozoic

PERIODS
- Permian
- Carboniferous
- Devonian
- Silurian
- Ordovician
- Cambrian

REVOLUTIONS
- Alpine
- Hercynian
- Caledonian
- Charnian
- Karelian

Mathematical Terms

6.159 Many mathematical terms are common nouns, hence they are set in lowercase:

equation, exponent, function, mantissa, power, progression, radical, relation, root, series, square, square root

These terms may be combined with other common terms to form a compound term. Set compound terms in lowercase:

arithmetic progression
binomial series
geometric progression
infinite series
power series

6.160 When a term involves a proper name, capitalize that name:

Bessel function
Fibanacci series
Laplace transform

While possessives rarely occur, one example is Maxwell's equations.

6.161 The abbreviated form of the primary trigonometric functions are

cos	cosine
sin	sine
tan	tangent

6.162 The abbreviated form of the complementary trigonometric functions are

cosec	cosecant
cot	cotangent
sec	secant

6.163 Express hyperbolic functions in abbreviated form. The terminal *h* indicates hyperbolic:

cosh	cosech
sinh	sech
tanh	

6.164 Inverse functions are indicated by the notation –1, which should be set as a superscript: $\sin^{-1}$ ß.

Astronomical Terms

Many astronomical terms come from Latin and Greek mythology. Also, many entities are named for their discoverers.

Planets

6.165 The term *planet* generally refers to the nine bodies within our solar system. However, the general definition of a planet as a "small cold body orbiting a star" means that other planets can exist.

6.166 Capitalize the names of the nine planets of the solar system:

> Mercury, Venus, Earth, Mars, Jupiter, Saturn, Uranus, Neptune, Pluto

Note the term *Earth* may be seen without an initial capital since it has become a common noun. However, for clarity and consistency, capitalization is recommended.

Capitalize the names of planet features:

> Nix Olympia, Pavonia Mons, Plain of Chryse, Valles Marineres

Moons

6.167 Moons are satellites. Earth's moon is referred to as *the Moon.*

All planets except Mercury and Venus have moons. In generic form, lowercase is used: The moons of Saturn are . . .

All moons are named and the names are capitalized:

Jupiter	Callistro
	Europa
	Gannymede
	Io
Mars	Phobos
	Deimos

Capitalize features—craters, seas, mountains, radial ridges, rills, and valleys—on the Moon's surface:

Alpine Valley, Hadley Rill, Jura Mountains, Rook Mountains, Teneriffe Range

Some names derive from Latin or are formed in Latin style. Note *Mare* (sea) and plural form *Maria* (seas), and *Oceanus:*

Mare Nubium, Maria Crisium, Maria Smythii

Craters often have single names such as

Copernicus, Lambert, Tycho

Comets

6.168 Comets are often named for their discoverers, hence treat the discoverer's name as a possessive adjective in the comet's name: Halley's comet, Tycho Brahe's comet.

A number of comets are not named for an astronomer. The usual form for these comets is the comet of 1843. Another form is the great comet of 1882. The term *great* is ascribed to a comet with a large tail that also approached close to Earth.

Meteors

6.169 Although individual meteors are unnamed, significant recurring meteor showers are named: the Perseid, the Leonids.

Note that the term *meteor shower* is omitted, it is understood from context.

Constellations

6.170 Capitalize the names of constellations (groups or configurations of stars). Many of these names come from mythology. The shape of the constellation suggested a particular mythological figure:

Scorpius, Sagittarius, Capricornus, Ursa Major, Ursa Minor, Ophiuchus, Orion, Corona Borealis

Note the vernacular forms for some constellations:

Big Dipper, Little Dipper

Stars

6.171 Capitalize the names of stars:

Alpha Centauri, Arcturus, Polaris, Rigel, Sirius

6.172 Many stars do not have individual names, but there are conventions that do uniquely identify them.

6.173 The Bayer method for "naked eye" stars is the Latin genitive form of the constellation name prefixed by a Greek identifier:

α-Centauri and ß-Centauri

These identify the first and second stars in the constellation of Centaurus.

6.174 Newer systems exist for the identification of fainter stars (not visible to the naked eye) and variable stars (which fluctuate in brightness). A common system is the use of roman capital letters. Historically, because the letters A and Q are already assigned to southern constellations, use R as the first letter:

1st star	R
2nd	S
. . . .	
9th	Z

Thus:

U Geminorium is the fourth star in Gemini.
R Scuti is the first star in Scutus.

This system can be extended by the use of double letters in the range R to Z:

10th star	RR
11th	RS
....	
18th	RZ
19th	SS
20th	ST
....	
54th	ZZ

Thus:

ST Cygni is the 20th star in Cygnus.

When further resolution is required, use the double letters in the range AA to QZ:

55th star	AA
56th	AB
....	
334th	QZ

6.175 For new variables a convention of the form V335 Cygni was adopted; this indicates the 335th variable star in Cygnus. This is an open (unlimited) system of identification.

6.176 Another common system of star identification is the Harvard observatory system. This system uses six digits to identify a star. The first four digits record the hours and minutes of ascension relative to star positions as of the year 1900. The next two digits describe the angle of declension in degrees. Roman digits are used to indicate north, italic digits indicate south. In the Harvard system, a designation such as α Leonis 094211 indicates a star at an ascension of 9 hours 42 minutes and declension 11° North relative to the star α Leonis (the first star in Leonis).

Classes of Stars

6.177 Stars are often classified in terms of their spectral radiation. Indicate this by the capital letters O, B, A, F, G, K, M, N, R, and S. A star may therefore be referred to as an M star or a K star.

Novas

6.178 Novas are named in the Latin genitive form. Thus, a name such as Nova Persei indicates a nova in Perseus.

Galaxies

6.179 Galaxies have a common naming system and alphanumeric naming systems that are cataloged. Galaxies in the common naming system include Draco system, Leo I system, and Leo II system.

6.180 Because of the immense number of objects in the universe, catalogs have been essential throughout the history of astronomy. The following catalogs provide the naming conventions in use:

Catalog	**Abbrev.**
Bonner Durchmusterung	BD
Cordoba Durchmusterung	CD
Flamsteed	
Henry Draper	HD
Index Catalog	IC
Messier	M
New General Catalog	NGC
Wolf	

Two catalogs of clusters, galaxies, and nebuli are still in common use: Messier (M) and the New General Catalog (NGC).

Johan Dreyer's New General Catalog (NGC) is generally preferred by professional astronomers. The NGC contains objects not cataloged by Messier. The Index Catalog (IC) supplements the NGC.

Exhibit 6–3:
Comparison of Galactic Cataloging Systems

Messier #	NGC #	Name	Object Type
1	1952	Crab Nebula	supernova remnant
8	6523	Lagoon Nebula	nebula
31	224	Andromeda Galaxy	spiral galaxy
44	2632	Praesepe	open cluster
57	6720	Ring Nebula	planetary nebula
88	4486	Virgo A	elliptical galaxy

Modern catalogs cite objects in both the Messier and NGC systems. Objects from other catalogs would appear as

66 Cygni, 72 Ophiuchi (66 and 72 represent Flamsteed numbers)

TECHNOLOGY TERMS

6.181 High technology creates new terminology to apply to new concepts, techniques, devices, and processes.

6.182 Major problems in technological terminology include

- Number of acronyms, abbreviations, and mnemonics;
- Obscurity of origins of terms;
- Borrowing terms from other disciplines; and
- Multiple usage of abbreviations and acronyms.

Each time a reader encounters poor terminology, comprehension declines and the reader's performance suffers.

As an example of multiple usage of abbreviations, consider the term *PCB,* which has at least three common meanings:

polychlorinated biphenyls
printed circuit board
process control block

Similarly, the term *spool* is widely used in computing. While we write about *spool files,* we also use the term as a verb:

The Print command spools the file to the printer.

The term *spool* was devised as an *initialism:* Slow Peripheral Operation On Line. But it became widely used and changed from an *initialism* to an acceptable word.

Computer Terms

Trends in Terminology

6.183 Trends in terminology are toward simplification. Such simplification combines words and uses lowercase letters.

6.184 Many of the terms used in computer technology are compound words; thus, they are commonly hyphenated:

log-in, log-on, log-off; on-line, off-line, in-line; sign-in, sign-on, sign-off; up-load, down-load, cross-load

However, continued change has made many of these hyphenated words into composite terms. This method works well for simple familiar terms such as those just given:

login, logon, logoff; online, offline, inline; signin, signon, signoff; upload, download, crossload

Be careful not to create awkward or incomprehensible terms. For instance, terms such as *cross-compilation* or *binary synchronous transmission* would not be combined.

Hardware Terminology

6.185 When used in a generic sense, set computer hardware terms in lowercase roman. Such equipment terms include

central processor	ink-jet printer
chain printer	laser printer
daisywheel printer	laser-jet printer
disk drive	line printer
dot-matrix printer	microcomputer
hard disk	minicomputer
high-speed laser printer	tape drive
impact printer	workstation

6.186 Identify specific computers by the manufacturer's name followed by the model designation:

IBM PC-XT™
Apple™ Macintosh™ II (or Macintosh II)
HP 3000
Apollo 4000
Sun 3/50 workstation
DEC LPS40™

Note that the abbreviation *PC* is registered by IBM. It may not be used as an abbreviation for the term *personal computer*. Also note that the terms *clone* and *PC-clone* are unacceptable terms. The correct term is *PC-compatible*, meaning *compatible with the IBM PC*.

Software Terminology

6.187 When writing in generic terms about computer software, use lowercase roman. Examples of such generic terms include

assembler	linking-loader
compiler	loader
cross-compiler	monitor

™ PC and PC-XT are trademarks of International Business Machines Corporation.
Apple and Macintosh are trademarks of Apple Computer, Inc.
DEC and LPS40 are trademarks of Digital Equipment Corporation.

database	operating system
decompiler	payroll program
executive	spreadsheet
interpreter	utility

Commercial Programs and Application Packages

6.188 Commercial software has legally registered names. The correct form for these names can often be found in trade journals. Manufacturers' literature offers a more definitive source. The manufacturer's name is not always an integral part of the software product name.

Some well-known names for operating systems, for instance, include

• CMS™	IBM Conversational Monitor System
• MS-DOS™	Microsoft Disk Operating System
• UNIX™	AT&T microcomputer operating system
• VMSP™	IBM Virtual Machine System Product

6.189 It is common to encounter constructions such as *UNIX-based.*

The XYZ system will be UNIX-based.

Avoid this by restructuring the sentence:

The XYZ system will be based on the UNIX operating system.

Personal Computer Software

6.190 Some of the better known business application packages for personal computers, and their orthographic conventions, include

- Excel™
- Lotus 1-2-3™
- PageMaker™

Mainframe Computer Software

6.191 Application software for mainframe computers uses many initialisms and acronyms as product names:

™ CMS is a trademark of International Business Machines Corporation.
MS-DOS and Excel are trademarks of Microsoft Corporation.
Lotus 1-2-3 is a trademark of Lotus Corporation.
PageMaker is a trademark of Aldus Corporation.
UNIX is trademark of AT&T Bell Laboratories.
VMSP is a trademark of International Business Machines Corporation.

- DCF™ Document Composition Facility
- GML™ Generalized Markup Language
- RPG™ Report Program Generator
- Script™ text processing language

Releases and Versions of Products

6.192 Release or version number of the software appears after the product name:

- MacDraw™ 1.9.6
- MacWrite™ 5.0
- MS Word™ 3.01
- QuarkXpress™ 1.01L
- WriteNow 2.0™

6.193 A common release or version convention for mainframe software is xx.yy.zz

Where

xx indicates the feature level,
yy indicates the enhancement level, and
zz indicates the maintenance level.

Thus MML Rel 4.03.05 would indicate for the Meta Markup Language product

Feature release 4,
Enhancement release 3, and
Maintenance release 5.

Feature releases deliver new features, enhancement releases introduce new functions, and maintenance releases fix software problems.

Computer Commands

6.194 Make command forms in documentation the same as in the interface. For instance, if commands must be entered in uppercase, show them in uppercase in the documentation.

™ DCF, GML, RPG, and Script are trademarks of International Business Machines Corporation.
MacDraw is a trademark of Claris Inc.
MacWrite is a trademark of Apple Computer, Inc.
MS Word is a trademark of Microsoft Corporation.
QuarkXpress is a trademark of Quark, Inc.
WriteNow is a trademark licensed to T/Maker Co.

When referring to commands in a generic sense, capitalize them:

> The Query Job command returns the status of the next job in the JCL system queue.

Computer Messages

6.195 Write system messages—error messages, prompts, and information messages—in the same way as commands. Both the screen and document text should be identical:

> PLEASE ENTER YOUR PIN NUMBER AND PRESS: 'OK'

Software Development Documentation

6.196 Software documentation—design specifications, system and program descriptions, flowcharts, and similar information—describes the software's structure. In such documentation, use specific conventions to refer to programs, subroutines, modules, data entities, and states.

6.197 Set all common terms in lowercase roman. Common terms include

buffer	operator	routine
code	process	stack
interrupt	program	subprogram
list	queue	subroutine
operand		task

6.198 When referring to these common terms by proper name, capitalize them:

> When an interrupt is received, control is transferred to the Clock subroutine. Bit 1 of the Program Status word is reset and Clock passes control to the Scan 1 program.

6.199 Use uppercase to refer to certain types of stacks:

FIFO stack	first-in first-out stack
LIFO stack	last-in first-out stack

Note also:

I/O buffer	input/output buffer

6.200 Use the solidus (/) in terms such as *input/output, read/write,* and *read/compare/write.*

6.201 Capitalize data names—record, field, and bit names:

Baud Rate field
Device Address field
Frames Received record
Reset bit
Status bit
X record

Programming Languages

6.202 There are many programming languages. Different languages work best for particular types of applications. Set names for programming languages in uppercase: COBOL, FORTRAN. Few of these names are true acronyms.

6.203 Modern trends are for initial caps only, although some names still insist on all caps. The following list shows mandatory capitals:

Algol	Algorithmic language
APL	A programming language (literal)
Basic	Beginners' all-purpose symbolic instruction code
C	Systems programming language originally developed for the implementation of the UNIX operating system
GML™	Generalized mark-up language
Lisp	List processing language
Pascal	High-level programming language named for the mathematician Blaise Pascal
PL/1	Programming language 1
REXX™	IBM executive programming language
Snobol	String-oriented symbolic language

6.204 Most programming languages have reserved words. These words have a specific predefined meaning. Reserved words are set in text based on the conventions for the programming language. In a language such as FORTRAN, all reserved words are uppercase:

LET, PRINT, FOR, WHILE, GOTO

In other languages, such as Algol and PostScript, reserved words are lowercase:

Algol—begin, end
PostScript—findfont, setfont, scale

™ GML and REXX are trademarks of International Business Machine Corporation.

Backus-Naur Form

6.205 For professional programmers, programming language reference manuals must specify the various language structures. Use metalanguages and syntactical notations for this purpose. A common standard notation is Backus-Naur Form (BNF). All legitimate forms allowed in a programming language are described by a series of BNF expressions. A typical BNF expression is

<digit> ::= 0 | 1 | 2 | 3 | 4 | 5 | 6 | 7 | 8 | 9

where | indicates *or*. An extended form of BNF is also in use.

Procedures

6.206 Precise terminology is particularly important in procedures. For example, consider the common term *login*. This term has several synonyms, *logon* and *signon* or *signin*. It is not too important which term is chosen, as long as consistency is preserved. If *login* is used, then *logout* should be used. If *signon* is used, then *signoff* should be used.

6.207 Be consistent when using imperative verb forms such as enter, press, key, type, hit.

6.208 When writing procedure titles, use action titles rather than nominalizations. Thus, do not use titles such as

Printer option settings
System software initialization

Rather, use procedure titles that contain an action verb. Use the imperative or gerundive (*ing*) form of the verb:

Imperative	**Gerundive**
Set printer options	Setting printer options
Initialize system software	Initializing system software

Use the friendlier gerundive form for documents such as user guides. Use the imperative form for procedural instruction documents such as installation and troubleshooting manuals.

Electronics

6.209 Since electronics advances rapidly, terms often come into common use before they appear in dictionaries. Useful sources for new terminology are trade journals and manufacturers' literature.

Circuit Technologies

6.210 The two genres of electronic circuits are analog (linear) and digital. Analog systems include audio systems, television and video systems, radar, and oscilloscopes. Digital systems include computers, microprocessors, computer terminals, and calculators. Digital technology has expanded into areas that have been traditionally analog, such as audio systems and telephone systems.

Passive Electronic Components

6.211 These electronic components do not perform active functions such as switching electronic signals. Passive components include:

capacitors[1]	inductors
coils	resistors
couplers	thermistors
filters[2]	transformers[3]

6.212 Capitalize names of passive components. If a qualifier precedes these terms, neither term is capitalized. Such terms include

auto-transformer	low-pass filter
band-pass filter	power transformer
ceramic capacitor	signal transformer
de-coupling capacitor	smoothing capacitor
electrolytic capacitor	tantulum capacitor
film capacitor	twin-T filter
high-pass filter	variable capacitor
intermediate frequency transformer	variable resistor

6.213 If a proper name precedes any of the terms in Section 6.212, capitalize that name: Tchebychev filter.

Active Electronic Components

6.214 Active components perform circuit functions such as amplifying and switching. Do not capitalize names for active components. If

[1] *Condenser* is an obsolete term for *capacitor.*

[2] Active filters also exist. These contain active components such as transistors and switching devices.

[3] Although a transformer may appear to be an amplifier, it can only step up either the voltage or current of an electrical or electronic signal. Transformers cannot give a power gain.

a qualifier that is not an acronym precedes these terms, capitalize neither term. Such terms include

avalanche diode	silicon controlled rectifier
clamping diode	switching diode
diode	thyristor
field-effect transistor	transistor
junction transistor	traveling-wave tube
klystron	tunnel diode
magnetron	varactor diode
point-contact transistor	

6.215 If a proper name precedes any of the terms in Section 6.214, capitalize that name: Zener diode, Schottky device, Gunn diode.

6.216 If a mnemonic precedes any of the terms in Section 6.214, capitalize the mnemonic: PNP transistor, NPN transistor.

6.217 Common initialisms for active devices include

Initialism	Device
CMOS	complementary metal-oxide-silicon field-effect transistor
FET	field-effect transistor
IGFET	insulated gate field-effect transistor
JFET	junction field-effect transistor
MESFET	metal-silicon field-effect transistor (no oxide)
MOSFET	metal-oxide-silicon field-effect transistor
NMOS	N-channel metal-oxide-silicon field-effect transistor
PIN diode	P-type intrinsic N-type diode
PMOS	P-channel metal-oxide-silicon field-effect transistor

6.218 There is a minor trend toward using lowercase for semiconductor material types, hence: p-type, n-channel.

Logic Circuits

6.219 Logic circuits are used extensively in digital systems such as microprocessors and computers. Acronyms identify a number of existing logic systems:

DTL	diode-transistor logic
ECL	emitter-coupled logic

I^2L	integrated-injection logic
RTL	resistor-transistor logic
TTL	transistor-transistor logic

6.220 Logic signals have two values based on symbolic logic: True and False. Capitalize these terms. Do not use full capitals (TRUE). Reserve this convention for acronyms and signal mnemonics. In circuit descriptions and logic dictionaries, a typical sentence or statement in this style would be

> When RST is True, the buffer contents are set to zero.

6.221 Another convention refers to logic signals as High and Low. Again, do not use full capitals. The abbreviations Hi and Lo are acceptable in laboratory notes, but not in formal writing. In circuit descriptions and logic dictionaries, a typical sentence or statement in this style would be

> When RST is High, the buffer contents are set to zero.

6.222 Most logic elements have specific terminologies. The following list shows the accepted terms (and their capitalization) for these elements:

BCD decoder
BCD-to-binary convertor
buffer
D flip-flop
EOR (Exclusive-OR) gate
flip-flop
full-adder
JK flip-flop
latch

6.223 Avoid cute or pedantic terms such as *flip-flip* and *flop-flip*.

Integrated Circuits

6.224 Integrated circuits (ICs) are complete circuits fabricated on a silicon chip. In contrast, circuits that use individual components are referred to as discrete component circuits. The terms used to describe the power and complexity of integrated circuits are

LSI	large-scale integration
VLSI	very-large-scale integrated circuits

The components fabricated in silicon integrated circuits include diodes, transistors, resistors, sometimes capacitors, and the logic elements mentioned in Section 6.219. The terminology for these devices is in Sections 6.214 and 6.219.

Telecommunication

6.225 Modern telecommunication systems are computer controlled. Many of the terms come from computer and electronic terminology.

Since it is necessary to connect telecommunications products and systems, the industry follows international standards developed by such organizations as CCITT (International Consultative Committee for Telephone and Telegraph) and ISO (International Standards Organization).

Abbreviations and Acronyms

6.226 Telecommunications uses many abbreviations and acronyms. The general rules for abbreviations and acronyms follow those given in Chapter 5.

6.227 The following list offers a sample of acronyms:

Bell operating company	BOC
Dynamic Network Architecture	DNA
Interexchange carrier	IEC
Inter local access and transport areas	Inter-LATA
integrated services digital network	ISDN
network control center	NCC
plain old telephone service	POTS
regional Bell operating company	RBOC
regional holding company	RHC

Transmission and Control Characters

6.228 The American Standard Code for Information Interchange (ASCII) Alphabet No. 5 defines the control characters for electronic transmission. Set the names for all such characters in uppercase:

ACK	Acknowledge
NAK	No acknowledge
WRU	Who are you?

Device States

6.229 Append the dynamic states of telecommunications devices to the device name:

DTE ready	Data circuit terminating equipment ready state
DTE wait	Data circuit terminating equipment wait state

Note the use of lowercase for the state (ready or wait).

6.230 If the state is used without the device name, capitalize the term:

Ready
Ready to Receive
Wait

SEVEN

Numbers and Symbols

7.1 Science writing and technical writing depend heavily on numbers and symbols, on measure and mathematics. Ideas and processes must be recorded quantitatively with accuracy and precision. Definitions must be accurate, operations mathematically rigorous, measurements appropriately precise.

USING NUMERALS

Measurements

7.2 Express units of measurement, time, and quantity as numerals.

7 meters	¼ pound
8 by 12 inches	3-½ quarts
1-½ miles	10 amperes
3 MeV	50 kG
1 Btu	17.5 seconds
8 cm	½ inch

But:

. . . a period of three years . . .

Mathematical Expressions

7.3 Use numerals instead of spelled-out numbers in mathematical expressions:

multiplied by 3	integrated from 0 to 1
divided by 14	*i*-values of 1, 2, and 3
a factor of 2	

Decimals

7.4 Put a zero before the decimal point in numbers less than one, and omit zeros after the decimal point unless they are significant digits: (See Section 7.33.)

0.25 inch
specific gravity 0.9547
gauge height 10.0 feet

But:

.30 or .50 caliber guns

Percentage

7.5 Use numerals to express percentage:

12 percent	25.5%
0.5 percent	5 percentage points

Proportion

7.6 Use numerals to express ratios and proportions:

ratio of 1 to 4   1:62,500
1/3/5 parts (by volume) of powder/alcohol/water

Time

7.7 Use numerals to report all time values (especially those reporting experimental results that happened over a specific time span):

6 hours 8 minutes 20 seconds
10 years 3 months 29 days

But:

four centuries, three decades, statistics of any one year, in a year or two

Clock Time and Dates

7.8 Use a consistent system for reporting time and date values that mark specific thresholds:

4:30 P.M., 4:30 in the afternoon, 16:30 hours (24-hour clock)
10 o'clock or 10 P.M.; not 10 o'clock P.M. (redundant)
12 M. (noon), 12 P.M. (midnight)

July 4, 1988; not July 4th, 1988
4th of July (the date); but Fourth of July (the holiday)
4 July 1988 (military form, no comma)

USING WORDS TO EXPRESS NUMBERS

7.9 Represent numbers not covered by the preceding rules by numerals if greater than ten, but spelled out if ten or less:

They bought 50 computers.

We performed seven experiments.

The ship housed 152 astronauts.

The target group had four members.

Always spell out round numbers:

thirty years ago
less than one million
one hundred people
about a thousand dollars

Always spell out a number at the beginning of a sentence:

Seventeen students completed the examination.
Twelve two-pound packages were delivered to the lab.

But:

The lab received 12 two-pound packages.

7.10 A spelled-out number should not be repeated as a numeral except in legal documents:

The grace period extends for three (3) months.

7.11 Treat related numbers close together at the beginning of a sentence alike:

Fifty or sixty miles away is snowclad Mt. Rainier.

Nine and eleven players make up the respective teams.

7.12 Simple fractions that do not express units of measurement are generally spelled out, but mixed numbers are written with numerals:

one-thousandth
one-half of the samples
one-tenth of the cases

But:

2-1/2 times as much, 3-1/2 cans of soup, 1/4 inch

USING HYPHENS WITH NUMBERS

7.13 Put a hyphen between numbers and words that form a unit modifier:

The rod has a 1-inch diameter (a diameter of 1 inch)
The 1-inch-diameter rod is heavy
6-foot-long board
five-member panel
6-percent-interest bonds (bonds yielding 6 percent interest)
11-fold, 150-fold

But:

threefold, tenfold

7.14 When two or more hyphenated compounds in a series have a common base element that is omitted in all but the first or last one, retain the hyphen to indicate suspension (the *suspense hyphen*):

Sections II-2, -3, and -7
Serial numbers 14685-122, -129, and -137
2- or 3-inch pipe (not 2 or 3-inch pipe)
2-to-3- and 4-to-5-ton trucks
8-, 10-, and 16-foot boards
2-by-4-inch boards

But:

boards 2 by 4 inches in cross section

7.15 Put a hyphen between the elements of spelled-out compound numbers from 21 through 99:

twenty-one, thirty-seven, eighty-two, ninety-nine

7.16 Put a hyphen between the numerator and denominator of a spelled-out fraction, except when one or the other already contains a hyphen:

two-thirds, one-thousandth, one three-thousandth
two one-thousandths (or two thousandths)

7.17 For mixed numbers in which the fraction is written with full-size numerals in slash or solidus form, use a hyphen to connect the whole number and the fraction:

16-3/4 pounds (not 16 3/4 pounds)
8-1/2 inches wide (not 8 1/2 inches wide)

7.18 A unit modifier that follows and refers back to the word or words modified (as in a parts list) takes a hyphen and is always written in the singular:

motor: ac, three-phase, 60-cycle, 115-volt
(not motor: ac, three phases, 60 cycles, 115 volts)
glass jars: 5-gallon, 2-gallon, 1-quart
belts: 2-inch, 1-1/4-inch, 1/4-inch

7.19 Do not hyphenate a modifier consisting of a number followed by a possessive noun:

two months' layoff (not two-months' layoff)
one week's pay
two hours' work

REPORTING LARGE NUMBERS

7.20 Use separating commas in numbers of five or more digits to mark off the thousands, millions, and the like.

26,575 3,775,000 $5,973,685,600

An exception to this rule is the *Système Internationale* (SI)—see Section 7.43—which recommends separating such numbers with spaces rather than commas.

26 575 3 775 000 $5 973 685 600

7.21 Do not use separating commas in four-digit numbers unless they appear with larger numbers that are separated:

Use	Not
7,560	7560
890	890
18,365	18,365
1,230	1230
28,045	28,045

7.22 Use the words *million, billion,* and so on, in large round numbers rather than writing out the zeros:

Sales of $6 billion (not $6,000,000,000 or six billion dollars)
Five million people were affected (not 5,000,000 or 5 million)
Pennsylvania's population is more than 11 million (not 11,000,000)

USING SCIENTIFIC NOTATION

7.23 Scientists and engineers think of numbers in terms of powers of ten. This standard representation enables them to express both large and small numbers as the product of some factor between one and ten multiplied by some power of ten. This representation simplifies calculations and saves space when very small or very large numbers must be recorded.

7.24 To represent a number as some factor times a power of ten:

- Place a decimal point to the right of the first nonzero digit in the number,
- Cut off all zeros after the last nonzero digit, and
- Let the resulting number be the factor.

Thus, the factor always represents a number whose first digit (to the left of the decimal point) ranges from 1 through 9. This factor is then multiplied by whatever power of ten (the *exponent* of ten) is required to make the product equal to the original number. For example, the standard representation of 795 is

7.95×10^2 (which is $7.95 \times 100 = 795$)

7.25 The exponent can be negative or positive, depending on whether the number being represented is smaller or larger than 1 (the number 1 is ten to the zero power). Various examples of small or large numbers and their equivalents in standard power-of-ten representation include

27.5	2.75×10 (not 2.75×10^1)
1.375	1.375 (not 1.375×10^0)
37,040,000	3.704×10^7
2,000,000	2×10^6 (not 2.0×10^6)
0.000715	7.15×10^{-4}
0.09004	9.004×10^{-2}
0.000000000145	1.45×10^{-10}
9.456	9.456
10.07	1.007×10

COMPARING POWERS OF TEN

7.26 When comparing the relative magnitudes of two numbers in the standard power of ten representation, remember that making an accurate comparison requires three steps

1. Compare their factors *and* their exponents,
2. Combine the two comparisons, and
3. Record resulting ratio.

7.27 To perform this process for the hypothetical numbers *a* and *b*, take the ratio of *b*'s factor to *a*'s factor and multiply it by the ratio of *b*'s exponent to *a*'s exponent. The result is the ratio of *b* to *a* (the desired comparison). For example, let *a* be 2.3×10^6 and let *b* be 6.9×10^3:

1. Divide the 6.9 factor of *b* by the 2.3 factor of *a* and get 3.

1a. Divide the 10^3 exponent of *b* by the 10^6 of *a* and get 10^{-3}. (Remember, division of such numbers is done by subtracting the exponent in the denominator from the exponent in the numerator, i.e., $3 - 6 = -3$).

2. Multiply the 3 by the 10^{-3} and get 3×10^{-3} or 0.003.
3. So we see that the ratio of *b* to *a* is only 0.003; in other words, *b* is only 3/1000 as large as *a*.

7.28 Below are a few more examples of this comparison process

Numbers	1. Factor Division	2. Exponent Division	Ratio
$b = 2.4$ $a = 1.6 \times 10^{-2}$	$2.4/1.6 = 1.5$	$1/10^{-2} = 10^2$	$b/a = 1.5 \times 10^2 = 150$

Numbers	1. Factor Division	2. Exponent Division	Ratio
$b = 1.9 \times 10^2$ $a = 9.5 \times 10$	$1.9/9.5 = 0.2$	$10^2/10 = 10$	$b/a = 0.2 \times 10 = 2$
$b = 7.1 \times 10^{-5}$ $a = 1.42 \times 10^{-2}$	$7.1/1.42 = 5$	$10^{-5}/10^{-2} = 10^{-3}$	$b/a = 5 \times 10^{-3} = 0.005$

7.29 Make the comparison obvious by putting the numbers to be compared in a *nonstandard* representation. Namely, express them in terms of the *same* power of ten, so that the factors alone reflect the comparative magnitudes of the numbers. Exhibit 7-1 illustrates how such a representation can make the relative magnitudes of the column entries immediately obvious.

The nonstandard scheme shown in Exhibit 7–1 is effective when the range of the numbers is only a few powers of ten. If the range is larger, say six or more powers of ten, use the standard representation. For such wide-ranging numbers, the nonstandard representation of Exhibit 7–1 would cause some digits to extend inconveniently far to the left and right of the decimal point.

Exhibit 7–1:
Comparison of Estimated Masses of Several Icebergs

	Standard Representation Mass (kg)	Nonstandard Representation Mass (kg)
Iceberg A	2.57×10^{10}	257×10^8
Iceberg B	8.43×10^8	8.43×10^8
Iceberg C	7.29×10^9	72.9×10^8
Iceberg D	1.87×10^{11}	1870×10^8

SOURCE: NOAA *Arctic Study,* 1980.

7.30 The idea used in the nonstandard representation in Exhibit 7–1—expressing all numbers to be compared in terms of the same power of ten—is commonly used in tables and graphs to avoid having to write out large numbers. For example, if a table contains a column of distances 3740, 3690, 3720, 3750, and 3710 expressed in meters, an editor might choose to represent the distances in kilometers instead—as 3.74, 3.69, 3.72, 3.75, and 3.71—to keep the numbers small.

NUMBERS IN TABLES AND GRAPHS

7.31 Tables and graphs in engineering and scientific reports often change the original unit by some factor of ten in order to have convenient-sized numbers in the table columns or on the graph scales. The method for expressing these units depends on whether a positive or a negative value for the power of ten is used. One approach is correct and useful; the other way is wrong, yet both are widely used. Exhibit 7–2 illustrates the difficulties:

Exhibit 7–2:
Tabular Data Presented as Powers of Ten

(Column 1) Yield Strength (psi)	(Column 2) Yield Strength (10^3 psi)	(Column 3) Yield Strength (psi $\times 10^{-3}$)
197,000	197	197
265,000	265	265
181,000	181	181
240,000	240	240

Column 1, the original values, cannot be misinterpreted. The column head tells what the numbers in the column represent (yield strength) and the units of measurement (psi, or pounds per square inch). But the numbers in column 1 are large, and space can be saved by using numbers with fewer digits, as has been done in the other two columns. The problem: how do we save space correctly and still insure accuracy?

The head in column 2 cites the unit of measurement as 10^3 psi, or 1000 psi. Since the unit of measurement is 1000 psi, and the first entry is 197 units, the first entry means 197,000 psi.

The head in column 3 forces the reader to conclude that the unit of measurement is 10^{-3} psi. If so, the first entry would mean 0.197 psi, which of course is incorrect. Why would someone write psi $\times 10^{-3}$? Instead of telling us what the numbers in the column must be multiplied by to get the actual measured values, the author is trying to tell us what the measured values were multiplied by to get the numbers in the column. However, this point of view is very misleading, because the column head is supposed to tell us the *unit of measurement.* And if we want to put 197 under the column head, the unit of measurement is obviously thousands of psi (10^3 psi), *not* thousand*ths* of psi (psi × 10^{-3}).

7.32 The same considerations apply when changing the measurement unit on a graph scale in order to use smaller numbers on the scale. A closely related idea—for example, changing meters (m) to millimeters (mm) or micrometers (μm) or kilometers (km), and so on—is discussed in Section 7.43.

REPORTING SIGNIFICANT DIGITS

7.33 Consider significant digits (or significant figures) when changing from one measurement unit to another. Significant digits in a number expressing a measurement or used in a calculation are an indication of the number's accuracy. Scientists and engineers follow a well-established system of rules for expressing this accuracy.

First, numbers that result from counting discrete objects have an infinite number of significant digits, since they can be completely accurate. Thus, the number 27 arrived at from carefully counting the number of ball bearings in a container could be thought of as 27.00000 . . . , with an infinite number of zeros possible. In contrast, the numbers obtained by measuring the weight of each of the ball bearings have a limited number of significant digits, depending on the accuracy of the weighing and recording process. A table of the weights of a few of the bearings might read:

Bearing Number	Weight (μg)
1	294.5
2	297.1
3	295
4	288.9
5	296.2

The measured weights of all these bearings except one are represented as having four significant digits, which in this case means they are considered accurate to the nearest 0.1 μg. The measured weight of bearing 3, on the other hand, is represented as having only three significant digits, implying accuracy only to the nearest 1 μg.

This difference should be taken seriously; it is not merely a typographical error. Whatever the reason for the implied tenfold reduction in the accuracy of bearing 3's weight as compared with that of the other bearings, it should be respected and allowances should be made for it in any calculation that includes bearing 3's weight.

For example, to calculate the mean weight of the five bearings, their individual weights would be added together and the result divided by 5. That average should be represented to only three significant digits, since that is the accuracy of the least accurate number in the calculation:

1. Adding the weights of the bearings produces 1471.7.

   294.5 + 297.1 + 295 + 288.9 + 296.2 = 1471.7

2. Dividing by 5 produces 294.34.

   1471.7/5 = 294.34

3. Rounding off to three significant digits produces 294.

If bearing 3's weight had also been to four significant digits, say 295.0, then it would have been appropriate to report the average to four significant digits also, as 294.3.

The example of the ball bearings just given suggests a few of the considerations involved in determining significant digits. For a more complete discussion of the concept, see *American National Standard Metric Practice* (ANSI/IEEE Std 268-1982, pp. 16–22) and *Suggestions to Authors of the Reports of the United States Geological Survey* ("Significant Figures," pp. 197–202).

7.34 Similar considerations arise in putting numbers on the scale of a graphical plot. Although data accuracy is not involved, appearance and consistency as well as avoiding excessive numerical precision are involved.

7.35 To create a scale for time, on the horizontal axis, start at zero on the left and proceed at 1-second intervals toward the right: 0, 1, 2, 3, 4.

Not

0.0, 1.0, 2.0, 3.0, 4.0.

Since all the numbers are integers, no decimal points are needed. In particular, a decimal point should never be used with the zero at the zero position of a scale. To create a time scale at 0.5-second intervals, use 0, 0.5, 1.0, 1.5, 2.0, 2.5.

Not

0, 0.5, 1, 1.5, 2, 2.5.

Since it is necessary to use a decimal point and a 5 to represent the numbers at the half-second points, maintain a consistent appearance by using a decimal point and a zero to represent those numbers at the one-second points.

ORDERS OF MAGNITUDE

7.36 When two quantities are the same order of magnitude, they are roughly the same size. One order of magnitude larger means ten times as large, but two orders of magnitude does not mean 20 times, and three orders of magnitude does not mean 30 times. Instead, when A is said to be three orders of magnitude larger than B, a reader might well infer that A is only *three times* the size of B, but the scientist means 10^3 or *one thousand times* as large.

When writing for a general audience, use familiar units of comparison—better to report A as 1000 times as large as B than to indicate that A is three orders of magnitude larger. In writing for a general audience and making comparisons in terms of orders of magnitude, define the concept.

MEASUREMENT UNITS

7.37 Although the United States government now has a preference for the metric system, the English system of weights and measures is still widely used. The English system does not conform well to decimal numbers:

12 inches make a foot
three feet a yard
1760 yards a mile
16 ounces make a pound
2000 pounds a short ton
2240 pounds a long ton

Calculating compound units that involve combinations of two or more basic measurement units such as those for mass and distance is almost overwhelming.

7.38 In contrast, the metric system takes full advantage of decimal numbers. The *CGS* system (Centimeter, Gram, and Second), for example, defines smaller and larger units of mass in 1000-fold increments:

The next larger unit than the gram is the kilogram (10^3 g).
The next larger unit is the megagram (10^6 g = 10^3 kg, a metric ton).

Similarly:

The next unit smaller than a gram is the milligram (10^{-3} g).
The next smaller unit is the microgram (10^{-6} g).

Even the second is divided into

thousandths (milliseconds)
millionths (microseconds)
billionths (nanoseconds)

Longer periods of time are often expressed in the familiar units of days, years, and the like.

7.39 The system of units called SI, established recently in a restructuring of the metric system to make it consistent, is discussed in Section 7.43.

ABBREVIATING UNITS

7.40 Abbreviating the names of measurement units in scientific writing saves space on the page and eases the strain on the reader's eyes. Such writing of necessity often contains many long words, such as *cubic centimeters, millimeters,* and *micrograms,* which fit in much more neatly in the form of their corresponding abbreviations (or symbols) cm^3, mm, and μg. (See Chapter 5.)

7.41 Define all but the most common unit abbreviations when they first appear in a text. Spell out the unit name in the text and follow it with the appropriate abbreviation in parentheses:

The sample weighed 35 micrograms (μg).

Thereafter, use the abbreviation alone.

When the first appearance of an abbreviated unit occurs in an illustration, the abbreviation should be defined in the caption. (See also Chapter 12.)

Note that the same abbreviation is used for both singular and plural of a measurement unit. No *s* is added for the plural. (See also Chapter 3.)

1 lb
5 lb—not 5 lbs.

7.42 A list of abbreviations for some of the more common measurement units follows. Section 7.43 discusses the SI units, the modern version of the metric system.

angstrom (10^{-10} m)	Å
atomic mass units	amu
bar (atmospheric pressure)	spell out
barn (10^{-24} cm^2)	b
British thermal unit	Btu
calorie (gram-calorie)	cal
centimeters per second	cm/s
cubic foot	ft^3
cubic inch	$in.^3$
curie	Ci
decibel (10^{-1} bel)	dB
degree (angle)	° or deg
degrees Celsius (centigrade)	°C
degrees Fahrenheit	°F
degrees Kelvin (absolute)	°K
degrees Rankine (absolute)	°R
electromagnetic units	emu
electron volt	eV
electrostatic units	esu
foot (feet)	ft
foot-pound	ft · lb
gallon	gal
gauss	G
giga-electron volt (10^9 electron volts)	GeV
horsepower	hp
hour	hr or h
inch	in.
kilobar (10^3 bars)	kbar
kilocalorie (kilogram-calorie)	kcal
kiloton (nuclear explosive yield)	kt
kilowatt-hour	kW · h
mass units	mu
megabar (10^6 bars)	Mbar
megaton (nuclear explosive yield)	Mt
millibar	mbar
minute (angular measure)	′
minute (time)	min
ounce	oz
pound	lb

pounds per square inch	psi
revolutions per minute	rpm
roentgen (x-ray dose)	R
second (angular measure)	″
square foot	ft^2
square inch	$in.^2$
thousand electron volts	keV
watt-hour	W · h
year	yr or y

SYSTÈME INTERNATIONALE (SI) UNITS

7.43 The *Système Internationale* (SI) units of measure, based on seven fundamental measurement units—metre, second, kilogram, ampere, kelvin, mole, and candela—are the preferred units in the international scientific literature. SI units use the French spellings—for example, metre, litre. (See also Chapters 4 and 6.)

This system, a modern version of the *MKSA* system (*M*eter, *K*ilogram, *S*econd, *A*mpere), is controlled by an international treaty organization. The objective of this organization is to see SI become the universally used measurement system in science.

7.44 Use of all-SI units will make calculations simpler for scientists and engineers. At present, because workers in different countries or different disciplines use various measurement systems, when they state a formula for a physical relationship they must also specify what measurement units each of the variables is expressed in. That will not be necessary when everyone uses SI units.

7.45 Although SI is a complete system of measurement units, covering all physical measurement needs, certain older, non-SI, metric units are so traditional that they are recognized as acceptable for use with SI units. For a thorough discussion of SI units and guidance in their use, including their relation to other measurement units, see *American National Standard Metric Practice* (ANSI/IEEE Std 268-1982).

Measurement Systems for Journals

7.46 Scientific journals and book publishers now ask their authors to use SI units. Authors can become familiar with SI units by consulting a reference such as *Metric Practice,* which contains conversion tables for changing from other units to SI units. But if authors use non-SI units in their work, they should report those values rather than converting them to SI.

For example, if their working units for length measurements are inches, they should report linear dimensions in inches, as measured, followed in parentheses by the equivalent SI value (millimeters or centimeters).

Or, if they measure magnetic flux density in gauss, they should report first the value in the gauss unit, the *working unit,* followed in parentheses by the equivalent value in the tesla unit.

7.47 These working unit values are the *real values,* with accuracy suitably indicated by the significant digits. The reader will recognize that the secondary value (the conversion to SI, in parentheses) is a rounded-off conversion and should not be depended on as an accurate figure. (See Section 7.33.)

7.48 If an editor insists that SI values only be used, then the author should be very careful to convert values to SI in a manner that preserves significant digits properly. Instructions for performing this operation are given on pages 16 to 22 of *Metric Practice.*

7.49 The SI prefixes for multiples and submultiples of units—for example, kilo- (k), mega- (M), milli- (m), and micro- (μ)—make it easy to select a proper-size unit for the data under consideration. In reporting measurements for very small objects whose weights are a few millionths of a gram, the prefixes replace such weights as

2.4×10^{-6} g (0.0000024 g)
1.6×10^{-6} g (0.0000016 g)

with the much simpler

2.4 μg and 1.6 μg

As another example, the energy of fast-moving particles in accelerator experiments can measure millions or billions of electron volts. Instead of expressing these energies in terms of 10^6 eV or 10^9 eV, scientists express them as MeV or GeV.

Abbreviations for SI Units

7.50 The standard abbreviations for SI units (given here in the preferred French spelling) include

ampere (base unit, electric current)	A
becquerel (disintegrations/s)	Bq

candela (base unit, luminous intensity)	cd
centi- (SI prefix, means 10^{-2})	c
centimetre (10^{-2} meter)	cm
coulomb (A · s)	C
cubic centimetre	cm^3
cubic metre	m^3
deci- (SI prefix, means 10^{-1})	d
farad (C/V)	F
femto- (SI prefix, means 10^{-15})	f
femtometre (10^{-15} metre)	fm
giga- (SI prefix, means 10^9)	G
gram	g
grams per cubic centimetre	g/cm^3
gray (J/kg absorbed radiation dose)	Gy
henry (Wb/A)	H
hertz (cycle per second)	Hz
joule (N/m)	J
kelvin (base unit, thermodynamic temperature)	K
kilo- (SI prefix, means 10^3)	k
kilogram (base unit, mass)	kg
kilohertz (kilocycles per second)	kHz
kilometre	km
kilometres per second	km/s
kilopascal	kPa
kilovolt	kV
kilowatt	kW
litre	L
lumen (cd · sr)	lm
lux (lm/m^2)	lx
mega- (SI prefix, means 10^6)	M
megagram (also called metric ton)	Mg
megahertz (megacycles per second)	MHz
megapascal	MPa
megawatt	MW
metre (base unit, length)	m
micro- (SI prefix, means 10^{-6})	μ
microfarad (10^{-6} farad)	μF
microgram	μg
microlitre	μL
micrometre	μm
micromole	μmol

microsecond	μs
milli- (SI prefix, means 10^{-3})	m
milliampere (10^{-3} ampere)	mA
milligram	mg
millimetre	mm
millilitre	mL
millisecond	ms
millivolt	mV
mole (base unit, amount of substance)	mol
nano- (SI prefix, means 10^{-9})	n
nanogram (10^{-9} gram)	ng
nanometre	nm
nanosecond	ns
newton (kg · m/s^2)	N
ohm (V/A)	Ω
pascal (N/m^2)	Pa
pico- (SI prefix, means 10^{-12})	p
picosecond (10^{-12} second)	ps
radian (plane angle measure)	rad
second (base unit, time)	s
siemens (A/V)	S
square centimetre	cm^2
square metre	m^2
steradian (solid angle measure)	sr
tera- (SI prefix, means 10^{12})	T
terajoule (10^{12} joules)	TJ
terawatt	TW
tesla (Wb/m^2)	T
volt (W/A)	V
watt (J/s)	W
weber (V · s)	Wb

CHOOSING APPROPRIATE SYMBOLS

7.51 Scientists and engineers represent unknown quantities, parameters, specific characteristics, and so on, with symbols that are usually letters, either roman (i.e., English) or Greek, and are almost always a *single* letter (except for any attached sub- or superscripts).

7.52 A few engineering symbols—for example, the Prandtl number (Pr) and the Nusselt number (Nu)—contain two letters, but there are not many such symbols.

7.53 In algebra, for instance, the unknown quantity in an equation is usually depicted as x. A scientist or engineer will label the same quantity with a letter that has a definite mnemonic connection to the name of the object or concept for which it stands. Velocity, for example, will probably be labeled v. If a problem involves the velocities of two objects, perhaps a train and an automobile, the scientist distinguishes them by adding appropriate subscripts, such as v_t and v_a.

7.54 In fact, the symbols writers use to represent quantities in their calculations can often give clues about their scientific sophistication. If a writer calculates vehicle weight and represents it in an equation as VW, we might have reason to doubt his or her familiarity with standard practices. This would be a bad symbol choice for two reasons:

1. To scientifically knowledgeable readers, VW in the equation represents two symbols, V and W, multiplied together.
2. Scientists normally choose the *quantity of interest,* in this case weight, as the mnemonic basis for a one-letter symbol, and use a subscript qualifier if need be.

A scientist, then, would more likely have chosen a symbol such as w_v for vehicle weight (w for weight, the quantity of interest, and v for the vehicle whose weight it is).

7.55 Use standard symbols where available. Often there is no need to invent a symbol, because most quantities of interest in science have long been associated with specific symbols that should be used to avoid confusion.

Mathematical Symbols

7.56 Use standard mathematical symbols for reporting mathematics. In the *Mathematics Dictionary,* the hundreds of entries under mathematical symbols are grouped into no fewer than six areas of mathematics:

- Arithmetic, Algebra, Number Theory;
- Trigonometry and Hyperbolic Functions;

- Elementary and Analytic Geometry;
- Calculus and Analysis;
- Logic and Set Theory; and
- Statistics.

Arithmetic, Algebra, Number Theory

$+$	Plus (positive)
$-$	Minus (negative)
$\pm$	Plus or minus (positive or negative)
ab	*a* times *b* (*a* multiplied by *b*)
a/b	*a* divided by *b* (ratio of *a* to *b*)
$=$	Equals
$\equiv$	Is identically equal to
$\neq$	Is not equal to
$\cong$	Is approximately equal to
$>$	Is greater than
$<$	Is less than
$\geq$	Is greater than or equal to
$\leq$	Is less than or equal to
a^n	$a \cdot a \cdot a \cdot a \ldots$ to *n* factors
$a^{1/2}$	The positive square root of *a*, for positive *a*
$a^{1/n}$	The *n*th root of *a*, usually the principal *n*th root
a^0	The number 1 (for $a \neq 0$); by definition
a^{-n}	The reciprocal of a^n ($1/a^n$)
$a^{m/n}$	The *n*th root of a^m
e	The base of the system of natural logarithms
$\log_a x$	Logarithm (to the base *a*) of *x*
$\log a$	Common logarithm (to the base 10) of *a*
$\ln a$	Natural logarithm (to the base *e*) of *a*
$\exp x$	e^x, where *e* is the base of the system of natural logarithms
$a \propto b$	*a* varies directly as *b*
i	square root of -1
$n!$	$1 \cdot 2 \cdot 3 \cdot \ldots n$ (*n* factorial)
$f(x)$	A function *f* whose argument is the variable *x*
$\lvert z \rvert$	Absolute value of *z* (numerical value of *z*, without regard to sign)
$\mathrm{Re}(z)$	Real part of *z* (where *z* is a complex number, i.e., has a real and an imaginary part)
$\mathrm{Im}(z)$	Imaginary part of *z* (where *z* is a complex number)
$\mathbf{i},\mathbf{j},\mathbf{k}$	Unit vectors along the coordinate axes
$\mathbf{a} \cdot \mathbf{b}$	Dot product (scalar product) of the vectors **a** and **b**
$\mathbf{a} \times \mathbf{b}$	Cross product (vector product) of the vectors **a** and **b**

Trigonometric and Hyperbolic Functions

sin	Sine
cos	Cosine
tan	Tangent (sin/cos)
cot	Cotangent (1/tan)
sec	Secant (1/cos)
csc	Cosecant (1/sin)
$\sin^{-1}x$	The principal value of an angle whose sine is x (for x real)
$\sin^2 x$	$(\sin x)^2$, and similarly for $\cos^2 x$, $\tan^2 x$, and so on
sinh	Hyperbolic sine
cosh	Hyperbolic cosine
tanh	Hyperbolic tangent

Elementary and Analytic Geometry

$\perp$	Perpendicular (is perpendicular to)
$\parallel$	Parallel (is parallel to)
π	Ratio of a circle's circumference to its diameter ($\pi \cong$ 3.14159 . . .)
(x,y)	Rectangular coordinates of a point in a plane
(x,y,z)	Rectangular coordinates of a point in space
(r,θ)	Polar coordinates
(r,θ,z)	Cylindrical coordinates

Calculus and Analysis

Σ	Summation sign, applied to an expression to indicate successive addition (requires specification of the addition to be done)
$\sum_1^n$ or $\sum_{i=1}^n$	Sum to n terms, one for each positive integer from 1 to n
$\sum_1^\infty x_i$	The sum of the infinite series $x_1 + x_2 + x_3 + \cdots$
Π	Product sign, applied to an expression to indicate successive multiplication (requires specification of the multiplication to be done)
$\prod_1^n$	Product of n terms, one for each positive integer from 1 to n
$\lim_{x \to a} y = b$	The limit of y as x approaches a is b, for $y = f(x)$

Δy	An increment of y
dy	Differential of y
dy/dx	Derivative of y with respect to x, where $y = f(x)$
$d^n y/dx^n$	The nth derivative of y with respect to x, where $y = f(x)$
D	The operator d/dx
$\int f(x)\, dx$	The indefinite integral of $f(x)$ with respect to x
$\int_a^b f(x)\, dx$	The definite integral of $f(x)$ between the limits a and b

Logic and Set Theory

$x \in M$	The point x belongs to the set M
$M = N$	The sets M and N coincide
$M \subset N$	Each point of M belongs to N (M is a subset of N)
$M \supset N$	Each point of N belongs to M (M contains N as a subset)
$M \cap N$	The intersection of M and N
$M \cup N$	The union of M and N

Statistics

χ^2	The chi-square distribution, $\chi^2(n) = \Sigma\, (u_i)^2$, where n is the number of independent observations $x_1, \ldots, x_n$ from a normal distribution $N(\varepsilon, \sigma^2)$, and the u_i are the standardized variables, $u_i = (x_i - \varepsilon)/\sigma$
P.E.	Probable error (same as probable deviation)
s	Standard deviation (from a sample)
σ_x	Standard deviation of the population of x
t	Student's t-test statistic
V	Coefficient of variation
Q_1	First quartile
Q_3	Third quartile
$E(x)$	Expectation of x
$P(x_i)$	Probability that x assumes the value x_i

HANDLING MATHEMATICAL EXPRESSIONS

7.57 This section explains the physical arrangement of mathematical expressions: horizontal spacing of elements, placement of equations, breaking equations that are too long to fit in a single line, and miscellaneous guidance on such tasks as choosing symbols of the appropriate size and putting subscripts and superscripts on the correct levels.

With the increasing availability of desktop publishing systems, many writers can now prepare mathematical material themselves in book-quality typography. The end of this section presents a summary of rules for mathematical typography, as well as some examples showing how it looks.

Horizontal Spacing of Elements Where a One-Space Separation Is Used

7.58 Use one-space separation on either side of mathematical symbols +, −, ±, and the like, between two quantities. Exceptions: Closed spacing—no space—is used in subscripts and superscripts, and may be used in other places where compactness is desired—as in tables.

$xy < p$	29 ± 0.4	$(x + y) + (u + v)$
$a + b = c - d$	$3.4 \times 10^{n-2}$	$r > r_0$

7.59 Use one-space separation to separate the abbreviations sin, cos, tan, sinh, cosh, tanh, log, ln, and so on, from characters on either side:

$\sin\ \omega t$	$2\ \cos\ \theta$	$i\ \cosh\ y$
$\tanh^{-1}\ y/x$	$r\ \tan\ \phi\ \sinh\ x$	$\log\ p$
$\sin^2\ p$	$2\pi\ \exp\ (x^2/2 + y)$	$\log_{10}\ R$

7.60 Use one-space separation to separate integral, summation, and continued product signs, the abbreviation lim, and the like, from characters on either side. When these signs carry limits extending to right or left, the limit (rather than the sign) is used as the starting point for counting the one-space separation:

$$\frac{1}{\pi}\int\ x\ dx \qquad p(x)\ \sum_{n=0}^{p+q+r}\ R_n\ X^{n-2} \qquad 2x\ \sum_{n=0}^{k}\ (y - y_n)$$

$$\int_{a+b}^{c}\ t^2\ dt \qquad 3\ N \lim_{x \to \infty}\ P_N(x) = Q$$

7.61 Use one-space separation between *unenclosed* fractions multiplied together:

$$\frac{pg}{xy}\ \frac{s}{t} \qquad \frac{1}{2}\ \frac{2}{3}\ \frac{3}{4} \qquad \left(\frac{x - y}{2}\right)\left(\frac{x + y}{2}\right)$$

But:

$$\left(\frac{x - y}{2}\right)\left(\frac{x + y}{2}\right)$$

7.62 Use one-space separation to separate differentials, derivatives, and so on, from other expressions by which they are multiplied:

$2x\ dy/dx$	$\iint (x^2 - y^2)\ dx\ dy$	$t^2\ \frac{d^2R}{dt^2}$
$\int (u - iv)\ dw$	$\int x^n\ dx$	$dA\ V(\rho,\theta)$

Where Closed Spacing Is Used

7.63 Use no spacing between quantities multiplied together when no multiplication sign is used:

$2ab$	$20pu$	$(p - q)(r - s)$	$p(q - r)$
$p(q - r)^{1/2}$	$x(1 - y^2)$	Y_aY_bZ	$7x^2y^2z^2$

7.64 Use no spacing for notation of function theory and calculus:

$\phi(x) + \theta(y) + \tau(z)$	$f(z) = 2z$	dx
$\frac{\partial t}{\partial x}$	d^2y/dx^2	$F^{(n+1)}(x) = \left[F^{(n)}(x)\right]$

7.65 Use no spacing between a symbol and its subscript or superscript, and between elements of the subscript or superscript:

B_k	$re^{i\omega t}$	F_{max}	R_{av}	x^2
$e^{-(x+iy)}$	L_{m+n+n}	$\log_{10} (a + b)$		

7.66 Use no spacing for signed quantities, between the sign and the quantity:

± 0.5	$+3$	>5	$-3/2$	$-ab$
$-\cos t$	$r = -ab^2$	~ 7.5	<3	

Placement of Equations

7.67 The space above and below displayed equations (those set apart from text) should be double the space used between lines in the text. This rule applies to the space between the equation and text and the space between one equation and another. The uppermost character in an equation is the starting point for counting space above; the lowermost character is the starting point for counting space below.

7.68 For horizontal placement of equations, centering is a fairly standard practice and is recommended if it can be done easily and conveniently. If centering is not convenient, however, choose some scheme, such as using a two-paragraph indent for all equations that will fit on a single line with that indent and no indent (flush left) for all longer equations.

7.69 Do not begin an equation to the left of the left margin. Instead, break the equation and use two or more lines as required.

7.70 When two or more similar short equations are written one after the other, align them on their equals signs, letting the longest one be the one that is centered (or given a two-paragraph indent). See the following examples. Notice Eqs. (1) and (2), which are examples of equations with their identifying numbers at the right and left margins, respectively:

$$x(y) = y^2 + 3y + 2. \qquad (1)$$

$$\begin{aligned} p(x,y) &= \sin (x + y), \\ p(x_0,y_0) &= 2 \sin x_0 \cos y_0, \\ q(x,y) &= \cos (x + y), \\ q(x_0,y_0) &= \cos^2 x_0 - \sin^2 y_0. \end{aligned}$$

$$(2) \quad l_2(z) = \frac{25}{\pi^4} l_0 \int_1^\infty \frac{n^3}{e^n - 1} dn$$
$$+ P_1(z) + P_2(z) + P_3(z) + P_4(z) + P_5(z) + P_6(z) + P_7(z).$$

Breaking Equations

7.71 Equations are commonly broken at the equals sign. The equals sign is brought down with the remainder of the equation:

$$\int_0^1 \left(f_n' - \frac{n}{r} f_n\right)^2 r\, dr + 2n \int_0^1 f_n f_n' \, dr$$
$$= \int_0^1 \left(f_n - \frac{n}{r} f_n\right)^2 r\, dr + n f_n^2(1). \qquad (3)$$

7.72 The second most preferred place (other than the equals sign) to break an equation is at a plus or minus sign not enclosed within parentheses, brackets, or the like. The plus or minus sign is moved to the next line with the remainder of the equation, which is placed so as to end near the right margin:

$$\begin{aligned}\phi(x,y,z) &= (x^2 + y^2 + z^2)(x - y + z)^2 \\ &\qquad + [f(x,y,z) - 3x^2]. \qquad (4)\end{aligned}$$

7.73 The next best place to break an equation is between parentheses or brackets indicating multiplication of two terms. A multiplication sign is brought down to the next line to precede the remainder of the equation. Thus, the previous equation may be written:

$$\begin{aligned}\phi(x,y,z) &= (x^2 + y^2 + z^2) \\ &\qquad \times (x - y + z)^2 + [f(x,y,z) - 3x^2]. \qquad (5)\end{aligned}$$

7.74 The same general rules apply to breaking equations in the line of text. When such an equation cannot be broken except at a place that leaves the line of text awkwardly short or long, it is often advisable to set the entire equation apart from the text as a displayed equation.

7.75 When an equation is so long that it requires three or more lines, observe the following rules:

1. Start the first line at the left margin.
2. End the last line at the right margin.
3. Keep the intermediate lines more or less centered and somewhat inside the margins:

$$\begin{aligned} & h\int_{u_k}^{u_{k+1}} \left[y_n + \Delta_1 y_n u + \frac{\Delta_2 y_n}{2}\left(u^2 + u\right) \right. \\ & \qquad\qquad \left. + \frac{\Delta_3 y_n}{6}\left(u^3 + 3u^2 + 2u\right)\right] du \\ & \qquad\qquad = h\left[y_n u + \Delta_1 y_n \frac{u^2}{2} + \frac{\Delta_2 y_n}{2}\left(\frac{u^3}{3} + \frac{u^2}{2}\right) \right. \\ & \qquad\qquad\qquad\qquad \left. + \frac{\Delta_3 y_n}{6}\left(\frac{u^4}{4} + u^3 + u^2\right)\right]. \qquad (6)\end{aligned}$$

Miscellaneous Typographic Details

7.76 Use a minus sign whose length matches the length of the plus sign. Depending on the typewriter, this may be the hyphen key or the underline key (raised to midline level). If your equipment has a special key for the minus sign, use that.

7.77 The large integral and summation signs are the ones normally used in displayed equations; the small signs are appropriate for use in the line of text.

7.78 Use a size for parentheses, brackets, braces, and similar marks of enclosure in keeping with the height of the expression enclosed. The small size is satisfactory for one-line expressions having either simple subscripts or superscripts (one hand-roll down or up on a typewriter). Built-up fractions and other tall expressions require larger sizes. Parentheses, brackets, braces, and the like should always be centered vertically on the main line of the expression they enclose.

7.79 The level of subscripts or superscripts outside marks of enclosure should be on the same level as the most extreme subscripts and superscripts inside—that is, as low as the lowest inside subscript and as high as the highest inside superscript. When no inside superscript exists, put the outside superscript on the level where the inside one would be if there was one:

$$(a^2+b^2+c^2)^3 \qquad (x+y)^3 \qquad [x_0+(a_1+b_1)\,x_1]_{av}$$

$$\left\{\left[\frac{x_1}{a_1}+\frac{x_2}{a_2}+\dots+\frac{x_n}{a_n}\right]_{x_i=y_i}\right\}^3$$

7.80 In general, put the superscript over the subscript, which gives a neat and compact appearance. If this arrangement is not possible with the equipment you have (for example, some word processing equipment cannot do it), then use other arrangements:

General Rule	Typical Exceptions
$a_1^2+b_1^2=c_1^2$	$a_1{}^2+b_1{}^2=c_1{}^2$
$L^{-2}\left(x+y\right)^2_{max}$	$\left(E_{max}\right)^2=\left[(x+y)_{max}\right]^2$
$x_i'+y_i'=z_i'$	$x_i{}'+y_i{}'=z_i{}'$

Punctuating Mathematics

7.81 When placing punctuation marks after mathematical expressions, place them immediately after the last character of the expression, on the main line.

7.82 Although a few publishers arbitrarily omit punctuation from displayed equations, the best mathematical writers depend on punctuation to add precision to their meaning, just as all writers do. Mathematical material, including displayed equations, should be punctuated according to the same rules used in any other writing. These principles can be illustrated by considering some specific questions.

7.83 Number the displayed equations—at least the more important ones and any that are referred to at other points—to provide a convenient way to identify them. The equation number is typically enclosed in parentheses and placed at the right margin on a level with the equation. Equation numbers in text are preceded by the abbreviation for equation (Eq.) followed by the equation number enclosed in parentheses. Treat the parentheses as an integral part of the equation number in written references to the number:

> . . . we note that Eq. (7) is a consequence of Green's theorem.

Punctuation for Introducing Displayed Equations

7.84 What punctuation should be used after the last word in the line of text introducing a displayed equation? Some writers invariably use a colon, but a colon is not always appropriate. In fact, if the sentence continues after the displayed equation and takes on any kind of grammatical complexity, the introductory colon can make the sentence as a whole impossible to punctuate properly:

> Thus we observe that the height, *h*, is given by:
>
> $$h = ut \text{ (meters)}, \tag{1}$$
>
> and the volume, *V*, by:
>
> $$V = \pi r^2 ut \text{ (cubic meters)}, \tag{2}$$
>
> where *u* is upward velocity (meters per second), *t* is time (seconds), and *r* is radius (meters).

In this example, the colons introducing the two equations only impede the flow of the sentence. An observant editor would remove both of them.

7.85 What punctuation mark then should be used to introduce an equation? The answer depends on how the equation fits into the sentence. In the preceding example, no punctuation is needed. In other cases a colon would be appropriate, in others a comma, in others a semicolon, in others a dash. Make the determination according to the usual punctuation rules for sentences; think of the equation itself as grammatically equivalent to the corresponding statement in words. (See also Chapter 3.)

Punctuation After Displayed Equations

7.86 What about punctuation after a displayed equation? If the equation ends the sentence, follow it with the appropriate end-of-sentence mark (usually a period, although a question mark is a possibility). If the equation is within a sentence, follow it with the appropriate in-sentence mark—which may be no punctuation at all, or a comma, or a semicolon, or a colon, or possibly a dash. Again, make the determination according to the usual punctuation rules governing sentences. (See also Chapter 3.)

The same considerations apply in punctuating equations and mathematical expressions that occur in the line of text.

Punctuating Mathematical Where *Lists*

7.87 How should one punctuate a *where* list? A common feature in mathematical developments is the presentation of an equation followed by a *where* list defining the symbols used in the equation:

> The force of gravitational attraction between two objects a and b is
>
> $$\mathrm{F} = k(m_{\mathrm{a}}m_{\mathrm{b}}/r^2), \qquad (3)$$
>
> where
>
> F = force (dynes),
>
> k = gravitational constant (6.670×10^{-8}),
>
> m_{a} = mass of object a (grams),
>
> m_{b} = mass of object b (grams),
>
> r = distance between a and b (centimeters).
>
> Several concepts should be noted in this example:
>
> 1. The *where* is placed below the equation and flush left.

2. The symbol definitions begin below the *where,* with the symbols indented based on the length of the longest definition that allows alignment of all equal signs. List the definitions in the same order in which they appear in the equation.
3. The equals signs in the list are aligned.
4. Units of measure for each item in the list are included, in parentheses, at the end of each line (note, however, that measurement units are not always needed in such a list).
5. The end of each line in the list is punctuated, with commas for the first four lines and a period for the final line, which in this case marks the end of the sentence. (If the definitions had contained internal punctuation, then a semicolon might have been used instead of a comma at the end of all but the last one.)

Include short and simple *where* lists in the line of text instead of displayed like this one. The example in Section 7.84, for instance, contains a list of only three definitions, which are simply run into the text.

Typographically Discriminable Zeros

7.88 One final note on mathematical writing style. Use of the correct symbols is very important to the quality of a mathematical presentation. Quality is degraded by use of the alphabetic character O (oh) for 0 (zero), or a hyphen for a minus sign, or the letter x for a multiplication sign (assuming the typeface you are using has the proper symbols available). The careful writer pays attention to all these details when producing a finished work of mathematical exposition.

BOOK-QUALITY MATHEMATICAL TYPOGRAPHY

7.89 Textbooks and journals on scientific subjects generally follow certain standard typographic conventions for presenting mathematical material. Each of the categories of mathematical symbols commonly used is represented by a distinctive typeface. This practice has two beneficial effects. It sets the mathematical material off from the accompanying text, thus improving the visual effect, and it speeds the experienced reader's understanding of the mathematics. A sample of such typography is given in Exhibit 7–3.

Exhibit 7–3: Mathematical Typography

Boldface roman for vector

Standard roman for mathematical abbreviations

4. The Dot Product. The *dot product*[1] of two vectors is defined to be the product of their lengths by the cosine of the angle between them. In symbols,

$$\mathbf{A} \bullet \mathbf{B} = |\mathbf{A}|\ |\mathbf{B}| \cos(\mathbf{A},\mathbf{B}), \tag{4-1}$$

where (**A**,**B**) is the angle from **A** to **B**. Thus, **A** • **B** is a *scalar*, not a vector. Geometrically,

$$\begin{aligned} \mathbf{A} \bullet \mathbf{B} &= |\mathbf{A}|\ (\text{projection of } \mathbf{B} \text{ on } \mathbf{A}) \\ &= |\mathbf{B}|\ (\text{projection of } \mathbf{A} \text{ on } \mathbf{B}). \end{aligned} \tag{4-2}$$

Evidently (**A**,**B**) can be measured in several ways. However, since $\cos\theta = \cos(-\theta) = \cos(2\pi - \theta)$, these different measures all yield the same value for **A** • **B**. The fact that $\cos\theta = \cos(-\theta)$ also yields

$$\mathbf{A} \bullet \mathbf{B} = \mathbf{B} \bullet \mathbf{A}, \quad \text{commutative law,} \tag{4-3}$$

Italic for scalar

and one easily verifies the additional properties

$$(t\mathbf{A}) \bullet \mathbf{B} = t(\mathbf{A} \bullet \mathbf{B}), \quad \text{associative law,} \tag{4-4}$$

$$\mathbf{A} \bullet (\mathbf{B} + \mathbf{C}) = \mathbf{A} \bullet \mathbf{B} + \mathbf{A} \bullet \mathbf{C}, \quad \text{distributive law.} \tag{4-5}$$

For proof of (4–5) use (4–1) to transform (4–5) into

$$|\mathbf{A}|\ |\mathbf{B} + \mathbf{C}| \cos\psi = |\mathbf{A}|\ |\mathbf{B}| \cos\phi + |\mathbf{A}|\ |\mathbf{C}| \cos\theta, \tag{4-6}$$

where the angles are defined in Fig. 12. Now (4–6) follows from

$$|\mathbf{B} + \mathbf{C}| \cos\psi = |\mathbf{B}| \cos\phi + |\mathbf{C}| \cos\theta \tag{4-7}$$

[1] The terms *scalar product* and *inner product* are often used.

SOURCE: Sokolnikoff, I. S. and R. M. Redheffer. *Mathematics of Physics and Modern Engineering* (New York: McGraw-Hill, 1958), p. 293.

Although most scientifically trained readers recognize this typographic coding of mathematical material, few of them are thoroughly familiar with the coding rules.

Character Sets for Mathematics

7.90 Mathematical expressions consist mainly of English and Greek letters plus various nonalphabetic characters—numerals (1, 2, 3, etc.) and special symbols (+, –, =, ∂, ∫, ∇). Only the different typefaces that are used for the letters will be considered in this discussion.

Typeface Conventions for Mathematics

7.91 Mathematical typography uses four different typefaces:

1. Standard typeface used for text, called roman, which is upright and unslanted (a, b, c, A, B, C),
2. Italic, which is distinctly slanted (*a, b, c, A, B, C*),
3. Boldface roman, whose letters are made in the roman shape but with much thicker lines (**a, b, c, A, B, C**), and
4. Boldface sans serif, which lacks the curlicues or serifs that decorate other typefaces (**a, b, c, A, B, C**).

Nonmathematical Character Typographic Conventions

7.92 Set nonmathematical characters in roman type. Letters of the alphabet represent not only mathematical quantities but also depict auxiliary, nonmathematical characters such as words, abbreviations, and identifying symbols. These nonmathematical characters are set in the same roman type as the rest of the text. Familiar examples (see Exhibit 7–4) include abbreviations for mathematical functions and measurement units, symbols for the chemical elements, prefix symbols for multiples of measurement units, and other standard abbreviations. Initial letters (e.g., those of words or people's names) are commonly used as identifying subscripts, as are arbitrary symbols such as a, b, c.

Exhibit 7–4:
Abbreviations and Other Nonmathematical Symbols in Mathematical Typography

(Set in roman type. These lists provide examples only and are not intended to be exhaustive.)

Mathematical abbreviations (see also Chapters 5 and 6).
arg, av, cos, cosh, cot, det, erf, exp, Im, lim, ln, log, max, min, mod, Re, Res, rms, sin, sinh, tan, tanh, tr

Measurement unit abbreviations (e.g., SI units)
A, C, cd, F, H, Hz, J, K, kg, lm, lx, m, N, rad, s, sr, T, V, W, Wb

Prefix symbols for multiples of SI units (see Section 7.43).
a, c, d, da, f, G, h, k, m, M, n, p, T

Other standard abbreviations
abs, approx, bcc, calc, cgs, const, diam, el, expt, fcc, inel, mks, obs, SI, theor, tot

Chemical symbols
B, Be, C, F, H, He, Li, N, O

Abbreviations and Nonmathematical Symbols: Typographic Conventions

Mathematical Quantities and Typography

7.93 For typographical purposes, mathematical quantities are divided into three categories: scalars, vectors, and tensors.

Scalar Typography

7.94 The most familiar of the three categories of mathematical quantities, and the most common in typical mathematical material, are the *scalars,* or ordinary numbers. Set scalars in italic type. In general, scalars are represented by single English or Greek letters. The familiar algebraic symbols x, y, and z represent scalars. The physical quantities mass, length, time, and temperature (commonly represented by m, l, t, and T) are scalars. Greek letters that represent scalars automatically take care of themselves, since the typical Greek typeface already has a slanted, italic appearance.

Vector Typography

7.95 Vectors, the next most familiar mathematical quantities, have both a scalar magnitude and a spatial direction. Set vectors in boldface roman type. (It is common to depict a vector as an arrow, whose length represents the vector's magnitude and whose direction represents the vector's direction.) The physical quantities velocity, acceleration, force, and momentum (commonly represented by the symbols **v**, **a**, **F**, and **p**) are vectors.

Vectors represented by English letters are set in boldface roman, as in the examples in the preceding paragraph. (Greek letters that represent vectors should be in a bold Greek typeface, unslanted if available.)

Tensor Typography

7.96 Tensors (also called *dyadics*) are the least familiar of the three quantities. Set tensors in boldface sans serif type. Tensors are mathematically different from scalars and vectors, so they are distinguished from them typographically. The inertia, stress, and strain tensors (**I**, **P**, and **S**) are examples.

7.97 In applying the foregoing rules, the primary difficulty is in distinguishing between scalars and nonmathematical symbols (vectors and tensors are usually identifiable by context). Exhibit 7–4 should provide assistance in making fine distinctions.

EIGHT

Addressing Nonnative Readers

- *Nouns*
- *Adjectives*
- *Prepositions*

Conditions—Not Definite

- *Adverbs*
- *Nouns*
- *Adjectives*

Location

- *Nouns*
- *Prepositions*

Function Words

- *Pronouns*
- *Articles*
- *Prepositions*

Joining Words

- *Adverbs*
- *Conjunctions*

Helping Words

- *Verbs*

Examples: Basic Usage for a Simplified English

8.1 Communications between two or more linguistic groups has always been a problem. Sign language, lingua franca, and pidgin have been the traditional solutions. Although useful in many instances, these solutions are limited for the communication of exact scientific and technical information.

Translation, as an alternative, presents other problems: the shortage of qualified translators; delays in publishing research; the need to develop multilingual scientists, engineers, and technicians.

A language develops to meet the communication needs of the users. Many languages have not developed the vocabulary required for science and technology. Translation of science and technical documents into these nontechnical languages requires the artificial development of terminology. In effect, the reader would be learning a new, and technical, language.

SIMPLIFIED ENGLISH

8.2 There have been various efforts to develop vocabularies and rules of grammar for controlled subsets of Standard English. These are called *Basic English, Fundamental English,* or *Simplified English.* Simplified English is defined as a subset of Standard English intended for science or technical communication.

8.3 Create a limited vocabulary for any Simplified English technique. Such a listing is not simply an approved word list. Each word must have one, and only one, meaning. Define technical terms in a glossary. If possible, illustrate them with example sentences.

8.4 Keep sentence construction as simple as possible. The preferred constructions are

- Present, past, and simple future tense;
- Active voice;
- Imperative mood.

8.5 Do not write sentences longer than 20 words in length, and no more than one sentence in 10 on a page should exceed 16 words in length. The intended audience's English-language skill should be your guide; the less skill and training, the shorter the sentences. (See also Chapter 2.)

8.6 Use illustrations, charts, graphs, and tables as often as possible to convey scientific and technical information. (See also Chapters 12 and 13.)

Simplified English Vocabulary

8.7 The lexicon (Section 8.27) includes the nontechnical words required for most technical documentation. This core vocabulary has fewer than 300 words. The words are grouped by function, as in a thesaurus, rather than simply alphabetically, as in a dictionary. These groups are

8.27	OPERATIONS—DEFINITE
8.30	OPERATIONS—NOT DEFINITE
8.32	THINGS—DEFINITE
8.36	THINGS—NOT DEFINITE
8.38	CONDITIONS—DEFINITE
8.44	CONDITIONS—NOT DEFINITE
8.48	LOCATION
8.51	FUNCTION WORDS
8.55	JOINING WORDS
8.58	HELPING WORDS

These groups are then subdivided into verbs, nouns, prepositions, adverbs, adjectives, pronouns, and conjunctions. The arrangement is alphabetical within the subdivisions.

8.8 In the development of a Simplified English vocabulary for a specific user, or group of users, it is probable that certain words in the core lexicon will be omitted or redefined and that other words will be added. The only rule that must not be broken is that each word has one, and only one, meaning in a specified *domain of discourse.*

For example, the word *tap* is not only a commonly verbalized noun as in

> *Tap* (n): A tubular plug for drawing liquid from a source and having a device for controlling the flow of liquid. A faucet. A tool for cutting internal screw threads.
> *Tap* (v): The insertion of a tap. To cut internal screw threads.

But the verb form:

> *Tap* (v): Strike a sharp light blow. Also provides: *Tap* (n): The sound of a sharp light blow.

Tap is also used as a prefix meaning a slender, tapered, cylinder, as in *Tap*-root.

There are many other usages, which, depending on the domain of discourse, refer to the drawing off of resources, energy, or information, and to the devices and procedures involved.

8.9 Select the one meaning of a word that is most useful for a given domain of discourse, and use that definition without variation throughout the text:

> Not: Use a standard *tap* and *tap* the container.
> Use: *Tap* the *tap* into the container.

Standard usage could include

> If the *tap* leaks, remove it.
> Use a pipe thread *tap* and *tap* the hole.
> Assemble a threaded *tap* with gasket to the container.

Lexicon Examples

8.10 Use the examples (Section 8.60) as guides to alternate constructions. Each example sentence offers common words and preferred sentence constructions.

TECHNICAL WORDS

8.11 Technical words are terms (1) specific to an item, process, or action; or (2) required for the description, operation, or maintenance of a system or equipment.

8.12 Develop the required technical word lexicon by

1. Preparing a complete parts list and a list of required tools and test equipment. Define these part names and add them to the lexicon. Illustrate the parts lists and the tools and test equipment:

    Bolt, Resistor, Hose, Tubing, Cover, Gear, Hammer, Torque-Wrench, Voltmeter, Micrometer

2. Listing the words that describe a process required for operation or maintenance. Define these words and add them to the lexicon:

    grind, drill, heat-treat, anodize, plate

3. Listing the words that describe an action required for operation or maintenance. Define these words and add them to the lexicon:

    boot (computer), fire (explosives)

The simplest method is to prepare a first draft of the document and catalog the process and action words. Review these terms, and replace them with words from the lexicon where possible. Add those words that must be used to the lexicon.

4. Listing and defining all technical terms in a glossary.

Jargon

8.13 Do not use jargon. Jargon as a verbal shorthand is useful for ease of communication between specialists, but it is a problem in communication between the specialist and the nonspecialist:

Not: The technician can *relate* to *user-friendly* documentation.
Use: Technicians with special training and experience can use this documentation.

Jargon, as the vernacular of a discipline, profession, or trade, does produce some technical words that are required for scientific and technical communications. These exceptions arise, particularly in developing technologies, when new words, or new meanings for existing words, must be used to describe items, actions, or processes.

Vernacular

8.14 Do not use vernacular:

Not: This is a *hassle-free* system.
Use: This system is *easy to use.*

Communication between groups that do not have a common general education and experience is difficult. Use only specified and defined words.

Abbreviations

8.15 Avoid abbreviations except for the standard international measurements such as mm, kHz, VA and kg. (See also Chapters 5, 6, and 7.)

If abbreviations are used, write the term(s) in full, followed by the abbreviation in parentheses, the first time it appears in each chapter or section.

Include all abbreviations, even the standard ones for units of measurement, in a list of abbreviations. (See also Chapters 5 and 6.)

Control and Instrument Markings

8.16 Always refer to markings on controls and indicators exactly as they appear:

Not: Set Auxiliary Switch to Panel Lights Bright.
Use: Set Auxiliary Switch to **panel brt.**

Homonyms

8.17 Do not use both forms of a homonym. That is, if you must use *lead* (n): Metal, chemical symbol Pb, do not use: *lead* (v): To go before or in front of.

There may be exceptions in the case of some technical terms:

Make sure *fire* protection equipment is in place.
Fire the charge.

Fire (n): Light and heat produced by combustion
Fire (v): Activate an explosive device

Verbalized Nouns

8.18 Do not verbalize nouns. Verbalized nouns are a special class of homonyms that is particularly difficult to differentiate for persons learning English. (See also Chapter 4.)

8.19 In some instances a word may be used more often as a verb than a noun. In these cases the verb form should be used in the lexicon. For example, if *tap* is selected as a verb that means "cut an internal thread," then all references to the tool for cutting the thread must be *tool, screw thread cutter, internal.*

Prefix of Negation

8.20 Do not use *un* as a prefix to indicate negation:

Not: It is UNNECESSARY to stop the motor.
Use: It is NOT NECESSARY to stop the motor.

Prefixes can cause other problems. *Flammable,* for instance, is used to label petroleum containers because the correct term—*inflammable*—is often confused with *unflammable.*

Gerunds

8.21 Do not use verbal nouns created with the *ing* suffix. Persons for whom English is a second language have difficulty with this construction: (See also Chapter 2.)

Not: If the temperature is *rising,* stop the motor.
Use: If the temperature *rises,* stop the motor.

Passive Voice

8.22 Do not use the passive voice in procedures or other instructions. Review all instances of passive constructions, and rewrite in the active voice when possible: (See also Chapter 2.)

Not: The exterior may be cleaned with a soft, lint-free cloth dampened with water or alcohol.
Use: Clean the exterior with a soft, lint-free cloth. Dampen the cloth with water or alcohol.

Stacked Nouns

8.23 If more than two related nouns are in sequence in a sentence, rewrite the sentence to separate them, or join two of the nouns with a hyphen to show relationship: (See also Chapters 2, 3, and 4.)

Not: Open the *water control valve.*
Use: Open the *water control-valve.*

LIST FORMAT

8.24 Arrange complex material in list format: (See also Chapter 2.)

Not: The system operates in four separate modes, Operation Cycle, Self Cycle, Make-up Cycle, and Aircraft Cycle.
Use: The system operates in four separate modes:

a. Operation Cycle,
b. Self Cycle,
c. Make-up Cycle, and
d. Aircraft Cycle.

This type of list format may be numbered or set as a bulleted list. The publication specifications and the purpose of the document will determine the appropriate style.

ILLUSTRATIONS

8.25 Use illustrations as often as possible. Illustrations are more easily understood than text. Often companies use wordless instructions for products destined for international markets. (See also Chapters 12 and 13.)

GUIDELINES FOR PREPARING DOCUMENTS FOR NONNATIVE READERS

8.26 Preparing documents for nonnative readers requires six basic steps:

1. Prepare a complete parts list for the apparatus, equipment, or system, including tool and test equipment, complete with illustrations.

2. Add the necessary technical terms to the lexicon.
3. Review the parts list illustrations. Rework and annotate to reduce the text. Review other data and arrange in data displays (charts, graphs, tables, and diagrams). A minimum of 50 percent of all pages should be data displays and illustrations. In order of preference use
   a. Illustrations,
   b. Charts,
   c. Graphs,
   d. Tables.
4. Write the text using
   a. Words from the lexicon;
   b. Active voice; and
   c. Present, past, or simple future tense.
5. Prepare the table of contents, glossary, and index.
6. Review the text for unnecessary use of
   a. Passive voice,
   b. Jargon,
   c. Abbreviations, and
   d. Homonyms.

LEXICON: BASIC VOCABULARY FOR A SIMPLIFIED ENGLISH

8.27 Operations—Definite

8.28 Verbs

add (*v*): to increase the number, dimension, quantity
adjust (*v*): to change to specified position or condition
apply (*v*): to put on
assemble (*v*): to attach items together
attach (*v*): to cause items to hold or stay together
bend (*v*): to change from straight
break (*v*): to separate by force
calculate (*v*): to find a result by using mathematics
clean (*v*): to remove unwanted materials
close (*v*): to move to a position that stops material from going in or out
compare (*v*): to examine for difference
connect (*v*): to put together to make one system
cut (*v*): to divide into parts with a sharp tool
disassemble (*v*): to take an assembly apart
disconnect (*v*): to cause a connection to come apart
do (*v*): imperative to perform an action
drain (*v*): to remove liquid
fill (*v*): to put into a container to a known level or pressure
find (*v*): to locate or discover
go (*v*): to move to or from something; to come into a condition
hold (*v*): to keep in a specified place or position
install (*v*): to attach an item to a second item, to connect an item in a system
lift (*v*): to move to a higher position
lock (*v*): to hold with a locking device
loosen (*v*): to cause something to become loose
make (*v*): to manufacture or fabricate
measure (*v*): to find the dimension or quantity
mix (*v*): to combine mechanically two or more materials to make one compound
move (*v*): to change the location
operate (*v*): to put or keep equipment in action
pull (*v*): to put a force on something in the direction of the power source

push (*v*): to put a force on something in a direction from the power source
put (*v*): to cause something to be in a specified position
read (*v*): to obtain data with the eyes
reject (*v*): to decide that an item is unserviceable
release (*v*): to make free
remove (*v*): to take from or move away
repair (*v*): to make an item serviceable
replace (*v*): to remove an item and to install another item in its place
rub (*v*): to move along a surface with pressure
seal (*v*): to close to prevent leaks
see (*v*): to get knowledge through the eyes
set (*v*): to put a control in a given condition or mode
start (*v*): to cause operation
stop (*v*): to make an end to operation
tighten (*v*): to make tight
turn (*v*): to cause to move about an axis or a point
use (*v*): to bring into action
weigh (*v*): to measure mass
write (*v*): to put data on a surface with a tool

8.29 Nouns

approval (*n*): permits specified action
correction (*n*): a change to make something correct
damage (*n*): a change in condition resulting from an unwanted occurrence
indication (*n*): something that shows
injury (*n*): damage to a person
inspection (*n*): a comparison of an item with its specification
installation (*n*): the procedure used to install an item
instruction (*n*): data for performing tasks
leak (*n*): a defect that lets fluid or light come out of or go into something
maintenance (*n*): the servicing and/or repair of an item or system
movement (*n*): a change of position
precaution (*n*): a procedure that tells you what to do to prevent accidents
preparation (*n*): the procedure to make something ready for use
protection (*n*): something to prevent damage
removal (*n*): a procedure that removes
result (*n*): what you get when you expend energy
selection (*n*): the result of choosing

8.30 Operations—Not Definite

8.31 Verbs

become (*v*): to come to be
continue (*v*): to not stop
decrease (*v*): to make or become less
increase (*v*): to make or become larger
lower (*v*): to move in a downward direction
occur (*v*): to take place
prepare (*v*): to make ready
prevent (*v*): to make sure something does not occur
recommend (*v*): to advise what is best
refer (*v*): tells you where to look for data
schedule (*v*): to plan for a specified time
supply (*v*): to give something

8.32 Things—Definite

8.33 Nouns

assembly (*n*): two or more items assembled for a function
component (*n*): part or subassembly
container (*n*): something that holds fluids or items
contents (*n*): something in a container
data (*n*): known facts
detail (*n*): a part of something
display (*n*): a visual indication
drop (*n*): a small sphere of fluid
flow (*n*): a continuous movement of liquid
group (*n*): a number of items that share a relationship
hole (*n*): an empty space in a solid object
item (*n*): part made for a product
joint (*n*): the place where two or more items are attached or two edges touch
mark (*n*): something to show location
mechanism (*n*): a system of parts that operate together
mixture (*n*): the compound you get when you mix materials
name (*n*): the identification of an item
number (*n*): a symbol showing how many or which one in a series
part (*n*): one piece or subassembly of a machine or other equipment
procedure (*n*): a number of steps that are in sequence
sample (*n*): a part of a group used for an inspection or test procedure
solution (*n*): a liquid that includes dissolved material

step (*n*): a specified part of a procedure
subassembly (*n*): an assembly that is part of a larger assembly
symbol (*n*): a sign that identifies an item or procedure
system (*n*): an assembly or connection of parts to do a specified operation
test (*n*): a procedure to examine an item or system
tolerance (*n*): permitted difference from a standard
type (*n*): of a specified group
unit (*n*): equipment with one function
value (*n*): a quantity that is given or calculated
work (*n*): effort exerted to do something

8.34 Adjectives

each (*adj, pn*): all of two or more considered separately

8.35 Pronouns

it (*pn*): that item
which (*pn*): the thing of things that
you (*pn*): the user
your (*pn*): of the user

8.36 Things—Not Definite

8.37 Nouns

output (*n*): power, energy, or quantity that comes out of a device
problem (*n*): something for which a solution must be found
something (*n*): an item that is not specified
source (*n*): something that supplies energy

8.38 Conditions—Definite

8.39 Verbs

are (*v*): refer to "be"
be; is; was; are; were (*v*): to occur, to have a property
can (*v*): to be able to
cause (*v*): to make occur
come (*v*): to move from "here" to "there"; to get into a condition
contain (*v*): to have or hold in
follow (*v*): to come after
have (*v*): to possess as a part or quality
include (*v*): to make or be a part of
is (*v*): refer to "be"
keep (*v*): to continue to do or hold
let (*v*): to permit an action
permit (*v*): to give approval
touch (*v*): to have one surface of an item against one surface of a second item
was (*v*): refer to "be"
were (*v*): refer to "be"

8.40 Adverbs

always (*adv*): at all times
here (*adv*): this position
how (*adv*): by what procedure
immediately (*adv*): without delay
mechanically (*adv*): controlled or operated by a mechanism
not (*adv*): adverb of negation
off (*adv*): not in operation
on (*adv*): in operation
out (*adv*): not in place or position
then (*adv*): subsequent in sequence
there (*adv*): that position
through (*adv*): in one side and out the other
together (*adv*): in one group at the same time
up (*adv*): to a higher position

8.41 Nouns

area (*n*): specified surface
capacity (*n*): the maximum quantity that can be held
defect (*n*): an imperfection or fault
difference (*n*): that which is different between two numbers or functions
distance (*n*): the dimension between two points
error (*n*): the difference from that which is correct
failure (*n*): the condition when an item can not operate satisfactorily
heat (*n*): energy measured as temperature rise
height (*n*): vertical distance
length (*n*): the distance from one end to the other end of an item
limit (*n*): a specified maximum or minimum distance, number, quantity, or time
mode (*n*): a specific set condition of operation
performance (*n*): the measured operation of a mechanism or system
position (*n*): the location of an item at a given time

quality (*n*): value or type
quantity (*n*): a specified amount
range (*n*): the distance an object or value can move
rate (*n*): a measurement of how frequently something occurs or how quickly a value changes
risk (*n*): a possible danger
strength (*n*): the ability to resist force
total (*n*): the full quantity
view (*n*): what can be seen
volume (*n*): the amount of space contained by the exterior of an item
weight (*n*): the force exerted on a mass by gravity

8.42 Adjectives

all (*adj*): total quantity
approved (*adj*): permitted by appropriate authorities
available (*adj*): that can be gotten for use
bad (*adj*): not acceptable
careful (*adj*): with care
constant (*adj*): without change
correct (*adj*): is in agreement with all requirements
dangerous (*adj*): that can cause injury or can kill
different (*adj*): not the same
dirty (*adj*): not clean
dry (*adj*): not wet
easy (*adj*): not difficult
empty (*adj*): all contents removed
equal (*adj*): the same
flat (*adj*): has a continuous surface in the same plane
free (*adj*): can move without limits
full (*adj*): at maximum quantity
good (*adj*): not bad
horizontal (*adj*): parallel to the horizon
initial (*adj*): of or at the start
left (*adj*): direction, opposite to right
level (*adj*): horizontal to the specified datum
loose (*adj*): not attached, without tension
maximum (*adj*): the largest quantity or value
mechanical (*adj*): controlled or operated by a mechanism
minimum (*adj*): the smallest quantity or value
movable (*adj*): that can be moved
necessary (*adj*): that must be
new (*adj*): not used before
no (*adj*): not any
only (*adj*): nothing more or different
open (*adj*): not closed
regular (*adj*): occurs at specified times or intervals
right (*adj*): direction, opposite to left
rigid (*adj*): can not bend
safe (*adj*): not dangerous
same (*adj*): alike in type, quality, amount
serviceable (*adj*): can be used
special (*adj*): not standard
specified (*adj*): identified, related to a specification
stable (*adj*): does not change or move
straight (*adj*): without bends
sufficient (*adj*): not less than required
tight (*adj*): permits no motion

8.43 Prepositions

without (*pre*): not with

8.44 Conditions—Not Definite

8.45 Adverbs

approximately (*adv*): sufficiently accurate, not precise
carefully (*adv*): with care
frequently (*adv*): occurs again and again in a limited time
heavy (*adv*): has a large weight
quickly (*adv*): at a fast speed

8.46 Nouns

condition (*n*): the state of an item
function (*n*): the operations that something must do

8.47 Adjectives

approximate (*adj*): sufficiently accurate, not precise
frequent (*adj*): that occurs again and again in a limited time
high (*adj*): relative greater value
hot (*adj*): at a high temperature
large (*adj*): more than average in size or capacity

long (*adj*): has greater length
low (*adj*): below a given value
moist (*adj*): moderately wet
possible (*adj*): that can occur
rough (*adj*): not smooth
short (*adj*): that has small length
slow (*adj*): at low speed
small (*adj*): below average in size
smooth (*adj*): has continuous finished surface
soft (*adj*): easily penetrated
some (*adj*): an undefined quantity, a part of a group

8.48 Location

8.49 Nouns

access (*n*): means of approach
bottom (*n*): the lowest place
end (*n*): the limit in distance or sequence
front (*n*): that part which normally faces the user (fixed equipment)
point (*n*): an accurate location
side (*n*): a specified surface
space (*n*): a distance or volume that has specified limits
surface (*n*): the external limits of an object
top (*n*): highest point of an item

8.50 Prepositions

above (*pre*): in a higher position
across (*pre*): from one side to another
after (*pre*): that follows specified operation or time
against (*pre*): in contact with
before (*pre*): preceding in time or sequence
between (*pre*): in the space or time that separates two items or occurrences
into (*pre*): movement with access to something, change
near (*pre*): closeness in space or condition

8.51 Function Words

8.52 Pronouns

that (*pn*): item referred to
these (*pn*): that shows things referred to
they (*pn*): that shows things referred to
this (*pn*): that shows thing referred to

8.53 Articles

a (*art*): indefinite article
an (*art*): indefinite article
the (*art*): definite article

8.54 Prepositions

at (*pre*): shows place, direction, or time
for (*pre*): shows purpose, time, result, or object of an action
from (*pre*): that shows point of departure
in (*pre*): that shows location, time, and limits
of (*pre*): that shows having
to (*pre*): that shows direction, limit
with (*pre*): that shows association, means, or instrument

8.55 Joining Words

8.56 Adverbs

also (*adv*): additionally

8.57 Conjunctions

and (*con*): connects words or clauses that are grammatically the same
after (*con*): that follows specified operation or time
because (*con*): as a result of
before (*con*): preceding in time or sequence
but (*con*): on the contrary
if (*con*): on the condition that
or (*con*): to show an alternative
than (*con*): used with comparative adjectives or adverbs
unless (*con*): but not if
until (*con*): sets limit for an occurrence
when (*con*): at the time that
where (*con*): in which place
while (*con*): at the same time

8.58 Helping Words

8.59 Verbs

must (*v*): helping verb that shows necessity
will (*v*): helping verb that shows simple future tense

8.60 EXAMPLES: BASIC USAGE FOR A SIMPLIFIED ENGLISH

Not	Use
Make sure you are *able* to remove the pin.	Make sure you *can* remove the pin.
Check for *abnormal* temperatures.	Check temperature is X degrees Celsius *plus or minus* Y degrees.
Apply heat to *accelerate* the curing.	Apply heat to *decrease* the curing time.
Check leakage is within *acceptable* limits.	Check leakage is *not more than* X liters in one day.
Special Tool XYZ is required to *accomplish* this procedure.	Special Tool XYZ is required to *do* this procedure.
Achieve access to the valve.	*Get* access to the valve.
Actuate the system.	*Start* the system.
Check there is *adequate* fuel in the tank.	Check there are X *liters of fuel* in the tank.
Advance the control two detents.	*Turn* the control two detents *to the right*.
This can *affect* the temperature.	This can *cause* the temperature *to rise*.
Do not *alter* the mixture.	Do not *change* the mixture.
ABC is an *alternate* manufacturer of this part.	ABC is *one more* manufacturer of this part.
X is the *amount* of spare parts.	X is the *quantity* of spare parts.
Anchor X to Y.	*Attach* X to Y.
Add *another* washer.	Add *one more* washer.
Do the *appropriate* procedures.	Do procedures *X, Y,* and *Z*.
This *arrangement* of items is system X.	This *configuration* of items is system X.
Ascertain the mixture is dry.	*Make sure* the mixture is dry.
Pump X can *assist* pump Y.	Pump X can *help* pump Y.
Make sure there is *at least* X liters in the tank.	Make sure there is *not less than* X liters in the tank.

Not	Use
Acid will *attack* this surface.	Acid will *damage* this surface.
Do not *attempt* this.	Do not *try* this.
Use *authorized* procedures.	Use *approved* procedures.
Use a rubber mallet to *avoid* surface damage.	Use a rubber mallet to *prevent* surface damage.
Back-off the retaining nut one turn.	*Loosen* the retaining nut one turn.
The system has a *backup* generator.	The system has an *emergency* generator.
Begin the test at step twelve.	*Start* the test at step 12.
Blank the spare connector.	*Put a cap on* the spare connector.
Blend two X with one Y.	*Mix* two X with one Y.
Bolt A to B.	*Attach* A to B *with bolts.*
Put *both* washers under the bolt head.	Put *the two* washers under the bolt head.
Bottom the clamp-ring bolt.	*Tighten* the clamp-ring bolt until the bolt shoulder touches.
This can cause a *breakdown.*	This can cause a *failure.*
Bring the item to the hangar.	*Move* the item to the hangar.
Brush the mixture on the surface.	*Apply* the mixture to the surface.
This *cannot* move.	This *can not* move.
Cap the connector.	*Use a cap and seal* the connector.
The system is *capable* of running for three days.	The system *can* run for three days.
Use *care* when releasing pressure.	Release pressure *carefully.*
Carry out steps a, b, and c.	*Do* steps a, b, and c.
Do not put item X in this *category.*	Do not put item X in this *group.*
Fill the *cavity* with amalgam.	Fill the *empty space* with amalgam.
Cease operation.	*Stop* operation.
Make *certain* the power is off.	Make *sure* the power is off.
Certain items are not used.	*Some* items are not used.
Check that the power is on.	*Make sure* the power is on.

Not	Use
Check the diameter of the hole.	*Measure* the diameter of the hole.
Clamp the wire to the pipe.	*Use a clamp* to *attach* the wire to the pipe.
Classify these items.	*Put these* items *into groups.*
Put item X *close to* item Y.	Put item X *near* item Y.
Coat the shaft with grease.	*Apply a layer* of grease to the shaft.
Commence operation.	*Start* the operation.
The items are connected to a *common* ground.	The items are connected to the *same* ground.
Use the interphone for *communication* with the operator.	Use the interphone to *speak with* the operator.
At the *conclusion* of the test, turn the power off.	At the *end* of the test, turn the power off.
Conduct the inspection inside the hangar.	*Do* the inspection inside the hangar.
Confirm the part is broken.	*Make sure* the part is broken.
Too much force was used, *consequently* the part failed.	Using too much force *caused* the part to fail.
Cover the tank.	*Put cover on* the tank.
Crack the bleed valve.	*Open* the bleed valve.
This will *damage* the item.	This will *cause damage to* the item.
Do not *delay* draining the radiator.	Drain the radiator *immediately.*
The second line *delivers* oil to the bearings.	The second line *supplies* oil to the bearings.
This will *deplete* the charge.	This will *decrease* the charge.
Depress the control.	*Push* the control *in.*
Do the inspection *described* in Section 2.	Do the inspection *specified* in Section 2.
Detach the hoses.	*Disconnect* the hoses.
Do a *detailed* inspection.	Do a *careful* inspection.

Not	Use
Use a water trough to *detect* leaks.	Use a water trough to *find* leaks.
Not-filtered fluid can be *detrimental.*	Not-filtered fluid can *cause damage.*
The test procedure is used to *diagnose* the fault.	Use the test procedure to *find* the fault.
Use colored connectors to *differentiate* between the lines.	Use colored connectors to *identify* the lines.
It is *difficult* to do.	It is *not easy* to do.
Dilute the solution with clean water.	*Add* clean water to the solution.
Remove the fuse to *disable* the equipment.	Remove the fuse so that the equipment *can not operate.*
The flag will *disappear.*	The flag will *go out of view.*
Discontinue the operation when the container is full.	*Stop* the operation when the container is full.
Display night signals.	*Show* night signals.
Dispose of the used parts.	*Discard* the used parts.
There are two *distinct* signals.	There are two *easy-to-identify* signals.
Do not *distort* the seal.	Do not *put* the seal *out of shape.*
Distribute the orders as scheduled.	*Give* the orders as scheduled.
Do not *disturb* the alignment.	Do not *change* the alignment.
There is a *drop* in pressure.	There is a *decrease* in pressure.
The decrease in pressure is *due to* bad seals.	The decrease in pressure is *because of* bad seals.
Check the temperature each hour for the *duration* of the test.	Check the temperature each hour until the *end* of the test.
It is the *duty* of the operator.	It is the *task* of the operator.
Ease the seal off the shaft.	*Carefully remove* the seal from the shaft.
Effect the repair today.	*Do* the repair today.
This will *eliminate* six steps from the procedure.	This will *remove* six steps from the procedure.

Not	Use
Employ a bar and move this rock.	*Use* a bar and move this rock.
Empty the bin.	*Remove all contents* from the bin.
If you *encounter* a problem, read the manual.	If *there is* a problem, read the manual.
End the test.	*Stop* the test.
Enlarge the opening.	*Make* the opening *larger.*
Ensure there is *enough* fuel.	*Make sure* there is *sufficient* fuel.
The item is *equipped* with lights.	The item *has* lights.
Make sure the *essential* tools are there.	Make sure the *necessary* tools are there.
Estimate the height.	*Make an estimate of* the height.
Every screw is plated.	*All* screws are plated. Or: *Each* screw is plated.
Make an *exact* measurement.	Make an *accurate* measurement.
This will *exceed* the standard weight.	This will be *more than* the standard weight.
There is *excess* material.	There is *too much* material.
Exercise care.	*Be careful.*
If you *experience* a problem, shut power off.	If you *have* a problem, shut power off.
Have *extra* fluid in a container.	Have *more* fluid in a container.
Extract the bearing.	*Remove* the bearing.
For use in *extreme* temperatures.	For use in temperatures *above* X degrees and *below* Y degrees.
A bad connection will cause a *false* indication.	A bad connection will cause an *incorrect* indication.
The cover is *fastened* with six screws.	The cover is *attached* with six screws.
The *fault* is in this assembly.	The *failure* is in this assembly.
Replace *faulty* items.	Replace *not serviceable* items.
Open the valve to *feed* more fluid into the system.	Open the valve to *supply* more fluid to the system.

Not	Use
Most items have only a *few* controls.	Items in this group have *three, four, or five* controls.
Fit the collar to the shaft.	*Install* the collar on the shaft.
Inspect the weld for *flaws.*	Inspect the weld for *defects.*
Assemble Item *A flush* with surface of Item B.	Assemble Item *A in line with* surface of Item B.
Assemble in accordance with the *following* procedures.	Use *procedures X, Y, and Z* and assemble the item.
Force the pin into the hole.	*Push* the pin into the hole.
Use a filter to remove *foreign* matter from the fluid.	Use a filter to remove (*dirt, water, metal particles*) from the fluid.
These instruments are *fragile.*	These instruments are *easily damaged.*
Make sure the threads are *free from* rust and dirt.	Make sure the threads are *clean.*
Fill the tank with *fresh* oil.	Fill the tank with *new* oil.
This item can *function* as a filter.	This item can *operate* as a filter.
The manufacturer will *furnish* this information.	The manufacturer will *supply* this information.
Look in the maintenance manual for *further* information.	Look in the maintenance manual for *more* information.
Gauge the *gap* between part A and part B.	*Measure* the *distance* between part A and part B.
There is a *gradual* increase in temperature.	There is a *slow and continuous* increase in temperature.
Halt the operation.	*Stop* the operation.
This will *happen.*	This will *occur.*
This gas is *harmless.*	This gas is *not dangerous.*
Heat the parts to be welded.	*Increase the temperature of* the parts to be welded.
Hoist the bale.	*Lift* the bale.
All A's are B, *however* not all B's are A.	All A's are B, *but* not all B's are A.

Not	Use
The two items are *identical.*	The two items are *the same.*
Bad connections will *impair* the systems operation.	Bad connections will cause the system to *operate incorrectly.*
It is *imperative* that this procedure is followed.	It is *necessary* that this procedure is followed.
Implement the modifications.	*Make* the modifications.
This is *impossible.*	This is *not possible.*
The test procedure is *incomplete.*	The test procedure is *not complete.*
Indicate the damaged parts.	*Show* the damaged parts.
This will *inform* the operator that the system has failed.	This will *tell* the operator that the system has failed.
Initiate procedure B.	*Start* procedure B.
Insert pins in the holes.	*Put* pins *into* the holes.
Instruct the operator in the correct procedures.	*Tell* the operator the correct procedures.
There is *insufficient* fuel.	There is *not sufficient* fuel.
This bearing has an *integral* seal.	A seal *is a part* of this bearing assembly.
Interconnect the two systems.	*Connect* the two systems.
Close the valve to *interrupt* the flow.	Close the valve to *stop* the flow.
This procedure does not *involve* the use of special tools.	This procedure does not *include* the use of special tools.
The solution will *issue* from the pipe.	The solution will *come out of* the pipe.
Jack the item into the correct position.	*Use a jack and lift* the item into the correct position.
It is the operator's *job* to get the data.	It is the operator's *task* to get the data.
Join the two hoses.	*Connect* the two hoses.
The illustration is *keyed* to the parts list.	The numbers on the illustration *refer to* items in the parts list.
There is a *lack* of spare parts.	There are *no* spare parts.

Not	Use
This item has the *latest* modification.	This item has the *last* modification.
The tank *leaks.*	The tank has a *leak.*
The assembly is shipped *less* the motor.	The assembly is shipped *without* the motor.
Cyanide is *lethal.*	Cyanide will *kill.*
The warning lamp will *light.*	The warning lamp will *come on.*
Limit the number of shims to three.	*Do not use more than* three shims.
Link the cables with the clevis.	*Attach* the cables with the clevis.
List the defects on the given form.	*Write* the defects on the given form.
The system will *lose* pressure.	The system pressure will *decrease.*
Maintain speed.	*Keep* speed constant.
The *major* cause of failure is bad maintenance.	The *primary* cause of failure is bad maintenance.
The system will *malfunction.*	The system will *not operate correctly.*
Make sure there is a *margin* of error.	Make sure there is an *allowance* for error.
Multiply the *measurement* in inches by 25.4.	Multiply the *dimension* in inches by 25.4.
Tighten the nut until the collar and hub *meet.*	Tighten the nut until the collar and hub *touch.*
Mend the leak.	*Repair* the leak.
This *method* is not safe.	This *procedure* is not safe.
Misalignment of the shaft will cause vibration.	If the shaft is *not aligned* it will cause vibration.
There is a *mistake* in the data.	There is an *error* in the data.
This is a *mobile* unit.	This is a *movable* unit.
Mount the bust on the stand.	*Put* the bust on the stand.
Do not use solvents of this *nature.*	Do not use solvents of this *type.*
Add fluid until the tank is *nearly* full.	Add fluid until the tank is *almost* full.

Not	Use
Never open the cover when power is on.	*Do not* open the cover when power is on.
If pressure is correct do the *next* step.	If pressure is correct do step X.
Do not use *non*standard parts.	Do not use a part *that is not* standard.
Note the time.	*Write* the time.
Notify the clerk.	*Tell* the clerk.
Obtain the parts from the manufacturer.	*Get* the parts from the manufacturer.
There are *often* delays.	Delays are *frequent.*
Replace the *old* parts.	Replace the *used* parts.
The cleaner can be used *once.*	The cleaner can be used *one time.*
Make sure the system is *operable.*	Make sure the system *can be operated.*
Use oil A if the temperature is between x and y; *otherwise* use oil B.	Use oil A if the temperature is between x and y; *if not,* use oil B.
If pressure is *over* x, stop the pump.	If pressure is *more than* x, stop the pump.
Do not *overfill* the tank.	Do not *fill* tank *above specified line.*
Make sure the fluid does not *overheat.*	Make sure the fluid does not *become too hot.*
Pack the bearing with grease.	*Fill* the bearing with grease.
Patch the leak.	*Repair* the leak *with a patch.*
Sharp objects can *penetrate* the skin.	Sharp objects can *go through* the skin.
Add one liter *per* cubic meter.	Add one liter *for each* cubic meter.
Periodically inspect the filters.	Inspect the filters *after each X hours of operation.*
This is a *poor* procedure.	This procedure is *not good.*
Position the collar on the shaft.	*Install* the collar on the shaft.
There is a *potential* for damage.	Damage *can occur.*

Not	Use
An inspection will *precede* the operation.	An inspection will be *done before* the operation.
Do *precise* measurements.	Do *accurate* measurements.
Brass tools are *preferable.*	Brass tools are *recommended.*
The gauge is *preset* to zero.	The gauge is *set* to zero.
Clean the equipment *prior to* inspection.	Clean the equipment *before* inspection.
Proceed with the test.	*Continue* the test.
This will *prohibit* operation.	This will *prevent* operation.
Use the *proper* procedure.	Use the *correct* procedure.
Use paint to *protect* the surface.	Use paint to *seal* the surface.
The supplier must *provide* the information.	The supplier must *supply* the information.
This is a *provisional* modification.	This is a *temporary* modification.
Use *qualified* parts.	Use *approved* parts.
Raise the handles.	*Lift* the handles.
See if there is a *rapid* rise in pressure.	See if the pressure rises *quickly.*
The temperature may *reach* 60 degrees.	The temperature may *rise to* 60 degrees.
Reactivate the control system.	*Set* the control system *to on.*
We did not get a *reading* from that meter.	We did not get a *value* from that meter.
After the part is installed, *readjust* the load control.	After the part is installed, *adjust* the load control.
Ready the part for shipment.	*Prepare* the part for shipment.
This is the *reason* for the failure.	This is the *cause* of the failure.
After the parts are cleaned, *reassemble* the item.	After the parts are cleaned, *assemble* the item.
Recharge the cylinders.	*Fill* the cylinders.
Recondition this assembly.	*Repair* this assembly.
There is a *reduction* in voltage.	There is a *decrease* in voltage.

Not	Use
Valves *regulate* the flow.	Valves *control* the flow.
Relieve the pressure.	*Release* the pressure.
Repeat procedure X.	*Do* procedure X.
Replenish the fuel.	*Fill* the fuel tanks.
You will *require* metric tools.	Metric tools are *necessary.*
A dirty filter will *restrict* the flow.	A dirty filter will *decrease* the flow.
This will *result* in damage.	This will *cause* damage.
Resume the test.	*Start* the test.
Retouch the paint as required.	*Apply* paint as required.
Return to procedure X.	*Go to* procedure X.
There is a temperature *rise.*	There is an *increase* in temperature.
Rotate the control.	*Turn* the control.
Sample the hydraulic fluid.	*Get* (a specified amount) of the hydraulic fluid.
Save the parts.	*Keep* the parts.
Scrap the parts.	*Discard* the parts.
Seat the bearing in the hub.	*Install* the bearing in the hub.
Secure the harness with clamps.	*Install* the harness with clamps.
Select the 0–10 range.	*Set the range control* to 0–10.
Separate the transmission from the engine.	*Remove* the transmission from the engine.
Service the vehicle.	*Clean, lubricate, and inspect* the vehicle.
Do not change the switch *setting.*	Do not change the switch *position.*
Dirt will *settle* in the tank.	Dirt will *go to the bottom of* the tank.
This procedure *shall* be done.	This procedure *must* be done.
The operator *should* have the manual.	The operator *must* have the manual.
There are *significant* changes in the procedure.	There are *important* changes in the procedure.

Not	Use
The parts are *similar.*	The parts are *almost the same.*
Situate the tank not less than 100 meters from the building.	*Install* the tank not less than 100 meters from the building.
Slip the ring of the shaft.	*Remove* the ring from the shaft.
Spread the compound on surface A.	*Apply* the compound to surface A.
Use the *stipulated* procedure.	Use the *given* procedure.
Store the film in a cooler.	*Keep* the film in a cooler.
Switch the power on.	*Set* the power *switch* to on.
Take the necessary precautions.	*Do* the safety procedures.
Tape the carton.	Use *tape* and seal the carton.
Test the warning system.	*Do a test* of the warning system.
Thin the paint.	*Add* solvent to the paint.
Clean the item *thoroughly.*	Clean the item *carefully.*
Transport the assembly to A.	*Move* the assembly to A.
Do not use *unauthorized* materials.	Do not use *not-approved* materials.
Uncap the connectors.	*Remove caps from* the connectors.
Keep the *undamaged* parts.	Keep the *serviceable* parts.
Do not operate when power is *under* X volts.	Do not operate when power is *less than* X volts.
Undo the spring clip.	*Release* the spring clip.
Keep a *uniform* temperature.	Keep a *constant* temperature.
One hour of operation will need *up to* 10 liters of fuel.	One hour of operation will need *as much as* 10 liters of fuel.
Utilize all systems.	*Use* all systems.
There are *various* defects in the system.	There is *more than one* defect in the system.
Vent the system.	*Remove pressure from* the system.
Verify there are no leaks.	*Make sure* there are no leaks.
The damage is *visible.*	The damage *can be seen.*
The technician will *want* the special tools.	The special tools are *necessary* for this procedure.

Not	Use
Wash the parts with water.	*Clean* the parts with water.
Put the *waste* in the bin.	Put the *discarded material* in the bin.
Watch for leaks.	*Look* for leaks.
This *way* is better.	This *procedure* is better.
Use the old parts *whenever* they are serviceable.	Use the old parts *when* they are serviceable.
Measure the material to see *whether* it can be used.	Measure the material to see *if* it can be used.
Do the *whole* procedure.	Do *all of* the procedure.
Withdraw the pins.	*Remove* the pins.

NINE

Quotations

9.1 Quotations are words, ideas, or data found in another source and used in your writing.

QUOTING

Why Use Quotations?

9.2 Use quotations for

- Credibility. Quoting an established source lends credibility to your position.
- Emphasis. Quotes stand out from the rest of the text.

When to Use Quotations

9.3 When using words, ideas, or data from another source, identify them as quotations. Use quotations to

- Add credibility to an idea,
- Include a well-constructed argument, or
- Add interest.

When Not to Use Quotations

9.4 Do not quote if the quotation offers no special information or if nothing is gained by directly quoting someone:

The President said he would campaign in all states.

Not:

The President said he would "campaign in all states."

9.5 Do not quote when a paraphrase or summary can provide the same information more concisely.

9.6 If you use so many quotes that the quotations, rather than your own writing, carry the message, you should be aware of the legal issues involved with overusing quotations. (See Section 9.82.)

Limit lengthy quotations, although you may need them on rare occasions, such as when the quotation source is unavailable to most readers. In these cases put the information in a separate appendix and use standard footnote techniques to refer to it. (See also Chapter 10.)

How to Use Quotations

9.7 Follow these guidelines when using quotations:

1. Make quotations as brief as possible.
2. Use only as much of the quotation as necessary to convey its sense.
3. Form sentences so that quotes fit naturally into the logical, grammatical, and syntactical flow of the writing.

Specific details on how to put quotations into texts appear in Section 9.17.

Why Cite Sources?

9.8 Cite the source of quotations or any data gathered from another source because:

1. It is a courtesy to other authors to attribute them as a source.
2. Citing sources allows readers to consult those sources for more information.
3. Citing also allows readers to assess the credibility of sources.

When using words, ideas, or data from another source, cite that source to avoid violating copyright law. In some instances failure to cite sources may lead to plagiarism or libel charges. (See Section 9.82 for more information on legal matters.)

When to Cite Sources

9.9 Cite the source of an idea or a particular expression of an idea:

1. Direct quotations occur when another's words, data, or ideas are used directly. Direct quotations must be attributed.

2. Indirect quotations, such as paraphrasing or summarizing another's words or ideas, must be attributed, especially if the source provided a first exposure to an idea.
3. Actual dialog, such as the exact words used by someone during an interview, must be cited.
4. Little-known facts, or facts or data (such as a report on new research) that are not commonly available from many sources must be cited.

When Citations Are Unnecessary

9.10 Do not cite commonly known facts:

George Washington was the first President.

9.11 Do not cite facts or beliefs commonly available from many sources:

Mount Everest is the highest mountain in the world.

9.12 Do not cite proverbial, biblical, or well-known literary expressions:

Love of money is the root of all evil.

TYPES OF QUOTATIONS

9.13 There are two main types of quotations, direct and indirect:

1. Direct quotations use another author's words, data, or ideas without changing them in any way.
2. Indirect quotations paraphrase or summarize another author's words or ideas.

9.14 Direct quotations faithfully reproduce the source's exact words or data. Use direct quotations when there might be a question about the original statement or a representation of it.

9.15 Indirect quotations are useful in representing the sense of a long quotation, when the original is unnecessarily long, or when the exact wording is not as important as the sense it conveys.

Direct Quotations

9.16 Use direct quotations to reproduce words or data exactly as they appear in their source. In deciding how to include a quotation, the basic consideration is length.

Types of Direct Quotations

9.17 Include quotations shorter than three or four lines within the text (run-in quotations):

> John Carroll tells us that readers often miss important information in manuals because they "come to the learning task with personal concerns. . . . They skip critical information if it doesn't address those concerns." The key, then, is finding a way to connect critical information with a reader's concerns.

9.18 Set off quotations longer than four lines (set-off quotations) from the text:

> John Carroll addresses the issue of why readers often miss important information.
>
> > *Skipping.* New users often seem to feel that what cannot be used can be skipped. One person we observed dismissed several pages of a training manual, commenting "This is just information" as she flipped past it. People come to the learning task with personal concerns that influence their use of training materials. They skip critical material if it doesn't address those concerns. They browse ahead until they find an interesting topic and ignore its prerequisites.

9.19 Nonstandard quotations—epigraphs, dialog, direct oral sources, question-and-answer format, and correspondence—are handled differently. (See Section 9.73.)

General Guidelines for Direct Quotes

9.20 Use the following guidelines when including direct quotes:

1. Make the quotations as brief as possible.
2. Use only as much of the quotation as necessary to convey the sense of the quotation.
3. Form sentences so that quotes fit naturally into the logical, grammatical, and syntactical flow of the text:

   > John Carroll tells us that users often miss important information in manuals because they "come to the learning task with personal concerns. . . . They skip critical information if it doesn't address those concerns."

   Not:

   > John Carroll tells us that users often miss important information in manuals because: "People come to the learning

> task with personal concerns. . . . They skip critical material if it doesn't address those concerns. They browse ahead until they find an interesting topic and ignore its prerequisites."

Indirect Quotations

9.21 Use indirect quotations to summarize or paraphrase material from another source. Use indirect quotations to convey only the main idea or ideas behind a lengthy external source. An indirect quotation calls less attention to itself.

9.22 Summarizing or paraphrasing, however, does not reduce the responsibility for accuracy. Any summary or paraphrase must accurately represent the source's ideas and arguments.

9.23 To paraphrase successfully, put the idea in your own words. Do not simply rearrange the words:
Original:

> *Skipping.* New users often seem to feel that what cannot be used can be skipped. One person we observed dismissed several pages of explanation in a training manual, commenting "This is just information" as she flipped past it. People come to the learning task with personal concerns that influence their use of training materials. They skip critical material if it doesn't address their concerns. They browse ahead until they find an interesting topic and ignore its prerequisites.

Successful paraphrase:

> Readers use documents the way they want, not the way the author intends.

Possible plagiarism:

> Users approach documentation based on their personal agendas, often skipping crucial material and ignoring prerequisite information.

RULES FOR PUNCTUATION AND STYLE

9.24 One general guideline summarizes the concern for good style and grammar when using quotations: The quotation should fit naturally, grammatically, and logically into the flow of the writing.

When using a quotation that is an incomplete sentence, either rewrite the quote to make the thought complete, or incorporate the quotation into an existing complete sentence.

Introducing Quotations

9.25 Always introduce quotations. The introduction identifies the source and tells the reader the value of the quote in relation to the subject. Note the difference in how the reader will approach the quotation based on the following two introductions:

> As Deterline wisely cautions, "Gas chambers work every time."

> Deterline shows his alarmist nature and knack for overstatement when he complains that "Gas chambers work every time."

The introduction does not need to be formal. In fact, it is usually best to avoid formality. Overly formal introductions, such as those that use "said" or "the following quotation," are poor style.

Introducing Run-in Quotations

9.26 Avoid unnecessary formality.
Use:

> As Deterline observes in his article on user friendliness, "Validity does not necessarily equal friendly. Gas chambers work every time."

Not:

> My point is confirmed by the following quotation from Deterline's article on user friendliness: "Validity does not necessarily equal friendly. Gas chambers work every time."

9.27 The introduction does not necessarily have to precede the quotation. Depending on the style used in the surrounding text, the introduction may be more appropriate in the middle of or after the quotation:

> As Carl Nagas once begged, "Stop, in the name of science."

Or:

> "Stop," Carl Nagas once begged, "in the name of science."

Or:

> "Stop, in the name of science," Carl Nagas once begged.

9.28 In cases where the quotation is longer than a sentence, it may be appropriate to use a more formal introduction. In this case the introduction is normally a complete sentence, and a colon rather than a comma may be used before the quote:

> The warning at the end of his article is clear: "In today's litigious society it is foreseeable that trainers could be sued. A lawyer wanted to get a trade article, which stated that inadequate training was being given to restaurant staff, admitted as evidence in a recent court case."

Introducing Set-off Quotations

9.29 As with run-in quotations, the introduction for a set-off quotation depends on the quotation's context. In general, use a complete sentence to introduce a set-off quotation that starts with a complete sentence, and use an incomplete sentence to introduce a set-off quotation that starts with an incomplete sentence.

Quotations introduced with a complete sentence can use either an informal sentence that ends with a period or a formal sentence that ends with a colon.

Informal Introduction

9.30 Use an informal introduction when the exact content of the passage is not important:

> In *The Mythical Man-Month,* Frederick Brooks describes the biggest fallacy of all when scheduling.
>
>> All programmers are optimists. Perhaps programming attracts those who believe in happy endings. Perhaps the nitty frustrations drive away all but those who habitually focus on the end goal. Perhaps it is merely that computers are young, programmers are younger, and the young are always optimists. But however the selection process works, the result is indisputable: "This time it will surely run," or "I just found the last bug."
>>
>> So the first false assumption that underlies the scheduling of systems programming is that all will go well.

Formal Introduction

9.31 Use a more formal introduction when the exact content of a passage is important:

> In *The Mythical Man-Month,* Frederick Brooks lists the following five factors that are the most common causes of scheduling disasters:

> First, our estimating techniques are poorly developed. More seriously, they reflect a false assumption, i.e., that all will go well.
>
> Second, our estimating techniques falsely confuse effort with progress, hiding the assumption that men and months are interchangeable.
>
> Third, because we are uncertain of our estimates, software managers often lack flexibility.
>
> Fourth, progress is poorly monitored. Proven monitoring techniques in engineering are considered radical innovations in software engineering.
>
> Fifth, when schedules slip, the natural response is to add personnel. Like dousing a fire with gasoline, this makes matters worse.

9.32 If the set-off quotation begins with an incomplete sentence, introduce it in the same manner as a run-in quotation—form the introduction so that the reader can flow smoothly from the introduction into the quotation. The incomplete sentence in the quotation should complete the initial sentence:

> In the third chapter, Brooks describes a method for doing large jobs efficiently. The method is based on a team that is
>
>> organized like a surgical team rather than a hog-butchering team. That is, instead of each member cutting away at the problem, one does the cutting and the others give support that enhances effectiveness and productivity.

Introducing Indirect Quotations

9.33 Usually an indirect quotation needs no specific introduction. The purpose of an introduction is chiefly to identify the source to the reader. This information should be naturally included within the summary or paraphrase:

> Brooks makes it obvious in Chapter 2 that the major reason for missing schedules is the lack of accurate scheduling.

CAPITALIZATION

9.34 Change the initial capitalization of the quotation to make it grammatically correct with the remainder of the text. Base that decision on how the quotation would appear if it were your own words rather than a quotation, and whether the quotation can be read as a part of the main text or as a separate sentence.

9.35 Changing capitalization from the original does not normally have to be indicated in any way. For legal documents, for some publishers, and in cases where the meaning may be affected if capitalization is changed, a change must be indicated by putting the changed letter in brackets. (See Section 9.67.)

Capitalizing Run-in Quotations

9.36 Capitalize the quotation based on the grammatical needs of the text. If the quote will be read as part of a sentence, treat it that way:

> Emerson might just as well have been talking about useless software interface features when he said that "a foolish consistency is the hobgoblin of little minds."

9.37 If the quotation should stand out as a separate sentence, then capitalize it that way:

> Some software interface features are created simply to be consistent with another feature. But, as Emerson once said, "A foolish consistency is the hobgoblin of little minds."

Capitalizing Set-off Quotations

9.38 Because set-off quotations are reproduced exactly as they appear in the original and generally are not part of the primary text, the capitalization of the original should not be changed.

9.39 The only exception would be set-off quotations that begin with an incomplete sentence. In such a case, follow the same rules as for run-in quotations; capitalize according to how the quotation fits into the sentence that introduces it:

> In the third chapter, Brooks describes a method for doing large jobs efficiently. The method is based on a team
>
>> that is organized like a surgical team rather than a hog-butchering team. That is, instead of each member cutting away at the problem, one does the cutting and the others give him every support that will enhance his effectiveness and productivity.

QUOTATION MARKS

9.40 Start and end a quotation with double quotation marks. For a quotation within a quotation, use single quotation marks for the nested quotation. While it may sometimes be necessary to punctuate several nested quotations, in which case you should alternate between double and single quotation marks, avoid such potentially confusing situations, whenever possible. If a quotation has more than one quotation nested within it, reconsider whether the quotation is obvious enough to be useful to the reader.

9.41 Do not use quotation marks for summaries or paraphrases.

Run-in Quotations

9.42 Always use quotation marks when including a direct run-in quotation. Start and end the quotation with double quotation marks. If the quotation includes a quotation, use single quotation marks for the nested quotation:

> As Fredd says, "The paper confirms that she said, 'I think the robber called it a cannon.' "

Set-off Quotations

9.43 Do not use quotation marks to start or end a set-off direct quotation, unless these marks are part of the style used by the publisher or the person for whom the writing is intended.

PUNCTUATION

9.44 Punctuation that is part of a quotation should be retained and included within the quotation marks. However, commas and periods at the end of a quotation should be changed to fit the sentence's grammar.

Commas and Periods

9.45 Punctuate the sentence as if the quote were your own words. Use a comma to introduce a quotation unless the quotation is a predicate nominative or a restrictive appositive:

> Franklin D. Roosevelt might just as well have been talking about the fear of using computers when he said that "all we have to fear is fear itself."
>
> The interface, the electronic means we use to talk to the computer, offers the user a map to an invisible territory. But, as Winston Niles Rumfoord once said, "Do not mistake the map for the territory!"

9.46 Commas and periods should always fall within the quotation marks. Change commas and periods at the end of a quote to match the context of the sentence containing the quote:

"I refuse to testify," he said.
He said, "I refuse to testify."

Other Punctuation and Quotations

9.47 Punctuation other than commas and periods go within the quotation marks if they are part of the quotation and outside if they are not:

"I refuse to testify!" he said.
He said, "I refuse to testify!"
He said, "Do I refuse to testify?"
Did she say, "I refuse to testify"?
Was "I refuse to testify based on the Fifth Amendment" all she was willing to say?

9.48 When the quotation ends a sentence, and the sentence has punctuation other than a period, put that punctuation outside of the quotation marks:

What does he mean by "I refuse to testify"?

9.49 Although the quotation in the previous example is a complete sentence, ending the quotation with a period would be redundant. If a quotation falls at the end of a complete sentence, and the quotation has special punctuation, however, use both the quotation's punctuation and the sentence's punctuation:

Did he say "Do I refuse to testify?"?
Did he say "I refuse to testify!"?

REVISING QUOTATIONS

9.50 Revise a direct quotation to

- Maintain the grammatical sense of the writing,
- Clarify the quote within the context of the writing,
- Keep the quotation as brief as possible, and
- Omit extra or distracting words or phrases.

9.51 In revising a quotation, remember the following guidelines:

- Do not change the meaning of the quotation, and
- Treat the source fairly.

9.52 Intentionally altering a quotation to change its meaning can be construed as libel.

9.53 Do not point out such revisions as

- Matching the text's capitalization by changing the first letter's case (unless writing a legal document or matching a publisher's style).
- Changing commas and periods at the end of a quotation to match the punctuation needs of the sentence that includes the quotation.
- Changing all caps to small caps, italic to underline, and similar typographical items, as long as the intended meaning is retained.

Omitting Parts of Run-in Quotations

9.54 The treatment of a run-in quotation depends on how much of the quotation is used and what part of it is omitted.

9.55 To indicate an omission, use ellipses points. Ellipses points are three periods with a space between the periods and the surrounding text.

When the quotation is an obviously incomplete sentence, no ellipses points are needed:

> Jackson warns that, partly due to "today's litigious society," trainers have to worry about being sued.

9.56 When the quotation is a sentence with the beginning omitted, do not use ellipses points. Simply treat the quotation's initial punctuation as part of your sentence: (See Section 9.34.)

> At the end of his article he warns that "it is foreseeable that trainers could be sued. A lawyer wanted to get a trade article, which stated that inadequate training was being given to restaurant staff, admitted as evidence in a recent court case."

9.57 When the quotation is a sentence with the end omitted, use ellipses points only if the omission makes the end of the sentence grammatically incorrect, or to intentionally call attention to the fact that the original text does not end there:

> The warning at the end of his article clearly places the blame: "In today's litigious society . . ."

9.58 Use ellipses points to indicate an omission when the quotation is a sentence with text removed from the middle:

> The warning at the end of his article is clear: "In today's litigious society it is foreseeable that trainers could be sued. A lawyer wanted to get a trade article . . . admitted as evidence in a recent court case."

9.59 When the quotation is longer than a sentence, and text is omitted between sentences or at the end of the first sentence, include enough of the text to form complete sentences and use four-dot ellipses points to indicate the missing text. In fact, four-dot ellipses points are actually a period followed by three ellipses points:

> Frederick Brooks identifies the biggest scheduling fallacy as the fact that "all programmers are optimists. . . . The result is "indisputable: 'This time it will surely run,' or 'I just found the last bug.'"

9.60 If the sentence preceding the omission in a multisentence quotation requires punctuation other than a period, use the punctuation and standard ellipses points:

> The article once again raised the question: "Should we certify technical writers? . . . We must answer the other questions before we can answer this one."

Omitting Parts of Set-off Quotations

9.61 When an incomplete sentence introduces the set-off quotation, the first sentence of the quotation, normally another incomplete sentence, completes the introduction. In this case, do not begin the quotation with ellipses points:

> Frederick Brooks says that the biggest scheduling fallacy is that
>
>> all programmers are optimists. Perhaps this modern sorcery attracts those who believe in happy endings and fairy godmothers.

9.62 When a complete sentence introduces the set-off quotation, the quotation normally begins with a complete sentence. In this case, do not use ellipses points:

> In *The Mythical Man-Month,* Frederick Brooks describes the biggest fallacy of all when scheduling.
>
>> All programmers are optimists. Perhaps this modern sorcery attracts those who believe in happy endings and fairy godmothers.

9.63 When a complete sentence introduces the set-off quotation and the quotation begins with an incomplete sentence, precede the first sentence with ellipses points:

> In *The Mythical Man-Month,* Frederick Brooks describes the biggest fallacy of all when scheduling.
>
> > . . . programmers are optimists. Perhaps this modern sorcery attracts those who believe in happy endings and fairy godmothers.

9.64 If text has been removed from within the quotation or at the end of the quotation, follow the same guidelines as for run-in quotations:

> In *The Mythical Man-Month,* Frederick Brooks describes the biggest fallacy of all when scheduling.
>
> > All programmers are optimists. Perhaps this modern sorcery attracts those who believe in happy endings. . . . Perhaps the nitty frustrations drive away all but those who habitually focus on the end goal. Perhaps it is merely that . . . the young are always optimists.

9.65 Setting off a quotation allows the writer to reproduce a long, continuous piece of text. However, if a set-off quotation needs considerable editing, either paraphrase or rewrite it to incorporate it into the primary text as a run-in quotation.

Adding Material to Run-in Quotations

9.66 Add material to a direct quote only to clarify something that might not be apparent to the reader without being able to see the entire quote in its original context. Enclose additions within a quotation in brackets ([]).

9.67 If a misspelling involves the omission of a letter, add the letter in the appropriate place within brackets. If the misspelling is obvious, either add the correct letter within brackets or use [*sic*]:

> He failed to use the cor[r]ect address.

9.68 *Sic* is Latin for “thus” or “so.” Use [*sic*] to indicate that an obvious error has been faithfully reproduced from the original. Misspellings, erroneous data, or apparent errors in logic can be indicated with [*sic*]. The latter error is probably the most important kind to identify in this manner:

> Jackson intended to orient [*sic*] the learner to the system.

Note that, stylistically, [*sic*] can convey more than the fact that a quote has been accurately reproduced. In some cases, especially where a quotation has been included so that it can be refuted, using [*sic*] has negative implications.

9.69 When the correct spelling, number, date, or other error is not obvious, put the correct information in brackets:

> Galacia, a region of central Europe, includes the north slope of the Carpathian Mountains and the valleys of the upper Vistula, Diester [Dniester], Bug, and Seret rivers.

9.70 Use brackets [] with nothing between them to indicate information missing from the source, such as when the original is unintelligible or incomplete. This practice often occurs when the source is damaged and unreadable.

9.71 Put clarifying notes in brackets. Explanatory text might include the source of italics, explanation of a pronoun reference, translations, and to add sense:

> According to several theorists, one of the most frequent errors made by instructors is expecting too much commitment *before* [emphasis in original] providing reinforcers. In volunteer organizations, consistent, frequent, timely, and *appropriate* [emphasis mine] reinforcement significantly influences commitment.

> When he [Tom Smithson] was a student, they say he was very wild.

> These questions [motivation planning questions] apply to planning for meetings, instructional sessions, projects, and other activities.

> Een schip op het strand is een baken in zee [A ship on the beach is a lighthouse to the sea].—Dutch Proverb

Adding Material to Set-off Quotations

9.72 The rules for adding to a set-off quotation are the same as those for run-in quotations. (See Section 9.66.)

OTHER DIRECT QUOTATIONS

Epigraphs

9.73 Epigraphs are interesting, relevant quotations placed at the beginning of a book or each chapter of a book.

9.74 Use the following guidelines for epigraphs:

- Do not use quotation marks.
- Use a "display" typeface. This can be as simple as using an italic font several points larger than the body text.
- If the source needs to be cited, the citation should be right-aligned below the quote, usually in a slightly smaller typeface.

An epigraph might look like this:

Small opportunities are often the beginning of great enterprise.
—Demosthenes

Quoting Oral Sources

9.75 Treat quotes from speeches or conversations in the same way as quotations from written sources. Cite the date, time, and location of the speech or conversation accurately. Also, get the speaker's permission to use quotations from the speech.

Dialog

9.76 Direct and indirect dialog are similar to direct and indirect quotes. (See Sections 9.16 and 9.21.) Direct dialog reproduces the dialog verbatim; indirect dialog paraphrases the speech.

9.77 Direct dialog does not require that the speaker be identified for each utterance. If only two people are speaking, then it is sufficient to indicate the speaker change by starting each speaker's words on a new line and by beginning and ending each speaker's words with double quotation marks:

Bob asked, "What are you doing?"
"Starting the experiment," Tom replied.
"But isn't the flask supposed to be empty?"
"No! It has to have distilled water in it."
"I think you better check that procedure list again."

Question and Answer

9.78 Question-and-answer format (Q & A) is sometimes used as an instructional device to convey technical information. For this kind of Q & A, use boldface, not quotation marks:

Why should I send in my registration card?
Registered owners receive immediate notice of updates, as well as a special price on upgrades.

Are there any other benefits?
Yes. You will also be notified of new products and receive other special offers from time to time.

Quoting From Correspondence

9.79 Correspondence includes letters or memoranda written by one person and sent to one or more other people. The rules for including quotations from correspondence are the same as for including quotations from other printed sources, such as books. If parts of the correspondence such as the salutation are included, reproduce them exactly, as when reproducing a set-off quotation.

Quoting Within Footnotes

9.80 To include a quotation within a footnote, follow the same rules for quoting and citing as if the quotation were within the body text. If the quotation is longer than two or three sentences, consider including it within the primary text as part of the discussion rather than as a footnote.

Foreign-Language Quotations

9.81 In general, treat foreign quotations in the same manner as other quotations:

- Accurately reproduce the quotation, including special typographic conventions such as non-English characters and punctuation.
- Except for rare circumstances, always provide an accurate translation. The entire point may be lost by including a foreign quotation that the reader does not understand.
- Put the translation within brackets like other quotation additions. Or, put the translation in a footnote, especially if additional explanation is needed.
- Never retranslate a quotation that has already been translated. Find the original and use it.
- Provide your own translation only if no other is available.
- Provide definitions of specialized or controversial terms.

LEGAL GUIDELINES

9.82 Legal issues for the scientific or technical writer can be summed up quite simply: If you are not certain whether to cite a source, then cite it.

9.83 If you have quoted so much from a source that you are not certain whether you need permission, get permission. In doubtful instances, check with company lawyers, the publisher, or supervisory personnel.

9.84 For every quotation, determine the

1. Need to cite. That is, determine whether a source must be cited, and if so, the correct attribution for the source.
2. Accuracy of the quotation. That is, determine whether the quote has been accurately reproduced or has faithfully retained the source's meaning when paraphrased or summarized.
3. Need for permission. That is, determine whether the copyright holder needs to grant permission, and if so, if you have gotten it.

Need to Cite Sources

9.85 Cite the source whenever using another's words or ideas. See also Chapter 10 for details on exactly how to cite sources.

9.86 Using another's words or ideas without citing the source, or incorrectly citing the source, constitutes plagiarism. Plagiarism is illegal and, depending on the circumstances, may involve liable suits and monetary penalties.

Accuracy of Quotations

9.87 The basic rule of accuracy of quotations is

Make sure the quote is absolutely accurate.

9.88 Accuracy includes not only the words but the ideas expressed in the source. As the *Associated Press Stylebook and Libel Manual* cautions: "Remember that you can misquote someone by giving a startling remark without its modifying passage or qualifiers" (French, Powell, Angione, the Associated Press, *The Associated Press Stylebook and Libel Manual.* Reading, Mass.: Addison-Wesley, 1984).

9.89 Worse, deliberately misquoting or misparaphrasing can be considered libel: "Libel is injury to reputation. Words, pictures, or cartoons that expose a person to public hatred, shame, disgrace or ridicule, or induce an ill opinion of a person are libelous" (French,

Powell, Angione, the Associated Press, *The Associated Press Stylebook and Libel Manual.* Reading, Mass.: Addison-Wesley, 1984).

9.90 Even if the quotation's author will not sue, and even if the misquote is very slight, it goes against the ethics of credible writing to be anything less than entirely faithful to the original.

9.91 When in doubt, follow this simple guideline: If the original quotation does not fit your needs without substantial changes, then don't use it.

Copyright Permissions

9.92 There is no specific limit on the amount of quoting you can do; the limits are on the amount you can quote from any one source before you need to get permission from the copyright holder. The Copyright Act of 1976 includes guidelines for copyrights and copyright permissions.

9.93 The Copyright Act of 1976 lists four factors that determine whether material can be used with or without permission:

1. The purpose and character of the use, including whether it is for commercial use or for nonprofit educational purposes;
2. The nature of the work;
3. The amount and substantiality of the portion used in relation to the copyrighted work as a whole;
4. The effect of the use upon the potential market for, or value of, the copyrighted work.

Fair Use of Copyrighted Material

9.94 Although the Copyright Act does not specifically define "fair use," the act does say that material quoted "for purposes such as criticism, comment, news reporting, teaching, scholarship, or research" does *not* require permission.

9.95 Fair use has a lenient interpretation for material used in an academic setting and in the news, for out-of-print and difficult-to-acquire material, or for short sections of large and expensive documents normally unavailable to readers.

9.96 An important test of fair use is the percentage of the source cited. Three paragraphs from a 30,000-page encyclopedia is clearly not the same quantity of citation as three paragraphs from a one-page article.

9.97 Many scattered quotes from a single source, or one or more long passages from a single source, probably require permission. Contact the publisher. Usually you need permission from both the publisher and the author.

9.98 Once you request permission, you *must* wait until you receive that permission before using the material.

9.99 List permissions received on the copyright page, in the preface, in the acknowledgments, in the notes, or in some other conspicuous place.

TEN

Referencing Source Material

REASONS FOR CITING SOURCES

10.1 Technical authors should explain how their work is related to supporting information. Acknowledging these sources is an important part of writing for scientific and technical audiences.

10.2 Properly citing sources satisfies legal requirements, strengthens the author's argument, maintains intellectual integrity, and provides a valuable resource for the reader.

10.3 When quoting other authors, either directly or indirectly (see Chapter 9), credit them by identifying the quotation's source.

WHEN NOT TO CITE SOURCES

10.4 Although authors should properly cite supporting material, sources need not be credited at all times. Writing that is too densely documented can be difficult to read. In such cases, rather than documenting every sentence in a passage, group references and document the entire passage as a unit:

> We are encouraged by reports that students using simulations have often learned faster and more thoroughly than students using the actual materials (Manning 1983, Alessi and Trollip 1985, Winthrow 1985, Abrams et al. 1986).

Ideas and information available from many sources are considered common knowledge and need not be documented unless the sources are quoted directly. (See Chapter 9.)

METHODS OF CITING SOURCES

10.5 Scientific and technical authors document their primary source material in many ways. In general, writers should use one of the following formats:

- Author and date in text, with reference list at end,
- Complete reference in endnotes,
- Complete reference in footnotes,
- Complete reference in text, or
- Bibliography (alone or in addition to any of the above).

10.6 In addition to these formats, authors can acknowledge sources with a single unnumbered note at the beginning of an article:

> The author is indebted to the work of Wallace Hurton and John Brantmann on interface design for many of the ideas expressed in this article.

Or with a list of unnumbered notes at the end:

> The thesis for the first study treatment reported here was suggested by Wallace Hurton's text *Interface Design* (Englewood Cliffs, NJ: Prentice-Hall, 1989).
>
> The second thesis was suggested by John Brantmann's book *Planning Electronic Texts* (Cambridge, MA: MIT Press, 1991).

Such notes may include page numbers that identify locations within the article.

10.7 The reference method depends on where the work will be published. Often professional organizations and technical publishers require specific styles for documenting sources. For example, many professional journals require a footnote format, while others require author and date citations within the text and reference lists at the end.

10.8 If publishers have rules for documenting sources, writers will either be provided a style sheet or referred to a standard reference work. If no referencing rules are supplied, writers should select a format and use it consistently throughout the work.

10.9 Acknowledgments of source material should provide enough information so interested readers can locate the original works. A proper citation of source material appearing in books should be based on information found on the full title and copyright pages:

- Author(s) names (first and middle initials are preferred to full names in technical documents),
- Book title (or, in the case of a compilation, the chapter or paper title, followed by the book title),
- Editor(s) or translator(s) name(s) (first and middle initials are preferred to full names in technical documents),
- Position in a series (volume number or the like),
- Edition number,
- Publication information (city and state [sometimes omitted for well-known cities] of publication, publisher, publication date),
- Relevant page numbers for reference.

Basic punctuation conventions for book citations are

> Citation number. Author's name, *Book title* (City, State: Publisher, Date), Relevant pages.

The following examples are typical citations:

Single author	1. J. P. Brantmann, *Planning Electronic Texts* (Cambridge, MA: MIT Press, 1991), 186.
Two authors	2. P. P. Snebur and K. R. King, *Complex Algorithms for Novel Data Structures* (New York: Lavosier Press, 1978), 211–215.
Three authors	3. W. Jackson, T. V. Quarel, and C. S. Watkins, *The Development of Computer Mediated Data Structures* (Dayton, OH: Littleton and Sons, 1984), 322–327.
More than three authors	4. T. V. Quarel et al., *Relevant Design Decisions for the Development of*

	LSVI Systems (Boston: Wentworth Publishing, 1988), 34–54.
Chapter of paper title in book	5. M. M. Wooldridge, "Recent Studies at Carverson Foundation in Avian Carcinogen Reactions," in *New Directions in Fauna Research*, C. V. Edwards, ed. (London, Ontario, Canada: Billingsworth Publishers, 1955), 444–450.
Editor, compiler, or translator	6. W. W. Wentworth, ed., *Research Methods for Migratory Avian Studies* (Madison, WI: University of Wisconsin Press, 1911), 19–28.
	7. P. P. Snebur and K. R. King, eds. and comps., *Complex Algorithms for Novel Data Structures* (New York: Lavosier Press, 1978), 254–257.
	8. J. Bertin, *The Semiology of Graphics,* trans. W. J. Berg (Madison, WI: University of Wisconsin Press, 1983), 19.
Volume in a series	9. A. G. Barrce, *The Development of Original Programming Languages,* vol. 11 of *A Complete History of the Rise of Automata and Other Thinking Machines* (London, England: Ticknor and Fields, 1904), 121–127.

10.10 A proper citation of source material appearing in journals and periodicals should include

- Author(s) names (first and middle initials are preferred to full names in technical documents),
- Title of the article or paper,
- Periodical's full title or an acceptable abbreviation,
- Volume and number of the issue (or publication date),
- Issue date,
- Relevant page numbers for reference.

Basic punctuation conventions for periodical citations are

Citation number. Author's name, "Article's title," *Periodical title* Volume (issue number or publication date): Relevant pages.

The following is an example of a typical periodical citation:

10. H. R. Billows, "Working with Multimedia in Restrictive Environments: The Problem of the Glass Cockpit," *Journal of Hypermedia Studies* 130 (June 1991): 117–152.

Dissertations and Ephemera Titles

10.11 Treat a dissertation title like an article title and follow it with the phrase "Doctoral dissertation" (or "Master's thesis") and the name of the university:

> C. A. Resnick, "Computational Models of Learners for Computer-Assisted Learning." Doctoral dissertation, University of Illinois at Urbana-Champaign, 1975.

Titles

10.12 List all titles exactly as they appear in the source, including upper- and lowercase letters. Articles and short prepositions (three letters or less) begin with lowercase letters, and the first word of a subtitle begins with a capital letter.

10.13 Set titles of books in *italics* using upper- and lowercase letters. Also set titles of magazines and journals in italics in upper- and lowercase and spell them out in full. In typewritten documents, indicate italics by <u>underlined</u> <u>roman</u> <u>text</u>.

10.14 Titles of articles appearing in periodicals or titles of chapters found in books are enclosed in quotation marks and are set in upper- and lowercase roman type.

10.15 Some journals require that all words in book, article, and dissertation titles begin with lowercase letters except for the first word and proper nouns:

> S. N. Singh, "A modified algorithm for invertibility in nonlinear systems." *IEEE Trans. Automat. Contr.* AC-26 (June 1984): 595–598.

Note that this rule usually does *not* extend to journal titles.

Abbreviating Journal Titles

10.16 Abbreviate journal titles in references, end- or footnotes, and bibliographies for scientific and technical journals. Examine relevant style guides to determine if an abbreviation practice already exists. However, be skeptical of abbreviations appearing in existing articles, unless the publishing agency has a recognized style. (See also Chapter 5.)

10.17 Stylistic differences among various professional style guides—notably the American Institute of Physics, the American Chemical Society, and the American Medical Society—include the use of periods and other punctuation, and the treatment of articles and prepositions. In addition, most short words are written out, as are some important terms (such as biochemistry). Most style guides italicize journal titles. Some sample journal title abbreviations include

Exhibit 10–1:
American Institute of Physics Journal Abbreviations

Abbreviation	Full Name
Acta Math. Acad. Sci. Hung.	Acta Mathematica Academiae Scientiarum Hungaricae
Adv. At. Mol. Phys.	Advances in Atomic and Molecular Physics
Ann. Chim. Phys.	Annales de Chimie et de Physique
Ann. Phys. (Leipzig)	Annalen der Physik (Leipzig)
Appl. Spectrosc.	Applied Spectroscopy
Comments Nucl. Part. Phys.	Comments on Nuclear and Particle Physics
IEEE J. Quantum Electron.	IEEE Journal of Quantum Electronics
Izv. Acad. Nauk SSSR. Ser. Fiz.	Izvestiya Akademii Nauk SSSR, Seriya Fizicheskaya
J. Mech. Phys. Solids	Journal of Mechanics and Physics of Solids
Kolloid Z. Z. Polym.	Kolloid Zeitschrift & Zeitschrift für Polymere
Rep. Prog. Phys.	Reports on Progress in Physics
Sov. J. Plasma Phys.	Soviet Journal of Plasma Physics
Trans. Am. Soc. Met.	Transactions of the American Society for Metals

Exhibit 10–2:
American Chemical Society Journal Abbreviations

Abbreviation	Full Name
Anal. Chem.	Analytic Chemistry
Appl. Phys. Lett.	Applied Physics Letters
Can. J. Chem.	Canadian Journal of Chemistry
Dokl. Acad. Nauk SSSR	Doklady Akademii Nauk SSSR
J. Am. Chem. Soc.	Journal of the American Chemical Society
J. Lipid Res.	Journal of Lipid Research
Nature (London)	Nature (London)
Naturwissenschaften	Naturwissenschaften

Abbreviation	Full Name
Polym. J.	Polymer Journal
Z. Anorg. Allg. Chem.	Zeitschrift der Anorganischen und Allgemeinen Chemie

Exhibit 10–3:
American Medical Association Journal Abbreviations

Abbreviation	Full Name
Acc Chem Res	Accounts of Chemical Research
Am J Cardiol	American Journal of Cardiology
Anaesthesia	Anaesthesia
Arch Dis Child	Archives of Disease in Childhood
Br Med J	British Medical Journal
J Toxicol Clin Toxicol	Journal of Toxicology, Clinical Toxicology
Lancet	Lancet
N Engl J Med	New England Journal of Medicine

10.18 In addition to normal abbreviation practices, some journals use superbrief abbreviations:

Exhibit 10–4:
Superbrief Journal Abbreviations

Superbrief Form	Expanded Form	Full Name
AIChE J.	Am. Inst. Chem. Eng. J.	American Institute of Chemical Engineers Journal
AIP Conf. Proc.	Am. Inst. Phys. Conf. Proc.	American Intitute of Physics Conference Proceedings
CA	Chem. Abstr.	Chemical Abstracts
Cancer J.	Clinicians	Cancer Journal for Clinicians
JAMA	J. Am. Med. Assoc.	Journal of the American Medical Association
JNCI	J. Natl. Cancer Inst.	Journal of the National Cancer Institute

10.19 If a journal does not have a recognized abbreviation, create one from the following listing used by the American Chemical Society and the American Institute for Physics.

Exhibit 10–5:
Frequent Journal Abbreviation Terms

Word	Abbreviation
Academy	Acad.
Accounts	Acc.
Advances	Adv.
Akademii (R: Academy)	Akad.
American	Am.
Analytical	Anal.
Angewandte (G: Applied)	Angew.
Annals	Ann.
Archives	Arch.
Astronomical, Astronomy	Astron.
Atomic, Atomnaya (R)	At.
Berichte (G: Reports)	Ber.
British	Br.
Bureau	Bur.
Chemical, Chemistry, Chemie (G: Chemistry)	Chem.
Chimie (F: Chemistry)	Chim.
Conference	Conf.
Discussions	Discuss.
Edition	Ed.
European	Eur.
Experimental	Exp.
Fizika (R: Physics), Fizicheskaya (R: Physical)	Fiz.
Industrial	Ind.
Inorganic	Inorg.
Institute, Institut (F)	Inst.
Internal	Intern.
Journal	J.
Japanese	Japan.
Japan	Jpn.
Khimii (R: [of] Chemistry), Khimicheskaya (R: Chemical)	Khim.
Letters	Lett.

Word	Abbreviation
Magazine	Mag.
Mathematical, Mathematics	Math.
Molecular	Mol.
Optical, Optiko (R), Optik (G), Optique (F)	Opt.
Physics, Physik (G), Physique (F)	Phys.
Proceedings	Proc.
Publications	Publ.
Royal	R.
Research	Res.
Review(s), Revue (F)	Rev.
Science(s), Scientific	Sci.
Section	Sect.
Series, Serie (F), Seriya (R)	Ser.
Society	Soc.
Soviet	Sov.
Special	Spec.
Standard(s)	Stand.
Symposium	Symp.
Technical	Tech.
Technology	Technol.
Zeitschrift (G: Journal)	Z.
Zhurnal (R: Journal)	Zh.

F — French
G — German
R — Russian

TEXTUAL NOTES AND REFERENCE LIST

10.20 A textual note, also called an "in-text citation" or "author-date citation," acknowledges a reference immediately after it has been stated. Place the note in parentheses. The note has two parts: the author(s) last name(s) and the publication date. If the work is a book or a long article, include a chapter or page number. The last number in parentheses below is a page reference:

> The structure of a computer tutorial must consider student motivation (Alessi and Trollip 1985, 281). Furthermore, the success rate for such programs . . .

10.21 Provide a reference list at the end of the article or chapter to supplement textual notes; the reference list gives complete citation for all works cited. (See also Sections 10.9 and 10.10.)

10.22 Textual notes are convenient because reference information appears with the text it supports. While the reader must look in the endnotes or bibliography for complete citations, in-text citations provide more information about sources and guidance to additional information than endnotes or footnotes. They are also less distracting than looking elsewhere in the text for references.

10.23 The use of many textual notes interrupts the flow of the argument more visibly than other reference methods. Textual notes are convenient for writers because they provide easier editing and text repagination. They are especially useful to writers who do not use computerized footnoting. The Modern Language Association, the University of Chicago Press, and many scientific and technical publications accept textual notes as a valid referencing method.

10.24 Give special attention to multiple authors, multiple publications in a given year, and multiple works in a single citation.

10.25 If a reference has three or more authors, cite only the first, followed by the abbreviation "et al." ("and others"):

> While it is difficult to assess the cost of comparable lessons, research suggests that interactive tutorials are educationally more effective than similar linear lessons (Abrams et al. 1986).

10.26 If an author has two or more cited publications in a given year, distinguish the sources by adding a letter to the date. Assign these letters only after ordering references alphabetically:

> A limited environment differs from an enriched environment in both number of personnel and amount of equipment (Kirsch 1988b).

10.27 List multiple citations for a single block of text in chronological order. Place commas only between citations:

We are encouraged by reports that students using simulations have often learned faster and more thoroughly than students using the actual materials (Manning 1983, Alessi and Trollip 1985, Withrow 1985, Abrams et al. 1986).

If page numbers are used in the citations, use semicolons to separate entries:

(Manning 1983, 154; Alessi and Trollip 1985, 18–22; Withrow 1985, 28; Abrams et al. 1986, 177).

If two works have the same year, place them in alphabetic order:

(Alessi and Trollip 1985, 18–22; Withrow 1985, 28).

10.28 Treat sources for tables and figures as textual notes by placing the reference at the foot of the table or figure (or the first page of multipage tables). In ruled tables or figures, enclose the reference in a box:

Table 7-1: Tests of Propellant Vent Ports

Propellant Vent Port	Flow		Pressure	
	Specified (psi)	Actual (psi)	Specified (psi)	Actual (psi)
	Model 635			
External	100.0	102.7	10.2	10.4
Internal				
Aft	28.0	28.0	3.0	2.9
Mid	14.0	14.7	1.5	1.6
Stern	28.0	28.1	2.5	2.5
	Model 635-AX			
External	120.0	122.7	12.0	11.4
Internal				
Aft	30.0	28.9	3.0	2.9
Mid	15.0	14.0	1.5	1.5
Stern	32.0	29.3	2.5	2.4

SOURCE: National Transportation Safety Board, Annual Report, 1992.

10.29 Introduce table references with SOURCE: or SOURCES: in emphatic type (conventionally small caps).

10.30 If the table or figure is reproduced directly from another publication, follow the citation with the words *reprinted with permission of the publisher* in parentheses:

SOURCE: J. R. Paulson, "Tensile Strength Loss due to Radiation," *Metallurgy Review* 128 (September 1986): 55 (reprinted with permission of the publisher).

If the publisher requires that the citation give credit with some other wording or format, honor the publisher's request.

10.31 If a source has no listed author, use the name of the sponsoring organization instead:

> The amount of information a viewer receives from the screen is not always predicted accurately by his apparent attention to the screen (Ohio University 1973).

10.32 At the end of the article or book, compile a single list of all sources cited in textual notes. Title this list "References," "Works Cited," "Works Consulted," or the like. Each entry in the list includes all the bibliographic information about the source:

> S. M. Alessi and S. R. Trollip. *Computer-Based Instruction: Methods and Development.* Englewood Cliffs, NJ: Prentice-Hall, 1985.

10.33 Organize the list in alphabetical order by author or, if no author or editor is listed, by sponsoring organization. (See Section 10.31.)

ENDNOTES

10.34 Endnotes are similar to footnotes, except endnotes are grouped together at the end of the entire work or at the end of individual sections, rather than at the bottom of the page.

10.35 Place endnote numbers at the end of a sentence (or clause), so that they do not interrupt the text. Use a small superscripted number (placed after any punctuation marks), or set the numeral normal and enclosed in brackets or parentheses (placed before any punctuation marks). This number refers to a specific endnote. Endnote numbers begin at 1 and continue sequentially within chapters:

> A computer tutorial's structure must also consider the students' motivation.[4]

Or:

> A computer tutorial's structure must also consider the students' motivation [4].

10.36 Create a numbered list, entitled "Notes," to provide the necessary bibliographic information. If superscripted endnote numbers appear in the text, punctuate them with a period—not superscripted—in the notes:

> 4. S. M. Alessi and S. R. Trollip, *Computer-Based Instruction: Methods and Development* (Englewood Cliffs, NJ: Prentice-Hall, 1985), 281.

If the endnote numbers appear bracketed in the text, bracket them in the notes:

> [4] S. M. Alessi and S. R. Trollip, *Computer-Based Instruction: Methods and Development* (Englewood Cliffs, NJ: Prentice-Hall, 1985), 281.

10.37 One of the more difficult citation decisions involves how to treat multiple references to the same work. The general practice has been to use "ibid." (meaning "in the same place") followed by a page number to refer to a work cited in an immediately preceding citation. In contemporary practice references to previous citations for books contain

- Author's last name(s),
- Key words from the title,
- Date (if there are multiple works in the same year),
- Relevant page numbers.

Basic punctuation conventions for shortened book citations are

> Citation number. Author's name(s), <u>Book</u> <u>title</u> (Date, if necessary), Relevant pages.

The following is a typical shortened book citation:

First reference	3. C. D. Perkins and R. E. Hage, *Airplane Performance, Stability, and Control* (New York: John Wiley & Sons, 1949), 148.
Shortened, subsequent reference	16. Perkins and Hage, *Airplane Performance,* 236.

Many readers still find this practice difficult to follow, especially if many references intervene between the full and shortened citation. One method that helps tie the original and subsequent versions together is to tell the reader how a work will be cited later within the original citation:

> 3. C. D. Perkins and R. E. Hage, *Airplane Performance, Stability, and Control* (New York: John Wiley & Sons, 1949), 148. Hereafter cited as: Perkins and Hage, *Airplane Performance.*

10.38 In contemporary practice references to previous citations for periodicals contain

- Author's last name(s),
- Key words from the title,
- Date (if there are multiple works in the same year),
- Relevant page numbers.

Basic punctuation conventions for shortened periodical citations are

Citation number. Author's name(s), "Article title" (Date, if necessary), Relevant pages.

The following is an example of a typical shortened periodical citation:

First reference	7. W. J. Hogan, "Positioning Data Displays in Technical Information," *Journal of Applied Psychology* 132 (10): 147.
Shortened, subsequent reference	22. Hogan, "Positioning Data Displays," 236.

Many readers still find this practice difficult to follow, especially if many references intervene between the full and shortened citation. One method that helps tie the original and subsequent versions together is to tell the reader how a work will be cited later within the original citation:

7. W. J. Hogan, "Positioning Data Displays in Technical Information," *Journal of Applied Psychology* 132 (10): 147. Hereafter cited as: Hogan, "Positioning Data Displays."

10.39 Endnotes group many references together; therefore they are a valuable resource to those readers who are interested in learning more about a subject. On the other hand, endnotes interrupt readers by forcing them to flip back and forth between the text and the end of the book or chapter.

Because endnotes are grouped together, editing the text and updating the notes list is easier than updating footnotes distributed throughout the work. Editing a work containing numbered notes, either footnotes or endnotes, often requires extensive renumbering. This task can be time consuming, although many modern word processing programs automatically renumber footnotes and endnotes during editing.

FOOTNOTES

10.40 Place footnotes at the bottom, or foot, of the page on which the reference occurs. A superscripted number (or less commonly, a bracketed number) follows the reference in the text. Also place the number at the bottom of the page, followed by the author's name and the rest of the necessary bibliographic information, as in an endnote. Use the same format (either superscript or bracket) for the note number at the foot of the page as for the note number in the text.

10.41 Footnotes are the traditional method for citing sources. In many disciplines and publications they are the accepted form for documenting sources.

10.42 Begin numbering footnotes with 1 for each chapter or section of the work. Although in many older works the first footnote on each page is assigned the number 1, this practice requires continual renumbering when the work is edited and paginated.

10.43 When citing the same work in two or more footnotes, follow the guidelines suggested in the section on endnotes. (See Section 10.34.)

10.44 Footnotes place references close to the text they support. This placement eliminates flipping from the text to a list of references at the end of the book or chapter.

10.45 However, footnotes have several disadvantages. They

- Distract readers by interrupting their reading to find the note at the bottom of the page,
- Can create typographically untidy pages,
- Do not provide a collected resource—as endnotes do—unless they are accompanied by a bibliography.

COMPLETE REFERENCE IN TEXT

10.46 In a work with only a few sources, it may be appropriate to include the entire reference in the text. Place the necessary information within parentheses and treat it as part of the text:

> Adults typically read texts to learn how to perform an action (S. T. Kerr, "Instructional Text: The Transition from Page to Screen," *Visible Language* 20 [1986], 368).

As the example shows, the reference overwhelms the text in which it appears. Therefore, this format should be used only when there are just a few references.

All of the conventions for creating short titles discussed in Sections 10.37–10.38 can be applied to parenthetical citations in text:

> While the ability to make screens function as text has been widely disputed, some successful designs have been used in the instructional setting (Kerr, "Instructional Text," 301).

BIBLIOGRAPHY

10.47 A bibliography is a list of works cited, consulted, or of interest to the reader. The reader usually assumes that the compiler of a bibliography has listed the most important works, if not all of them. Thus, a bibliography has a broader purpose than a reference list (which is a simple list of works cited).

10.48 A bibliography can be used alone or with footnotes or endnotes. Since footnotes deprive the reader of a collected group of references (see Section 10.45), a bibliography is especially convenient with footnotes.

10.49 A bibliographic citation for books should include

- Author(s) names (first and middle initials are preferred to full names in technical documents),
- Book title (or, in the case of a compilation, the chapter or paper title, followed by the book title),
- Editor(s), translator(s), or compiler(s) name(s) (first and middle initials are preferred to full names in technical documents),
- Position in a series (volume number and the like),
- Edition number,
- Publication information (city and state [sometimes omitted for well-known cities] of publication, publisher, publication date (citations in the natural and social sciences often place this after the author).

Basic punctuation conventions for book citations are

Author(s) last name, first and middle initials. *Book Title.* City, State of publication: Publisher, Date (could be placed after the author's name in some styles).

Brantmann, J. P. *Planning Electronic Texts.* Cambridge, MA: MIT Press, 1991.

Brantmann, J. P. 1991. *Planning Electronic Texts.* Cambridge, MA: MIT Press.

Sometimes the first line of a bibliographic entry is outdented, so that the author's name stands out:

Brantmann, J. P. *Planning Electronic Texts.* Cambridge, MA:
  MIT Press, 1991.

Typically, bibliographies are alphabetized. In some fields they are treated as both bibliography and reference list. This latter kind of bibliography is still alphabetic, but each entry is also numbered. These numbers are used in the text to credit sources. The following examples are typical bibliographic citations for books:

Barrce, A. G. *The Development of Original Programming Languages.* Vol. 3 of *A Complete History of the Rise of Automata and Other Thinking Machines.* London, England: Ticknor and Fields, 1904.

Bertin, J. *The Semiology of Graphics.* Translated by W. J. Berg. Madison, WI: University of Wisconsin Press, 1983.

Brantmann, J. P. *Planning Electronic Texts.* Cambridge, MA: MIT Press, 1991.

Jackson, W., T. V. Quarel, and C. S. Watkins. *The Development of Computer Mediated Data Structures.* Dayton, OH: Littleton and Sons, 1984.

Snebur, P. P., and K. R. King, eds. *Complex Algorithms for Novel Data Structures.* New York: Lavosier Press, 1978.

Wentworth, W. W., ed. *Research Methods for Migratory Avian Studies.* Madison, WI: University of Wisconsin Press, 1911.

Wooldridge, M. M. "Recent Studies at Carverson Foundation in Avian Carcinogen Reactions." In *New Directions in Fauna Research.* Edited by C. J. Connolly. London, Ontario, Canada: Billingworth Publishers, 1955.

10.50 A bibliographic citation for periodicals should include

- Author(s) names (first and middle initials are preferred to full names in technical documents),

- Title of the article or paper,
- Periodical's full title or an acceptable abbreviation,
- Volume and/or issue number,
- Issue date (citations in the natural and social sciences often place this after the author),
- Relevant page numbers.

Basic punctuation conventions for periodical citations are

Author(s) last name, first and middle initials. "Article title." *Periodical title* Volume or issue number (issue number or date): Page numbers.

Thus, a typical bibliographic entry for a periodical would be

Billows, H. R. "Working with Multimedia in Restrictive Environments: The Problem of the Glass Cockpit." *Journal of Hypermedia Studies* 130 (June 1991): 107–164.

Sometimes the first line of a bibliographic entry is outdented, so that the author's name stands out:

Billows, H. R. "Working with Multimedia in Restrictive Environments: The Problem of the Glass Cockpit." *Journal of Hypermedia Studies* 130 (June 1991): 107–164.

10.51 While most bibliographies combine all document types—books, periodicals, reports, and the like—into one alphabetic list, a long bibliography should be divided into sections by subject matter or by document type. Provide headings that describe the organizational method. Dividing a bibliography by subject matter, for instance, produces

Computer-assisted Training

Keller, J. M. "Motivational Design of Instructions." In *Instructional Design Theories and Models: An Overview of the Current Status.* Edited by C. M. Reigeluth. Hillsdale, NJ: Lawrence Erlbaum, 1983.

Macklin, V. C. "Computer Mediated Instruction in Rural Schools." *Academic Computing* 11 (June 1987): 64–84.

Stein, J. S., and M. C. Linn. "Capitalizing on Computer-based Interactive Feedback: An Investigation of Rocky's Boots." In *Children and Microcomputers.* Edited by M. Chen and W. Paisley. Beverly Hills, CA: Sage Publications, 1985.

Video Disk Training

Manning, D. P., P. Balson, D. Ebner, and F. Brooks. "Student Acceptance of Videodisc-Based Programs for Paramedical Training." *T.H.E. Journal* 11 (November 1983): 105–108.

Thorkildsen, R. "Using an Interactive Videodisc Program to Teach Skills to Handicapped Children." *American Annals of the Deaf* 130 (11): 383–385.

10.52 Organization by document type is a characteristic of dissertations and of personal bibliographies that support academic résumés. Here is an example organized by document type:

Books

Bertin, J. *The Semiology of Graphics.* Translated by W. J. Berg. Madison, WI: University of Wisconsin Press, 1983.

Brantmann, J. P. *Planning Electronic Texts.* Cambridge, MA: MIT Press, 1991.

Jackson, W., T. V. Quarel, and C. S. Watkins. *The Development of Computer Mediated Data Structures.* Dayton, OH: Littleton and Sons, 1984.

Snebur, P. P., and K. R. King, eds. *Complex Algorithms for Novel Data Structures.* New York: Lavosier Press, 1978.

Wentworth, W. W., ed. *Research Methods for Migratory Avian Studies.* Madison, WI: University of Wisconsin Press, 1911.

Journals

Macklin, V. C. "Computer Mediated Instruction in Rural Schools." *Academic Computing* 11 (June 1987): 64–84.

Manning, D. P., P. Balson, D. Ebner, and F. Brooks. "Student Acceptance of Videodisc-Based Programs for Paramedical Training." *T.H.E. Journal* 11 (November 1983): 105–108.

Thorkildsen, R. "Using an Interactive Videodisc Program to Teach Skills to Handicapped Children." *American Annals of the Deaf* 130 (11): 383–385.

10.53 If several works by the same author are listed, replace the primary author's name (or duplicate authors' names) with a dash in the second and subsequent citations:

Jackson, W. "Computer Mediated Instruction in Rural Schools." *Academic Computing* 11 (June 1987): 64–84.

———. "Student Acceptance of Videodisc-Based Programs for Paramedical Training." *T.H.E. Journal* 11 (November 1983): 105–108.

———, T. V. Quarel, and C. S. Watkins "Using an Interactive Videodisc Program to Teach Skills to Handicapped Children." *American Annals of the Deaf* 130 (11): 383–385.

ELEVEN

Creating Indexes

11.1 Indexing scientific and technical documents has three phases:

- Planning—selecting an indexer, scheduling, determining index length, and preparing a keyword list;
- Preparation—determining the format, the organization method, the alphabetic arrangement, and page design;
- Procedure—marking entries, preparing index cards or an index data file, filing and sorting with or without a computer, and writing and editing index entries.

11.2 This chapter will assist individuals with little or no indexing experience, especially the author indexing her own writing. In addition, it can serve as a useful refresher for the experienced indexer.

KEY INDEXING TERMS

11.3 Five terms are important in discussing indexes:

- Entry—an index item, consisting of a heading, a locator and/or a cross-reference, and any subheadings together with their locators and references.
- Heading—a word or phrase based on some material in the text, arranged in alphabetical order.
- Subheading—a word or phrase subordinate to the heading, under which specific references to the text are located.
- Locator—the identifier leading to information. Although the locator is generally a page number, it might also be a section

or paragraph number. List locators in ascending order in technical matter.

- Cross-reference—a pointer from one heading or any of its subheadings to another heading. A *see* cross-reference points to a heading that collects all relevant information. A *see also* cross-reference points to another heading that contains additional information.

NEED FOR INDEXES IN TECHNICAL MATERIAL

11.4 Technical literature collects and organizes facts, but these facts are worthless without reference points to guide the reader to them. An index offers readers a way to discover relevant material. The index is a subject finder; it leads the reader to information by gathering the references related to that subject in one place.

11.5 The index is particularly important in scientific and technical documents because it saves the reader time. The index simplifies the search. By making relevant information readily available, a good index makes the reader more efficient.

The Goals of a Good Index

11.6 The best indexes help the reader find information with the least effort, in the shortest time. Good indexes have four characteristics:

- Complete—If the author emphasizes a particular fact in the text, that fact should also be emphasized in the index. Depth is important; every important point in the text should be indexed. One can, however, overindex by including trivial entries.
- Accurate—The information should be where the index indicates it will be. Names, dates, and sequences must be spelled correctly, be in the correct order, and be cross-referenced properly. Edit the index with the same care given to the primary text.
- Concise—Entries should be clear and logical. The more concise the index entries, the more usable the index will be:

  "Steel" is general.
  "Steel, annual production of" is more exact.

  Avoid duplicate headings, entries indexed by something other than the keyword, and circular references.

- Consistent: Spell the word "color" or "colour," but never use both. If the manager's name is Jane Smyth, do not index her as "Smythe" or "Smith."

PLANNING AN INDEX

11.7 In planning an index, consider

- Who should create the index,
- How long it should be,
- How much time it should take,
- What it should look like, and
- What it should contain.

11.8 As in all written works, an index should consider an audience's needs, age, education, reader expectations, and usage patterns. While preparing the index, ask "Is this important? Will readers want to look it up?" (See Section 1.8.)

11.9 Keep in mind that the broader the subject, the more difficult it is to index. A broad subject encompasses more terms, more concepts, more cross-references. Such complexity makes it all the more important to be concise and accurate in index entries.

11.10 Consider the need to revise the index along with the primary text. Plan the index not only for today's readers, but also for tomorrow's.

11.11 Examine indexes in similar books, to see if your index includes material that someone searching through more than one source covering the same subject would expect to find.

The Indexer

11.12 The good indexer has

- A knowledge of index preparation,
- An awareness of reader expectations, and
- The ability to judge technical content.

11.13 Typically, the author knows both the material and the audience, and probably allocates some writing time to indexing. However, the author's closeness to the topic might mean a loss of objectivity. Also, she may not know enough about indexing to prepare a good index. Following the guidelines in this section will simplify the procedure for someone with little indexing experience.

11.14 An indexer should work with an unabridged dictionary, a technical dictionary for the subject being indexed, and the house style guide.

11.15 If more than one indexer works on a project—for example, on the index to a multivolume set—one person should manage the project to maintain consistency. A style guide (see Section 11.21) is valuable for this activity.

Planning the Size of an Index

11.16 The length of the index varies with the book's length and complexity. It is difficult to offer an ideal index length. A short index—two percent of the total number of pages—might be adequate. On the other hand, in an encyclopedia, the index can consist of 10 to 20 percent of the primary text. In general, the index length should be two to five percent of the primary text.

11.17 A publication's size limits might restrict an index's completeness. If index length is limited, the indexer must estimate the number of entries per primary text page that can be created.

Generally, five or fewer references per primary text page produce an index that occupies two percent of the total number of pages available.

Twelve or more references per page produce an index that occupies five percent of the total number of pages. The density of the primary text determines the actual number of entries, and the index format and selection conventions determine its length.

Preparation Time and Indexing

11.18 A major difficulty in indexing is assuring that the right locator occurs in the final index entry. For example, in indexing technical manuals, there is not enough time to prepare the index prior to the book's release because page proofs arrive late in the publication cycle.

11.19 Computerized writing and indexing has an advantage over manual indexing. Any computerized word processor allows an author to embed a control character or text string (a "flag") to denote an index entry. These embedded flags can be collected electronically at any time. Indexes can thus be made available to reviewers while the book is still in draft form.

11.20 No firm rules exist regarding the amount of time needed to prepare an index, but an indexer should be able to index even the most complex technical single-volume work in 80 hours.

The primary difficulty is that indexes for technical material often need an additional edit for technical accuracy by personnel familiar with the product's operation that the document supports. While indexing, spend about 40 percent of the time reading, marking copy, and sorting entries and 60 percent analyzing and editing the index. Analyzing and editing is so labor-intensive because of the number of decisions made in that stage: dividing material into headings and subheadings, controlling synonyms, and cross-referencing terms.

Indexer's Style Guide

11.21 The indexer's style guide defines the in-house rules governing indexing preparation and production. As such, it should be an integral part of the organization's style guide. Indexing rules might detail

- Format style,
- Typographic considerations,
- Listing of indexable terms,
- Cross-referencing conventions, and
- Construction rules for headings and record entries.

INDEX ORGANIZATION

11.22 Three decisions must be made in planning an index:

- Whether to present the index in indented or paragraph format,
- Whether to organize the index alphabetically or by subject,
- Whether an alphabetic index should be arranged in word-by-word or letter-by-letter order.

Indented or Paragraph Format

11.23 The two basic index formats are indented and paragraph. The indented format provides a well-organized vertical display of entries that can be scanned quickly and easily, and it is recommended for detailed technical indexes. This style begins with a primary entry that lists the major pages for an entry. All subentries in that section are treated as a vertical list and serve as a table of contents for the primary entry. Although the paragraph format begins with the same listing for the major pages, all subentries are treated as phrases and punctuated accordingly. This format often allows both subentries and all or part of their page references to break across lines. Thus, the paragraph format is more difficult to scan for specific subentries, and it does not have the same organizational structure—the table of contents—found in the indented format. The following examples compare the indented and paragraph-entry style:

Indented (Acceptable)

Maps 136–148
  blowups of, 142
  deleting, 143, 145
  display options of, 136–139, 141
  exposing and burying, 141, 143–144

Paragraph (Unacceptable)

Maps 136–148; blowups of, 142; deleting, 143, 145; display options of, 136–139, 141; exposing and burying, 141, 143–144

11.24 The following sections consider only the indented format, which is more common in technical and scientific works.

Alphabetic Versus Subject Indexes

11.25 Organize indexes in alphabetical order. Alphabetic indexes are easy to create, and the organization scheme is simple, predictable, and easy to use.

11.26 Although alphabetic indexing is recommended, there are some other useful alternatives. The subject-oriented classified index (such as the categorical organization of *Roget's International Thesaurus*) is often more useful for technical material. For example, the index in a collection of scientific biographies might use scientific fields as subject headings and list scientists working in each field as subheadings.

11.27 However, two problems must be overcome in creating and using the subject index:

First, an expert on the topic must create a subject index. Few writers have the knowledge needed to create a good subject index.

Second, the use and organization of a subject index must be explained. A complicated organization might discourage the less knowledgeable reader.

11.28 The alphabetized, indented index offers three essential usability features:

- The vertical, visible orientation allows rapid scanning.
- The alphabet provides a predictable organization method.
- The alphabet lends itself to cross-referencing and makes it easy for the user to move around within the index.

The reader can find a search term, or its synonym, quickly in an alphabetized, indented index.

Alphabetic Arrangements

11.29 An index may be alphabetized in two ways: the word-by-word and the letter-by-letter arrangements.

11.30 In the word-by-word system, alphabetize by the entry's first word. If two entries have identical first words, alphabetize by the second word, then the third, and so on to the first comma, semicolon, or period. In the word-by-word system, "nothing comes before something"; that is, letters followed by a space precede the same letters followed by more letters. New York, for example, precedes Newfoundland.

11.31 In the letter-by-letter system, consider every letter up to the first comma, semicolon, or period. Ignore spaces, so Newfoundland precedes New York. Most dictionaries use the letter-by-letter style.

The letter-by-letter has an advantage over the word-by-word style because the reader knows where to look for compound expressions that sometimes appear as one word and sometimes are split, depending on the author's preference.

The following lists compare the word-by-word and letter-by-letter schemes:

Word-by-word	**Letter-by-letter**
Liquid helium	Liquidambar
Liquid lattices	Liquid-fueled rockets
Liquidambar	Liquid helium
Liquid-fueled rockets	Liquid lattices

Both alphabetic systems are in use and are acceptable. However, be consistent after choosing a system. Tell the reader which system has been chosen in an introductory note to the index. (See Section 11.89.)

General Rules for Alphabetizing

11.32 Both the word-by-word and letter-by-letter arrangements follow these general rules. Section 11.63 and following applies these general rules to specific cases:

- Ignore nonalphabetic and nonnumeric characters such as hyphens, quotation marks, apostrophes, and diacritical marks when alphabetizing. Thus, for example, ignore both the hyphen in "Gell-Mann" and the apostrophe in "Pickett's Charge."
- Stop alphabetizing at the first significant punctuation—that is, a period, comma, or semicolon.
- Alphabetize words that precede a comma or period before any identical words followed by additional words with no intervening punctuation. "Copper, oxidation of" precedes "Copper oxides."
- Ignore relational words such as articles, conjunctions, or prepositions, and alphabetize by the keyword following the relational word.

INDEX PAGE DESIGN

11.33 Layout and typographic considerations include indentation, punctuation, and capitalization. Include these rules in an organization's style guide. Following them will simplify index preparation, editing, and production.

Layout Considerations

11.34 Use the indented style for technical indexes because it provides an easy-to-scan and well-organized vertical display of entries. Set indented indexes in hanging indent style (about four characters for the second and all subsequent lines in an entry), in which

- The heading is set flush against the left margin.
- Each subheading begins on a new line, indented (usually) one em. Sub-subheadings are indented two ems. Use no more than three heading levels.
- Indent all runover lines one em deeper than the largest subheading indentation.

11.35 Set indexes in two columns (sometimes three or even four), with the right margins unjustified. In general, typeset the index in a typesize two sizes smaller than the text. Thus, if the text was set 12-on-14, the index would be set 10-on-12. Use the same typeface as the text for the index.

Typographic Considerations

11.36 In an indented index, capitalize the first word of a heading and set other entry words and the initial word in subheadings in lowercase. Some indexes also capitalize the initial word in subheadings. Any word or abbreviation normally printed in lowercase letters keeps its form as a heading or subheading, such as "p-n junctions" or "f-stop." Similarly, any term normally capitalized, such as a proper name, keeps its form.

11.37 Italicize the cross-referencing terms *see* and *see also*. Capitalize them when they begin a new line, or when a period separates them from the rest of the entry. Also italicize a phrase representing an inclusive class of terms: *See also names of participating institutions*. If an entry lists a series of cross-references, the most significant reference—for example, the definition—may appear in italicized or boldface type.

11.38 Use parentheses to add alternative endings to words: "Europe(an);" to differentiate between homonyms: "Filter (electrical)" and "Filter (mechanical);" or as a qualifier, providing further information.

Punctuation Considerations

11.39 Punctuation is straightforward in an indented index. Use a comma to

- Mark an inverted phrase so that the key word comes first,
- Separate entries from their locators, and
- Separate page locators within an entry from one another.

11.40 Use a period to separate an entry from a cross-reference following it.

11.41 Use a semicolon to separate the parts of a multiple cross-reference.

11.42 Do not punctuate the end of an entry.

The following sample index illustrates the indentation, capitalization, and punctuation conventions:

Cameras, electronic, 492, 499
Carbon flash, 559
Cassiopeia A (3C461), *See* Radio sources; Supernovae
Catalogs
  of bright nebulae, 74
  of dark nebulae, 74, 120
    Lundmark, 121
    Lynds, 123
    Schoenberg, 123
  Herschel's (of nebulae), 119
  of planetary nebulae, 484–85, 563
    Perek-Kohoutek, 484, 563
    Voronstov-Velyaminov, 484
  of reflection nebulae, 74
  3C catalog of radio sources, revised, 630
Central stars. *See* Planetary nebulae
Cerenkov radiation, 668, 709
Chemical composition, 71. *See also* Abundances; *names of individual elements*
  of stars and nebulae, 405
Clark effect, 756, 758, 765

INDEXING PROCEDURE

11.43 For the most part indexes are still compiled by hand. That is, an indexer manually writes out entries and combines them into coherent groups. However, many new electronic methods for creating indexes have recently come into common use, particularly in industry. This section discusses both manual and computerized indexing procedures.

An Overview of the Indexing Procedure

11.44 Whether indexing with or without a computer, the steps are essentially the same:

1. Determine the length of the index, based on time, space, and any other constraints. Assign a number of pages to the index and create the page specifications. Determine how many entries can fit in that space. Estimate three to seven entries per primary text page.
2. Select a format (indented or paragraph), organization (alphabetic or subject), alphabetic arrangement (word-by-word or letter-by-letter), and page specifications for the index.

3. Skim through the manuscript and study the table of contents to identify important topics, and thus what to index. The proportion of content allotted to each topic determines approximately how much content that topic will contribute to the index.
4. Mark the index entries. Authors can note entries as they write. Indexers working from a finished draft can mark the entries on the page proofs.
5. Prepare index cards (if indexing manually) or an index data file (if indexing with a computer).
6. Alphabetize cards or sort the data file according to either the letter-by-letter or word-by-word arrangement.
7. Edit the cards or the data file by categorizing headings and subheadings, consolidating synonyms, determining cross-references, and deleting trivial entries.
8. Write an introductory note that explains how to use the index (see Section 11.89).
9. Prepare and proofread a final copy for the editor and typesetter.

The Preparatory Steps

11.45 The preparatory steps for indexing include the first three steps found in Section 11.44. They are the same whether you index manually or by computer. These three steps are

1. Determine how many pages can be assigned to the index and how many entries will fit in that space. Find out if the publisher or editor has placed any constraints on size.
2. Select a format, organization, alphabetic arrangement, and page specifications for the index.
3. Skim through the manuscript and determine what to index.

Determining What to Index

11.46 Compile a list of words and phrases (also called a keyword list or a thesaurus) from the text. List all their abbreviations and acronyms, and show the synonymous or hierarchical relationships and dependencies among them. List each term with its related terms and cross-references. This list resolves questions about standard meanings, usage, and synonyms. The following example shows how a keyword list for types of insulating materials might be fashioned. An "R" denotes a root term; a "B" denotes a branch term:

Batt insulation
  R: Insulation
  Use instead of: Flexible insulation
Beadboard
  R: Rigid insulation
  Use instead of: Expanded polystyrene insulation
Block insulation: Use Rigid insulation
Expanded polystyrene insulation: Use Beadboard
Fiberglass
  R: Loose-fill insulation
Flexible insulation: Use Batt insulation
Insulation
  R: Construction
  R: Energy conservation
  B: Batt insulation
  B: Loose-fill insulation
  B: Rigid insulation
Loose-fill insulation
  R: Insulation
  B: Fiberglass
  B: Vermiculite
Rigid insulation
  R: Insulation
  B: Beadboard
  B: Urethane
  Use instead of: Block insulation
Urethane
  B: Rigid insulation
Vermiculite
  R: Loose-fill insulation

Keyword lists are usually flexible and can be changed to fit an organization's needs and specific objects or functions.

11.47 In order to know where to look for indexing terms, examine the manuscript's organization. Use the table of contents to determine the importance of each topic. The relative size of each topic in terms of total content should determine roughly how much space should be given to each index topic. Performing this overview does not require a finished manuscript; it can be started before the manuscript is finished.

11.48 Establish the overall scope of the index. Include:

- Introductions,
- Prefaces,

- Executive summaries,
- Conclusions,
- Appendixes, and
- Footnotes.

11.49 Index data displays (tables, charts, and diagrams) and illustrations, but not their content: Parameter values, table of.

11.50 Use an abbreviation for data displays and illustrations in the index. Explain the abbreviation practices in the index introduction (see Section 11.89).

Matter to Omit

11.51 Do not index

- Items listed in the table of contents or in lists of figures or tables,
- Passing mentions used to illustrate a point ("large companies such as A and B"),
- Authors or titles cited as references and listed in the bibliography,
- Names of people or places mentioned in the preface ("the author would like to thank . . ."),
- Section summaries,
- Review questions,
- Title page material,
- Bibliographic entries, or
- Glossary terms.

Manual Indexing

11.52 Before beginning, resolve all the questions raised in the sections on planning and organizing indexes. Prepare a large work space for handling the cards and the page proofs. Use the following procedure when indexing manually:

1. Mark the index entries on primary text pages. Highlight or underline the entries in the text, writing the appropriate modifier in the margin, preceded by a colon or some other indicative punctuation mark. Be consistent throughout.

   For example, to index this paragraph, highlight "index entries" in the first sentence, and note ":marking in proofs" as a modifier in the margin beside it.
2. Prepare index cards, with one entry per card. Include the entry, the modifier, and the locator or locators (if the entry

is longer than one page) on each card. In the previous example, the card would read "index entries :marking in proofs," along with the page number.

Keep the cards in page order; do not sort them yet. After completing this preliminary index card set, verify all spellings, capitalization, and locators by comparing them with the primary text pages.

3. Alphabetize the cards. Divide the cards into one stack per each letter of the alphabet. Then sort each stack on the second letter, the third, and so on based on the letter-by-letter or word-by-word method. (See Section 11.30.) Sort until all cards are in strict alphabetical order.
4. Edit the cards. Identify headings and subheadings, and main and subordinate terms for groups of synonyms. Alphabetize all subheadings under their headings. Prepare cross-references, and eliminate duplicate and trivial entries. Resolve any questions by checking against the manuscript. Number the completed card set.
5. Prepare the final index. Write an index introduction. Type the index in a double-spaced column (using the column width of the planned index) on 8-½ × 11-inch sheets. Insert blank lines to separate letters of the alphabet. Number the manuscript pages consecutively. Proofread the manuscript and send it to the editor or typesetter.

Using the Computer in Index Preparation

11.53 Creating an index with a computer has three steps:

1. Determine the document's topic and purpose—what's important and what to index.
2. Mark, collect, and sort the entries.
3. Analyze and edit the entries to make a cohesive index.

The first and third steps above require intellectual decisions. The computer provides a powerful tool for completing the second. Sections 11.54, 11.58, 11.61, and 11.62 examine three types of computer-assisted indexing: indexing with word processors, markup languages, and index composition programs.

Word Processors and Indexing

11.54 All word processors with an indexing function follow a similar pattern. To use this function, the author places a special character (flag) in the text near a term to be indexed. These flagged entries are recorded in a data file. When the data file is processed, the program collects the flagged entries and creates an index in a preplanned format. The following example uses the caret "^" and brackets "^[" and "]^" to create an entry

^[Indexing, computerized^]

in the text file at the point where computerized indexing is discussed. After placing the flag, the text may be moved, but as long as the flag stays with the primary text, the index function will put the correct locator in the index.

11.55 In some cases flagging an index entry creates an index database file with an onscreen data entry form. This form contains heading, subheading, cross-referencing, and typographic instructions, and controls the index's content and appearance. The index database file stores this information, for future editing, alphabetic sorting, and printing.

11.56 Word processor indexing provides many advantages:

- Locators correspond to the primary text regardless of the number or types of edits,
- Printing is based on a preplanned format,
- Alphabetizing of entries is correct,
- Information is transferred correctly from marked pages to index cards to typesetter's copy, and
- The index is constantly verified and proofread.

11.57 Word processor indexing has some disadvantages:

- Possible limit to the number of index entries allowed by the computer program, and
- Computer memory limits might prohibit some users from using these indexing programs.

Markup Languages and Indexes

11.58 A markup language, or computerized typesetting, creates an index entry by embedding a typesetting command (rather than a flag) in the text.

For example, the typesetting command "\index" could indicate an index entry, and brackets could indicate the term(s) to be indexed. Typing "\index[Typesetting, computerized,]" in the primary text tells the markup language to create an index entry—"Typesetting, computerized"—and put it in an index data file with its locator.

11.59 Markup languages have the same benefits as the word processor for indexing:

- Primary text and index items remain together regardless of editing,
- Locators are always correct, and
- The transition from marked proofs to cards to typesetter's copy is eliminated.

11.60 The markup language has several minor difficulties for indexing:

- Data files must be sorted separately, and
- Markup languages require "programming" and take time away from writing.

Index Composition Programs

11.61 Index composition programs automate the indexing procedure after the markup step. The author marks index entries in the primary text file as in manual indexing and types the index information into a preformatted data file. The composition program sorts the index information and applies typographic, layout, and punctuation rules.

The most serious drawback to composition programs is that page proofs and the locators are not automatically updated. Since the index data and primary text files are independent of each other, the index data file must be manually updated.

Computer Indexing Conclusions

11.62 Computer indexing provides some useful capabilities:

- In the most effective applications of computer indexing, the text is available electronically. In fact, indexing with word processing or markup languages is most efficient when the index terms are marked while the text is being written.
- A word processor with an indexing function saves time for an author, who can index while writing. Also, writers only need to learn how to use a word processor, which makes the learning time shorter and eliminates using two programs—word processor and indexer—that might be incompatible.
- Unlike word processors, markup language commands are the same regardless of the device—computer, word processor, or typesetter—used to embed them in text. The same commands produce the same output whether Brand X or Y device created the commands.

- An indexer other than the author might perform better with an index composition program, since it does not matter if the primary text exists as electronic data. These programs are more accurate than manual indexing.
- As with all indexes, any computerized approach still requires that the author learn how to write and edit index entries. All three approaches—word processor indexing features, markup languages, and index composition programs—produce a page-by-page listing of index entries that can be edited. This editing process can resolve such accidental indexings as reporting "John Smythe" as "John Smyth."

WRITING AND EDITING INDEX ENTRIES

11.63 Index entries should be concise and meaningful. This section shows how to prepare good headings and subheadings, work with synonyms and cross-references, and create a good entry style and language. Indexing language consists of both vocabulary and syntax—"information science" is quite different from "information, science." Index terms should be based on concrete nouns and noun phrases; vague and trivial terms should be avoided. While natural word order should be used, noun phrases should be inverted to emphasize the keyword (the one readers are most likely to see first). Indexers should also take an active role in determining the index terms that are incorporated into an organization's style guide.

Headings and Subheadings

11.64 Make headings general, subheadings specific. For instance, the heading "Printing processes" is a general classification. The subordinate subheadings divide this general heading into its specific subclasses: gravure, letterpress, lithography, and so on. Similarly, both "Diesel engine" and "Gasoline engine" are subheadings of "Internal combustion engine."

11.65 Make each entry a heading rather than a subheading, unless it

- Begins with the same word or words as the heading, thus denoting a close relationship, and
- Is ambiguous unless read in the context of the main heading. "Annual production of" cannot stand alone without a heading—"Steel"—to provide context.

11.66 Use no more than two subheading levels. More are confusing, and the level of depth might indicate a flaw in heading choice.

11.67 In addition to a logical relationship, headings and subheadings have a grammatical relationship. Subheadings can include a relational word such as an article, preposition, or conjunction for clarity, although such words are ignored in alphabetizing:

Semilogarithmic charts, 85–110
  characteristics of, 85–87
  interpretation of, 87–91
  for portraying spatial data, 108–110
  and special applications, 91–95

11.68 Index an object or concept that has no formal name by function or material. While terminology is well defined and more standardized in technical than in nontechnical fields, new and unnamed items are common.

11.69 Avoid under- and overanalyzing headings. Typically, a long string of undifferentiated locators follows an underanalyzed heading. Divide such a heading into subheadings. While using seven undifferentiated locators with an entry is acceptable, for clarity, try to include no more than four.

11.70 In contrast, a long list of subheadings, each with only one locator, might indicate overanalysis:

Maps
  borders for, 137
  icons used in, 138
  projection of, 136
  styles of, 136
  varying size of, 137

Consolidate some of these subheadings to make the heading appear less elaborate:

Maps
  display characteristics of, 136–138

Synonyms

11.71 Classify synonyms to link related references in two steps:

1. Group those terms that have nearly the same meaning, and
2. Decide which synonyms should be main entries and which should be cross-referenced to it.

For instance, in indexing a book on printing technology, choose one term—Flying, Misting, Spitting, or Spraying—as the heading and cross-reference the remaining terms to it.

11.72 Follow the author's preference and choose as the main entry the form used most often in the book. If the author does not show a preference, use the most familiar form as the main entry.

11.73 The synonym problem occurs with common terms such as "Conscription," "Draft," and "Selective service," in which the accepted term changes over time. Similarly, seemingly opposite terms can have the same meaning: "developing countries" and "underdeveloped countries." In cases such as these, use the current, and not a superseded, term. If necessary, index any outdated terms and cross-reference to the current one.

Cross-References

11.74 Use cross-references to provide guides to related information, not just the same information under a different heading. There are two types of cross-references—*see* and *see also:*

- *See* cross-references all the information related to a term within another term. Use this when referencing from a

  Popular to the official term,
  Table salt. *See* Sodium chloride

  Synonym to the entry with the information,
  IMP. *See* Packet Switching Node

  Replaced name to the current name,
  Leningrad. *See* St. Petersburg

  and

  Pseudonym to the given name.
  Saki. *See* H. H. Munro

- In the *see also* cross-reference, direct the reader to an entry that contains additional information:

  Access devices, 224. *See also* Packet Assembler/Disassembler

  Place the cross-reference after the heading or subheading it augments, rather than after the last subentry.

11.75 In cross-referencing, avoid making a bad logical relationship between entries. Avoid, in particular, circular searches. For instance, save readers from the frustration of this circular reference:

> Inert gases, 150, 177. *See also* Neon
> Neon. *See* Inert gases

Index Entry Style

11.76 Indexing language consists of both vocabulary and syntax. Write the typical index entry with a noun or a noun phrase supported by prepositions (e.g., "Indexes, preparation of, by computer"), which the reader can put into a normal sentencelike structure—"Preparation of indexes by computer." Apply the following guidelines in styling index entries:

1. Use the author's preference in terminology to determine synonym choice, spelling, and capitalization. Use the forms that appear most often in the text.
2. Use natural word order, but always invert noun phrases so that the keyword—the one the readers are most likely to seek—comes first. "Formats, table of" is more meaningful than "Table of formats." Index "Michigan, Lake" and not "Lake Michigan," "Copyright laws" and not "Laws on copyright."
3. Homonyms present an indexing challenge. The same term may have different meanings in different technical fields. "Nesting," for instance, means something quite different to an ornithologist and a computer programmer. To differentiate between homonyms, use separate headings with qualifiers in parentheses:

Filters (electrical)	Condensers (electrical)
Filters (mechanical)	Condensers (steam)

11.77 Two rules govern the use of singular versus plural number in index entries:

- Use the singular for processes "Fermentation," properties "Conductivity," concepts "Relativity," and objects "Oxygen."
- Use the plural for common objects and classes of objects "Trees," "Planets."

Rules Governing Index Locators

11.78 When recording locators on cards or in a data file, be accurate, and verify everything. This rule is especially important in manual indexing. Other guidelines

- Supply comprehensive page numbers. Do not use the outdated convention of indicating a range of pages with a lowercase f; record actual page numbers—116–118, not 116f or 116–18. Record each page: 73, 75–76. Type en dashes, not hyphens, in specifying locator ranges, or specify to the production staff to use en dashes.
- Collect entry repetitions that are in close proximity as a single page range for the entry: 116–120. Do not use the outdated convention—116 passim—since it does not specify an end-of-range.
- Index footnotes with a lowercase n: 54n; if the footnote is on a page with other footnotes, specify the indexed footnote by number: 54n2. Specify the note number in italics.
- Avoid circular searches with *see* and *see also*. Make sure that the locator is correct and that the headings are identical.

Rules Governing Special Cases

11.79 This section describes some special cases often encountered in scientific and technical indexing, and offers guidelines for dealing with them.

Abbreviations and Acronyms in Indexes

11.80 Avoid indexing abbreviations and acronyms directly. Use the full spelling whenever possible, with the abbreviation or acronym in parentheses, for example, "Seasonal Affective Disorder (SAD)." In scientific and technical writing, however, the rule is not quite so strict because of the nature of the material. If an abbreviation is more familiar than the full name to an audience, use it. For example, a general audience might be more familiar with "AZT drugs," but a professional audience might prefer "Azidothymidine (AZT) drugs." To index by abbreviation, follow strict letter-by-letter alphabetic order and cross-reference the term if necessary.

Accents, Apostrophes, and Diacritical Marks in Indexing

11.81 Ignore accents, apostrophes, and diacritical marks in alphabetizing. Do not use the Americanized spellings of ä, ö, ü, and so on (ae, oe, ue); keep the markings in place and merge them with the unmarked letters. If there are many such terms, follow the custom of the native language. For example, in Swedish the å, ä, and ö follow the z in the dictionary. Be sure to include a description of any unusual alphabetizing practices in the index introduction. (See Sections 4.9–4.12, 4.75, 6.26–6.54, 11.89.)

Entities and Alphabetic Ordering

11.82 Alphabetize businesses, book and periodical titles, and other "literals" by the first keyword; for example, "John Hancock Insurance Company, the" and "*Wall Street Journal,* the." Invert society or agency names so that the keyword comes first: "Standards, National Bureau of (NBS)."

Foreign Terms in Indexes

11.83 Do not translate foreign phrases, especially in specialties such as legal or medical indexing. Translations have no meaning to index users, who will not know where to look. Index the term as readers expect to see it.

Names in Indexes

11.84 To index names use the following guidelines (see Sections 6.26–6.54):

- Use the personal name that is best known: such as novelist "Eliot, George" and not "Evans, Mary Ann," pseudonym.
- Use the familiar form of a personal name: "Poe, Edgar Allen," and not "Poe, E. A."
- Index pseudonyms by proper name, with the pseudonym in parentheses: "Clemens, Samuel (Mark Twain)."
- Index by the first part of name in a compound name: "McGraw-Hill," not "Hill," and "Bulwer-Lytton," not "Lytton."
- Use the full form of a place name. To index a "long-established entity," index by the apparent modifier, since that is how the place is known: "West Virginia" instead of "Virginia, West." Index less known places by the noun and not the modifier: "Korea, North."
- Use the last or most current name in the event of a name change: "Istanbul," not "Constantinople."

Numerals in Indexes

11.85 Treat numerals as if they are spelled out: index "4th" as "fourth," "2, 5-AM" under "two," and so on. Some cases might dictate the use of serial order over alphabetic order: An example is within a series of related items incorporating numbers or numerals in their names. For example, index a series of aircraft as F3D, F4D, F5D, and so on. Index "Mark V" before "Mark IX."

For other nonalphabetic orders, use chronology for time order and numeric order for related patents or related chemical substances such as the B vitamins.

Possessives in Indexes

11.86 Place the possessive form of a noun after the noun itself, even in a letter-by-letter arrangement:

Bessell, Friedrich W.
Bessell functions
Bessell's equation
Bessell's formulas

Symbols in Indexes

11.87 Treat symbols as if they are spelled out: "=" as equal, "%" as percent, "+" as plus, and so on.

Identical Headings in Indexes

11.88 Alphabetize identical headings in the following order:

1. Person's name,
2. Place name,
3. Common noun, and
4. Titles of printed or audiovisual works.

For example, "Cicero" precedes "Cicero, Illinois." The topic of "Industrial dynamics" precedes the book title "*Industrial Dynamics*."

Including an Introductory Note

11.89 Introduce the index in order to help the reader use it better:

- Note the alphabetic arrangement used ("terms in this index are arranged in letter-by-letter alphabetical order").
- Include instructions for interpreting special type ("page numbers for definitions are in italics; page numbers for illustrations are in boldface type").
- Indicate abbreviations used ("the abbreviation 'il.' indicates that the entry contains an illustration of the indexed item").

- Describe how to interpret foreign characters ("German vowels with umlauts are merged with vowels having no umlaut").

FINAL PROOFREADING

11.90 When the index has been completed, make one last proofreading pass before sending it to the editor or typesetter:

1. Verify the alphabetic order of headings and subheadings.
2. Check for consistent spelling, capitalization, and punctuation.
3. Indent all headings and subheadings correctly.
4. Determine that no single heading or subheading has more than seven locators, and try to use no more than four.
5. Verify the accuracy of all locators.
6. Cross-reference all synonyms adequately.
7. Make sure all cross-references identify headings correctly.
8. Delete repetitious, extraneous, or trivial entries.
9. Write a preliminary note to explain how to use the index.

TWELVE

Creating Useful Illustrations

Selecting Illustrations
Photographs
- *General Guidelines for Photographic Sessions*
  - *Planning Photographic Sessions*
  - *Preparing for Photographic Sessions*
  - *Directing Photographic Sessions*
  - *Selecting Photographs*

Renderings
- *Special Considerations for Producing Renderings*

Line Drawings
Icons
Constructing Illustrations
- *Assessing the Reader's Needs*
- *Gathering Source Information*
- *Selecting Appropriate Information*
- *Constructing Clean Hand Sketches*
- *Working With the Artist*
- *Printing Final Illustrations*
- *Special Considerations for Producing Illustrations*
  - *Selecting Views*
  - *Including a Hand*
  - *Numbering Illustrations*
- *Labeling Illustrations*
  - *Captioning Illustrations*
  - *Subtitling Illustrations*
  - *Callouts for Illustrations*
  - *Illustration Notes*
    - *Source Notes*
    - *General Notes*
    - *Specific Notes*
  - *Type Legibility for Illustration Text*
- *Positioning Illustrations*

Highlighting
Color Highlighting

12.1 Typically, illustrations can be used in various types of documents, from job training materials to proposals, brochures, and pamphlets. Illustrations can

- Add information,
- Help explain content,
- Visually confirm information in text,
- Motivate readers and make reading more interesting,
- Provide visual relief for a page heavy with text, and
- Sometimes replace text.

12.2 Using illustrations will help readers

- Perform procedures,
- Visualize mechanisms and processes, and
- Remember what they have read.

12.3 Illustrations will be especially helpful when users

- Have limited reading skills,
- May not have on-the-job time to read, or
- Speak a language other than English (see Chapter 8).

12.4 This section provides guidelines on

- Selecting illustration types—line drawings, photographs, renderings; and
- Constructing, labeling, highlighting, and placing illustrations.

SELECTING ILLUSTRATIONS

12.5 When selecting an illustration technique,

To Show	Use
Exactly how something looks to help readers recognize it or prove it is real	Photographs
Something that does not exist now but will in the future, or something that is difficult or impossible to photograph	Renderings
Specific features or characteristics of a complex object How to assemble, install, use, or maintain a product How something works	Line drawings
A picture that represents objects, actions, or locations in a form that can be quickly understood	Icons

PHOTOGRAPHS

12.6 Use photographs to show exactly how something looks to help readers recognize it or to prove its authenticity. Photographs also can motivate readers by making documents more interesting, attractive, and accessible.

General Guidelines for Photographic Sessions

12.7 Work with a professional photographer who can provide expert guidance. Plan and direct the photo session and select the photographs to use. In some cases an artist may assist in this process.

Planning Photographic Sessions

12.8 Plan photographic sessions by

- Determining what photos would best support the text,
- Detailing in writing the elements that should appear in each photo,
- Determining when the object or event to be photographed will be available,
- Deciding if the object or event can be transported or must be photographed at a specific location,

- Contacting the photography studio or in-house photographer and scheduling a photo session,
- Arranging for a model, if required. (Sometimes employees can be used as models.)

Preparing for Photographic Sessions

12.9 Prepare for the photo session by

- Inspecting the item(s) to be photographed to insure it is in good condition (no obvious scratches, proper labels, etc.), and
- Briefing the photographer on the photos required.

Directing Photographic Sessions

12.10 Direct the photographic session by

- Setting up the item(s) to be photographed for each picture, and
- Posing the model properly (if applicable).

Selecting Photographs

12.11 To select photos

- Examine proofs, select the best photos, and have prints made.
- Examine the prints with the artist to decide on retouching and cropping.
- Mark callouts (text that appears within the illustration itself, to explain its features; see Sections 12.52 and 12.53) on photocopies of the prints. To do so, first make a photocopy of each print. Leave enough border on each photocopy to write on. Mark the callouts legibly near the features to be labeled. Using a ruler, draw a line from each callout to the appropriate feature.

RENDERINGS

12.12 Use renderings to show something that does not exist now but will in the future, or something that is difficult or impossible to photograph. Some renderings are so realistic that they can be mistaken for photographs.

Special Considerations for Producing Renderings

12.13 Provide the artist with as much information as possible to create a rendering. Engineering design drawings are helpful. In addition, arrange a meeting among the artist, marketing, and engineering personnel.

LINE DRAWINGS

12.14 Use line drawings to show

- Specific features or characteristics of a complex object. For instance, a one-dimensional line drawing might be included in a product description to illustrate the size of an object:

Exhibit 12–1:
Dimensioned Line Drawing

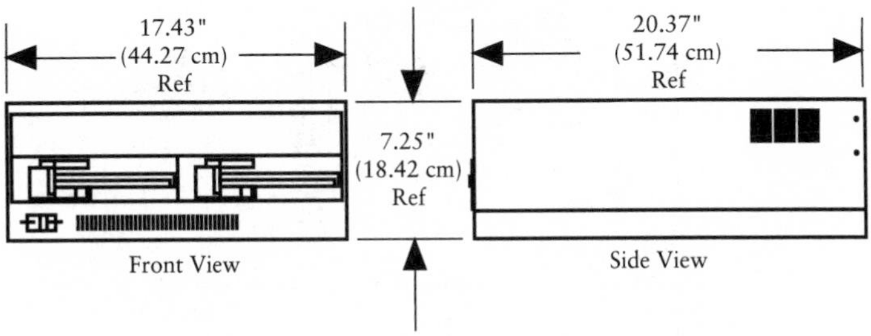

- How to assemble, install, use, or maintain a product. This illustration type also shows how something works. The following figure, for example, integrates text and a drawing so that readers can verify they are performing the procedure properly:

Exhibit 12–2:
Installation Drawing

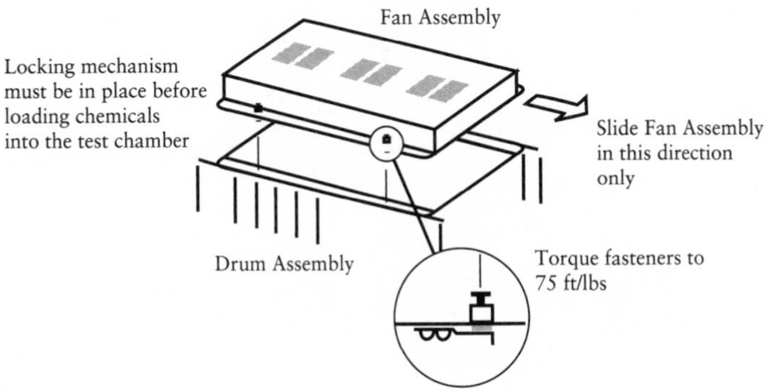

12.15 Do not mix illustrations types in procedures. Using one style (photos or line drawings) allows the reader to recognize the object and its parts from step to step and between procedures.

Exhibit 12–3:
Single Style Illustration

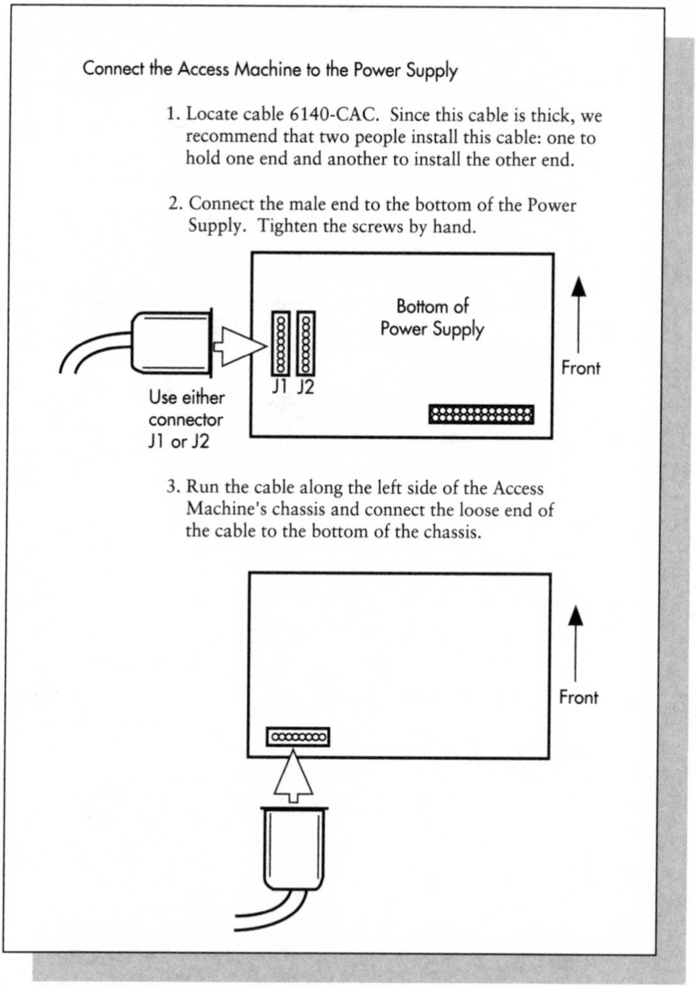
Connect the Access Machine to the Power Supply

1. Locate cable 6140-CAC. Since this cable is thick, we recommend that two people install this cable: one to hold one end and another to install the other end.

2. Connect the male end to the bottom of the Power Supply. Tighten the screws by hand.

3. Run the cable along the left side of the Access Machine's chassis and connect the loose end of the cable to the bottom of the chassis.

12.16 Exclude needless detail from line drawings. This allows readers to focus on important parts or features they must identify or manipulate. In Exhibit 12–3, for instance, only the cable and its connector are detailed; the fasteners needed to complete the assembly have been excluded.

12.17 Consider wordless instructions for machinery set-up tasks (connecting a cable to a computer) and emergency procedures (aircraft emergency instruction cards) based on spatial relationships.

ICONS

12.18 Use icons, stylized pictures of actual objects, in an area where text would be too small to be legible.

12.19 Use color to intensify an icon's meaning. Color attracts attention. On the other hand, color could mislead or be less appealing to a color-blind person. Also, color is more expensive to print and, if only icons use color, it may be difficult to maintain color alignment (registration) and provide easily identifiable colors for such small, and perhaps numerous, objects.

12.20 Use standard icons for specific disciplines (chemistry, electrical engineering, etc.) or those appearing in the public domain (international signage for personal services). The most commonly known icons in scientific and technical documentation are those approved by such established standards organizations as the International Organization for Standardization (ISO) and the American National Standard Institute (ANSI).

Exhibit 12–4:
International Icons

CONSTRUCTING ILLUSTRATIONS

12.21 Effective illustrations provide information without forcing readers to search through unnecessary detail. To be effective, illustrations must be of high quality and relevant to the text.

12.22 The following sections discuss the aspects of the process important to providing functional, technically accurate illustrations:

- Assessing the reader's needs,
- Gathering source information,
- Selecting appropriate information,
- Preparing clean hand sketches,
- Working with the artist, and
- Printing final illustrations.

Assessing the Reader's Needs

12.23 Before creating illustrations, understand the needs of the particular audience as clearly as possible. (See also Chapter 1.)

12.24 Help readers understand a product, device, or concept by presenting a view that best orients them to the subject. It is not always easy to understand the reader; doing so requires some intelligent guesswork. In writing an installation manual, for instance, walk through all tasks step by step. To tell users how to perform the installation

- Perform the installation personally (and, ideally, alone),
- Interview experts on the tasks,
- Ask experts what type of information the user needs to know, and
- Talk to the people who will make the installation.

Gathering Source Information

12.25 To use artist-created illustrations, such as drawings, provide the artist with good source material. Technical accuracy begins with reliable source material. The kind of source material varies with local work relationships and the subject of the illustration. Generally, appropriate source information can include documents from engineering or marketing departments:

- Engineering drawings (schematics or blueprints),
- Prototype object or product,
- Artist's rendering,
- Hand sketch of the actual device,

- Photograph of the device,
- Engineering reports,
- Specifications,
- Engineering design drawings, and/or
- The device itself.

12.26 To use photographs in a document, plan and direct the activities of the photographer to insure that the photos meet their intended purpose. (See Section 12.6.)

Selecting Appropriate Information

12.27 Select material from the source information that readers need to perform the task or understand the concept. Base this selection on an understanding of

- Who the audiences are,
- What they need to know, and
- How they will use the information.

12.28 Use the product, device, or software. If possible, install the hardware and learn the basics of maintaining it. Note the kind of information needed to perform the task, and provide that information to the reader. Consider whether the readers have more or less experience than you do.

Constructing Clean Hand Sketches

12.29 Provide rough sketches of the intended art for a graphic artist or technical illustrator. Include such information as desired views, the use of screens or color, the general size and orientation, and references to comparable company artwork.

12.30 Include one or more carefully drawn sketches, tracings, xerographic samples, or prints (electronic or printed) of the intended final illustrations.

12.31 After gathering source material and selecting appropriate information, construct hand (or electronic) sketches for the artist or technical illustrator. Make these sketches neat and clean. Use a ruler, templates, a compass, and other appropriate drawing equipment.

12.32 Label each illustration, with either a self-stick removable note or a nonreproducible pencil, so the artist can identify its subject matter. Provide a copy of the text so the artist can see the relationship between the text and the supporting illustrations.

12.33 In addition to hand drawings, give the artist such source material as blueprints or schematics. If possible, provide the artist with the actual device; if it is not transportable, arrange for the artist to create a working hand sketch from the actual device.

Working With the Artist

12.34 If the artist needs additional information, the author must obtain it. For instance, if the illustrations depict hardware the author, or perhaps a product designer, must provide the actual device for the artist as reference material.

12.35 Review the artist's work at least once for

- Technical accuracy,
- Typographic errors,
- Correct typographic conventions,
- Accurate text references,
- Accurate use of abbreviations,
- Correct specialized symbols and signs,
- Appropriate line weights for specific uses,
- Leader arrows correctly positioned,
- Obvious physical size cues, and
- Accurate object orientation.

12.36 Consider any suggestions a knowledgeable artist makes for presenting information, keeping the readers' needs in mind. If both the author and illustrator understand these needs, they can create usable and appealing illustrations.

Printing Final Illustrations

12.37 In printing illustrations on an offset press, review a blueline (also known as blueprint) with the illustrator. (Since bluelines increase the printing cost somewhat, request them in advance.) Do not make substantive changes at this point, however. Check

- Illustration accuracy,
- Proper screen placement,
- Consistency of screen percentages, and
- Missing type.

Special Considerations for Producing Illustrations

Selecting Views

12.38 In documenting hardware, the view of the device depends on the illustration's purpose and the readers' needs. Alternatives to the typical front view include top view, side view, bottom view, partial view, and cut-away view. When constructing the initial print package, check with an artist and consider the view (or views) that will be most useful to readers.

Including a Hand

12.39 Include a hand in the illustration only when it is important to show the user exactly how to hold or manipulate a device. This practice may be particularly useful in wordless instructions, in assembly and set-up instructions, and wherever the risk of personal injury exists.

Numbering Illustrations

12.40 Number illustrations sequentially with arabic numerals in a manner similar to the document numbering system. Sequential numbering may continue through the entire document (1, 2, 3) or be contained within a major section or chapter. For instance, illustrations in Chapter 10 would appear as Figure 10.1 (or 10–1), Figure 10.2 (or 10–2), and so forth. Illustrations should also be described by using a single term. For instance, all nontext elements in a manuscript might be called *displays* or *exhibits*. If there is no doubt as to what part of the text the illustration relates to, the caption may not need a figure number:

Unacceptable	**Acceptable**
Drawing 7–1	Exhibit 7–1
Photo 7–1	Exhibit 7–2
Rendering 7–1	Exhibit 7–3

12.41 If the text refers to illustrations, number

- Every illustration and refer to each by number in the text;
- Illustrations in the order they appear and are cited;
- Illustrations and data displays in a single series, calling them all figures or exhibits;
- Illustrations separately from data displays: tables, charts, and diagrams.

While it is possible to number and name illustrations separately from data displays, that technique is generally not typical of technical information practice.

12.42 If appendixes contain illustrations, number the illustrations using the same method as the rest of the work. For example, if illustrations are numbered sequentially throughout the text and the last illustration in the text is Exhibit 42, then make the first illustration in the appendix Exhibit 43. If illustrations are numbered by chapter or section, the first illustration in Appendix B would be Exhibit B–1 (or B.1).

Labeling Illustrations

Captioning Illustrations

12.43 Give the illustration a unique caption that clearly distinguishes it from other illustrations. Feature the distinction; do not obscure it in the caption. Avoid a series of captions that all begin with the same words. Such long captions obscure the most important and useful part of the caption:

Unacceptable

Sodium Concentrations of
Mississippi
Sodium Concentrations of Amazon

Acceptable

Mississippi Sodium Concentrations
Amazon Sodium Concentrations

12.44 The caption should not

- Repeat information found in the text;
- Provide unnecessary background information;
- Attempt humor (Mental Illness in Illinois: Nonstandard Deviants).

12.45 Use longer and more descriptive captions for a nontechnical audience than those for a purely technical audience. Such captions should contain complete sentences. They typically tell the reader what the illustration contains and how to interpret it:

For Technical Readers

Groundwater Contaminants in
Chicago Metropolitan Area

For Lay Readers

Groundwater Contamination Increases in Chicago Area.
Arsenic Shows Largest Increase.

12.46 Phrase the caption as a noun or noun phrase. Avoid relative clauses; use participles instead:

Unacceptable	**Acceptable**
Number of Patients who Showed Advanced Symptoms	Patients Showing Advanced Symptoms

12.47 Center one- or two-line captions horizontally over the illustration. Centering the caption helps the reader visually associate it with the illustration and distinguish it from the text.

12.48 Left-justify captions over three lines.

12.49 Illustration captions and numbers may be omitted altogether when the illustration is simple and appears right where it is mentioned in the text.

Subtitling Illustrations

12.50 Use a subtitle for essential, but secondary, information that applies to the illustration as a whole. Use subtitles for

- Scope of a study (Madison County, Alabama),
- Units of measurement (in thousands of dollars),
- Units of analysis (by month),
- Conditions of an experiment (T = 450°F),
- Statistical criteria (s = 6.223), and
- Number of test subjects (n = 104).

12.51 Enclose subtitles in parentheses. Center the subtitle below the caption:

Depth Gauge
(metric dimensions)

Callouts for Illustrations

12.52 Use callouts to explain specific illustration features. This text attracts the readers' attention to selected features and adds specific details. Leader lines, directional lines without arrowheads, or arrows often connect callouts to illustration features.

12.53 Position callouts to the left, right, above, or below the subject of the illustration. If the leader line or arrow runs through a dark portion of the illustration or through general text, use white space around the arrow to make it visible. Where possible, keep the leader lines or arrows straight, rather than broken at an angle.

12.54 For illustrations that will be translated, use numbered callouts in the illustration and provide a numbered list to define each numbered callout in the illustration. This practice allows the same illustration to be used for many languages.

Also be aware that many cultures have attitudes about professional roles and workplace conditions. Select props and people for photographs based on a careful review of their cultural implications.

Illustration Notes

12.55 Illustrations can have three types of notes:

- Source—documents, much like bibliographic citations for texts, the source of the information presented in the illustration,
- General—explains, clarifies, expands, or elaborates on information applicable to the entire illustration, and
- Specific—supplies critical information applicable to a particular feature.

12.56 Place illustration notes in three positions:

- Enclosed in a box at the bottom of boxed illustrations,
- At least two rows or 2 picas below all artwork in an unboxed illustration,
- Collected on a note page immediately after a multipage illustration.

12.57 If all illustration notes appear on a note page at the end of a multipage illustration, cross-reference this page with this note on the bottom of each page that contains a note reference: *See notes at end of illustration* (*display, exhibit,* or whatever term describes illustration).

Source Notes

12.58 Source notes go at the foot of the illustration (or the first page of a multipage illustration) before any other notes.

12.59 Introduce source notes with SOURCE: or SOURCES: in emphatic type (conventionally caps and small caps). Use source notes to acknowledge the source of the data, if other than the author's original research. Use the same form as for footnotes in text.

12.60 If the illustration is reproduced directly from another publication, follow the citation with the words *reprinted with permission of the publisher* in parentheses:

SOURCE: J. Paulson, "Tensile Strength Loss Due to Radiation," *Metallurgy Review* (September 1986): 55 (reprinted with permission of the publisher).

If the publisher requires that the citation give credit with some other wording or format, honor the publisher's request.

GENERAL NOTES

12.61 General notes include information about the entire illustration or large parts of it. Information appropriate for general notes includes

- How the illustration information was gathered.

  NOTE: Meteorological features common to global warming were provided by the National Oceanographic and Atmospheric Administration and are based on their studies of Antarctic conditions during the years 1974 through 1979.

- How accurate or reliable the callouts are.

  NOTE: Equipment dimensions are accurate to two decimal places.

- How information has been manipulated.

  NOTE: The cutaway views in this exhibit have been generated from preliminary design drawings and may be inaccurate.

- Other related information or cross-references.

  NOTE: For a similar view see Exhibit 14.

12.62 Place general notes below source notes at the foot of the first page of the illustration. Introduce general notes with NOTE: or NOTES: set in emphatic type (conventionally capitals and small capitals).

SPECIFIC NOTES

12.63 Specific notes give more information about a specific illustration feature.

12.64 Mark specific notes with raised (superscripted) symbols above and to the right of the item they refer to. Separate note references that occur together with spaces, not commas. If a note occurs in a blank cell, put the note reference in parentheses:

Single note[1]	Double note[2 3]	(4)

12.65 Use special symbols when there are many specific notes in an illustration. Special symbols include the asterisk, dagger, double dagger, section, parallel, and number mark. To create even more special symbols, double or even triple them (**, ***, ##, ###). Do not mix special symbols (*#, #*@). In statistics * and ** are abbreviations for probabilities of .05 and .01 respectively.

12.66 Use the same type of symbols throughout the illustration.

12.67 Place specific notes in three positions:

- Enclosed in a box at the bottom of boxed illustrations,
- At least two rows or 2 picas below all artwork in an unboxed illustration,
- Collected on a note page immediately after a multipage illustration.

See also Sections 12.56 and 12.57.

12.68 Place specific notes in the order they occur in the illustration: starting in the upper left corner and proceeding clockwise.

Type Legibility for Illustration Text

12.69 Use the same size and style typeface as the body text for illustration captions, subtitles, and callouts, but practical constraints may require that they be smaller.

12.70 Use upper- and lowercase in figure captions. All caps are harder to read and should be used only occasionally for emphasis or for acronyms.

12.71 Set illustration notes in smaller or less emphatic type than captions, subtitles, or callouts.

Positioning Illustrations

12.72 Position illustrations to integrate them with the text in a manner most usable to the reader. When placing illustrations on the page, keep them

- Within the page's grid structure, and
- Close to their reference in the text.

12.73 Many companies use a standard grid structure to control documentation. Placing the illustration within the grid structure enhances the document's organization and contributes to clean copy:

- If positioned improperly, the illustration will appear detached or randomly placed ("float") on the page; and
- If too little or too much space surrounds the illustration, the page will look cluttered or the illustration will float.

12.74 Always place the illustration as close as possible to the first reference to it. Illustration or figure *reference* means the place in the text where the reader is told to look at the figure.

12.75 In procedures, place the figure immediately following (or beside) the verbal instruction related to that illustration. See Exhibit 12–3, for instance.

HIGHLIGHTING

12.76 Shading is a useful way to highlight illustrations. A shaded area is known as a screen. A screen is a block of color or gray covering a portion of an illustration or text to be given the reader's special attention.

12.77 When used carefully, a screen can

- Draw the reader's attention to a portion of an illustration;
- Keep an illustration from floating on the page; and
- Highlight featured text, such as a title or list of promotional features.

12.78 Screen, or shade, areas by percentage. The lightest useful screen—least dense—would probably be about 10 percent, the darkest perhaps 40 to 60 percent (a 100 percent screen means that the screen is the pure color; 100 percent black literally would mean no screen). Plan tonal values so that they are used consistently throughout a work. In his *Semiology of Graphics* Jacque Bertin, for instance, suggests the following values for tones:

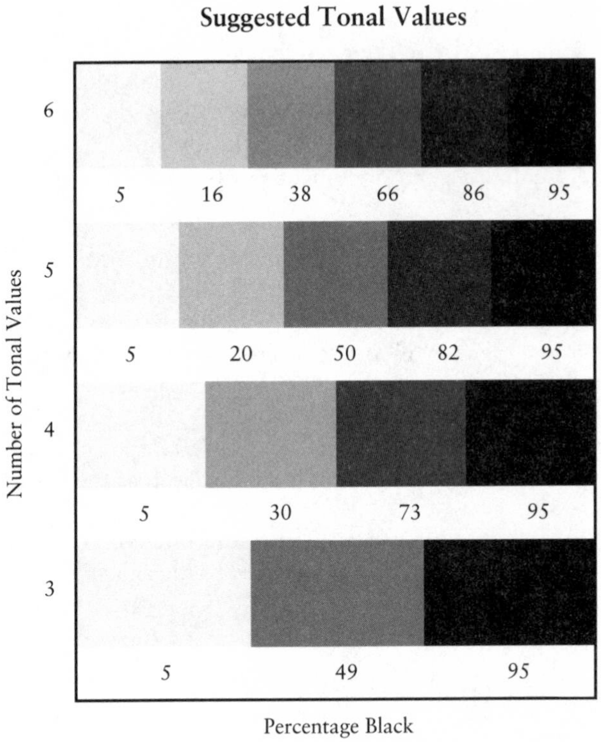

12.79 The total value should highlight the illustration or text while allowing for maximum readability. Useful screen percentages vary with the type of paper on which the document will be printed and the color of the ink used. For example, since glossy (coated) paper is less absorbent than dull or matte (uncoated) paper, a screen will appear slightly darker on glossy paper. Similarly, a 50 percent brown screen would probably be too dark over black ink, making the document difficult to read. The goal is to make the screens attractive and functional, and the screened illustration readable. To select screens for illustrations, discuss percentages with the artist.

Color Highlighting

12.80 Screens can be in color or shades of gray. While color adds appeal, it

- Varies with the qualities of the paper stock,
- Varies with the ambient lighting,
- May be ineffective for color-blind people,
- May have adverse cultural implications for foreign markets,

- May be reproduced by a monochromatic photocopying device,
- Is more time consuming to prepare, and
- Costs more to print.

12.81 Consult with an artist or illustrator about alternatives and special considerations for using color.

THIRTEEN

Creating Usable Data Displays: Tables, Charts, and Diagrams

13.1 Tables, charts, and diagrams—data displays—present technical information in ways that text alone cannot. These data displays condense, summarize, and organize complex details. They can reveal relationships and allow reference to specific facts.

WHAT ARE DATA DISPLAYS: TABLES, CHARTS, AND DIAGRAMS

13.2 Various authorities do not even agree on what to call a table, a chart, a graph, or a diagram. In this chapter the terms mean

Table

Column 1	Column 2	Column 3	Column 4
Item 1	Item 1	Item 1	Item 1
Item 2	Item 2	Item 2	Item 2
Item 3	Item 3	Item 3	Item 3
Item 4	Item 4	Item 4	Item 4
Item 5	Item 5	Item 5	Item 5
Item 6	Item 6	Item 6	Item 6
Item 7	Item 7	Item 7	Item 7
Item 8	Item 8	Item 8	Item 8
Item 9	Item 9	Item 9	Item 9
Item 10	Item 10	Item 10	Item 10

Collection of individual pieces of information arranged in rows and columns.

Chart

Visual display in which scaled dimensions represent numeric values. The term chart *includes* graphs.

Diagram

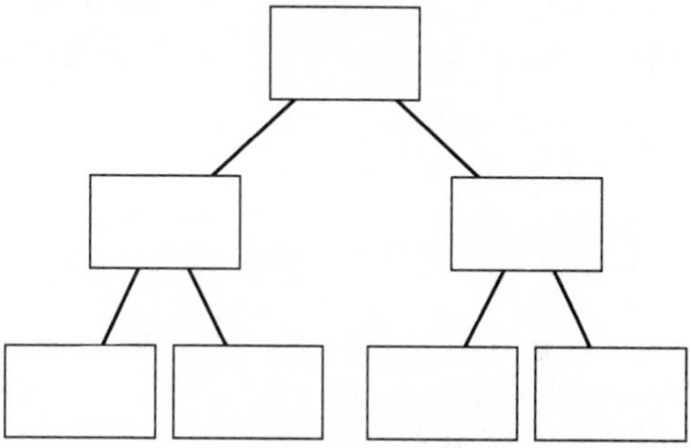

Visual representation of the relationships among the components of a system process, organization, or procedure.

Which Should I Use?

13.3 Charts, tables, diagrams, and text can all show relationships among numerical data or other forms of information. Use the right form for a particular situation:

If relationships in data are . . .	And . . .	Then use . . .
Qualitative	Simple	Text
	Complex	Diagram
Quantitative	Exact values most important	Table
	Patterns and trends most important	Chart

For more detailed recommendations, see the individual sections on tables, charts, and diagrams.

CHARACTERISTICS OF DATA DISPLAYS

Legible

13.4 Data displays should be compatible with surrounding text. Lines and lettering must be large and heavy enough so they do not fade out when reduced, printed, or photocopied.

13.5 It is impossible to specify type sizes and styles in a general guide like this, which must cover publications produced on typewriters, desktop publishing systems, and commercial typesetters. Individual presses and publishing organizations should set up standards for type and line weights used in data displays. Emphasis, for instance, can be achieved by varying typographic elements:

Bold vs. normal weight,
Italic vs. roman style,
Underlined vs. not underlined, and
Larger vs. smaller type.

13.6 Use upper- and lowercase for titles, headings, and text in tables, charts, and diagrams. ALL CAPS IS HARDER TO READ AND TAKES UP MORE SPACE.

13.7 Do **not** mix too many sizes and *styles* of LETTERING. Use one basic typeface for all text in the data display. Vary the size and weight for emphasis. Avoid elaborate, decorative fonts. Use italic or slanted lettering sparingly.

13.8 Size the data display to present the data legibly, not to fill up space on the page. Never spread out the rows and columns of a table merely to fill out a column or page.

Integrated With Text

13.9 Place data displays in the normal flow of text where possible. Avoid clumsy cross-references and unnecessary page flipping. Do not, however, place a data display in the middle of a paragraph.

13.10 Data displays that are not essential to the text's message, present large amounts of raw data, or are used solely for occasional reference may fit better in an appendix.

13.11 Tables, charts, and diagrams should use the same margins and typefaces as the text. Many organizations still use serif type for text and sans serif for tables, charts, diagrams, and other illustrative displays. If the work is to be typeset, there seems little aesthetic reason to use a different typeface for displays than the one used for text. Likewise the same type should be used in the body of data displays as in the body of text. Using smaller type implies that information in the data display is less important.

13.12 Do not use unnecessary boxes or rules to segregate text from graphics.

13.13 Leave at least the equivalent of two blank text lines above and below tables. Center or left-justify the table in the available space.

Easy to Locate Information

13.14 Because data displays are used for quick retrieval of an individual fact or reading a single data value, the scanning reader must be able to recognize them on the page. When integrated with the text, data displays must be visually distinct. Normally the extra margins and rectangular layout of tables provide sufficient clues for the reader. If not, data displays can be announced visually by

- Placing bold rules before and after the display. This draws attention to the display but separates it from the text.
- Printing the display on a light-colored background. The background should be no more than 10 percent black.

13.15 Avoid using a dot pattern to create a gray background, especially if the display contains numbers with decimal points.

Complete Data Displays

13.16 Design displays to be complete and self-contained. The essential meaning of the display should be obvious, even when presented without the surrounding text. The reader should never have to search the text to discover a display's meaning.

Simple Data Displays

13.17 If a display seems too complex, consider whether it offers too much information or too many types of information. If a display requires more than two levels of headings, it is probably too complex and should be divided into two or more simpler displays.

Accessible Data Displays

13.18 Emphasize the similarities and differences of displays. Group similar items and present them in the same way. Conversely, separate dissimilar items and present them differently.

Numbering Data Displays

13.19 If the text refers to displays, number

- Every display and refer to each by number in the text,
- Displays in the order they appear and are cited,
- Displays and figures in a single series, calling them all figures or exhibits,
- Displays separately from illustrations (Illustrations usually include line drawings and photographs).

While it is possible to number displays separately from illustrations, that technique is generally not typical of technical information practice.

13.20 Do not create separate numbering series for different types of figures:

Unacceptable	Acceptable
Display 7–1	Figure 7–1
Photo 7–1	Figure 7–2
Chart 7–1	Figure 7–3
Diagram 7–1	Figure 7–4

13.21 Number displays with arabic numbers. Follow the same numbering scheme for displays as for pages. Number displays by chapter or section—for example, Figure 5–7 for the seventh figure of Chapter 5.

13.22 If appendixes contain displays, number the displays using the same method as the rest of the work. For example, if displays are numbered sequentially throughout the text and the last display in the text is Display 42, then make the first display in the appendix Display 43. If displays are numbered by chapter or section, the first display in Appendix B would be Display B–1 (or B.1).

Primary Title for Data Displays

13.23 Give the display a unique title that clearly distinguishes it from other displays. Feature the distinction; do not obscure it in the title. Avoid a series of titles that all begin with the same words. Such long titles obscure the most important and useful part of the title:

Unacceptable

Sodium concentrations of Mississippi
Sodium concentrations of Amazon

Acceptable

Mississippi Sodium concentrations
Amazon Sodium concentrations

13.24 The title should not

- Repeat information found in the text,
- Provide unnecessary background information,
- Draw conclusions from the data (Smoking hinders prenatal development),
- Express editorial comments (Incompetence costs the University badly needed research grants), or
- Attempt humor (Mental Illness in Illinois: Nonstandard Deviants).

13.25 Titles for a nontechnical audience tend to be longer and more descriptive than those for a purely technical audience. Such titles should contain complete sentences. They typically tell the reader what the display contains and how to interpret it:

For Technical Readers	For Lay Readers
Groundwater Contaminants in Chicago Metropolitan Area	Groundwater Contamination Increases in Chicago area. Arsenic shows largest increase.

13.26 Phrase the title as a noun or noun phrase. Avoid relative clauses; use participles instead:

Unacceptable	**Acceptable**
Number of Patients who Showed Advanced Symptoms	Patients Showing Advanced Symptoms

13.27 Center one- or two-line titles horizontally over the display. Centering the title helps visually associate it with the display and distinguish it from the text.

13.28 Left-justify titles over three lines.

13.29 Display titles and numbers may be omitted altogether when the display is simple and appears right where it is mentioned in the text.

Subtitling Data Displays

13.30 Use a subtitle for essential, but secondary, information that applies to the display as a whole. Use subtitles for:

- Scope of the study (Madison County, Alabama),
- Units of measurement (in thousands of dollars),
- Units of analysis (by month),
- Conditions of an experiment (T = 450°F),
- Statistical criteria (s = 6.223), and
- Number of test subjects (n = 104).

13.31 Enclose subtitles in parentheses. Center the subtitle below the main title:

Starting Salaries of Electrical Engineers

(n = 104, mean = $30,608, s = $994)

13.32 Put less important and less concise information in display notes or in the text.

TABLES

13.33 Tables organize information into rows and columns by category and type, and simplify access to individual pieces of information. They accurately present many quantitative values at one time.

When to Use Tables

13.34 Although text can express numerical relationships, use a table to present more than a few precise numerical values. Use tables to

- Present a large amount of detailed information in a small space;
- Support detailed, item-to-item comparisons;
- Show individual data values precisely; and
- Simplify access to individual data values.

Parts of Tables

13.35 Tables conventionally contain these main parts:

Exhibit 13–1:
Parts of a Table

Stub head · Row head · Title · Column spanner · Column head · Field spanner · Notes

Table 7-1: Tests of Propellant Vent Ports

Propellant Vent Port	Flow		Pressure	
	Specified (psi)	Actual (psi)	Specified (psi)	Actual (psi)
	Model 635			
External	100.0	102.7	10.2	10.4
Internal				
Aft	28.0	28.0	3.0	2.9
Mid	14.0	14.7	1.5	1.6
Stern	28.0	28.1	2.5	2.5
	Model 635-AX			
External	120.0	122.7	12.0	11.4
Internal				
Aft	30.0	28.9	3.0	2.9
Mid	15.0	14.0	1.5	1.5
Stern	32.0	29.3	2.5	2.4

NOTE: Tests performed at 72°F and 65% relative humidity.

Not all tables contain all these parts, nor should they. This example table is more typical than ideal.

Column Headings

13.36 Give each column a short heading to describe information in that column. While independent variables go in row headings, column headings label dependent variables:

Table 7-1: Tests of Propellant Vent Ports

Propellant Vent Port	Flow		Pressure	
	Specified (psi)	Actual (psi)	Specified (psi)	Actual (psi)
	Model 635			
External	100.0	102.7	10.2	10.4
Internal				
Aft	28.0	28.0	3.0	2.9
Mid	14.0	14.7	1.5	1.6
Stern	28.0	28.1	2.5	2.5
	Model 635-AX			
External	120.0	122.7	12.0	11.4
Internal				
Aft	30.0	28.9	3.0	2.9
Mid	15.0	14.0	1.5	1.5
Stern	32.0	29.3	2.5	2.4
NOTE: Tests performed at 72°F and 65% relative humidity.				

13.37 Place information in the column heading that

- Applies to all members of that column, and
- Is unique to that column.

13.38 If information is too long or detailed to go in the heading, put it into a table note and cite the note in the column heading:

Thickness	Dielectric Strength[1]	Volume	Surface Resistivity
NOTE: Explain the content of note 1 here because it is very long and would not fit in the column head.			

13.39 Capitalize column headings the same as the table title. Set column headings in smaller or less emphatic type than the table title. Column headings may be singular or plural as appropriate. Center one-line column headings. Left-justify longer column headings.

13.40 All text on the table should be set horizontal:

Thickness	Dielectric Strength	Volume	Surface Resistivity

13.41 Turn column headings that are too long to fit over their columns counterclockwise so they read from bottom to top. Turn headings only if they cannot be shortened to make them fit. Avoid diagonal heads:

Unacceptable

Power Generation in Southeast States

Alabama	Arkansas	Florida	Georgia	Louisiana	Mississippi	North Carolina	South Carolina	Tennessee	Virginia

Acceptable

Power Generation in Southeast States

Alabama	Arkansas	Florida	Georgia	Louisiana	Mississippi	North Carolina	South Carolina	Tennessee	Virginia

13.42 If a heading is a lead-in phrase completed by a table entry, end the heading with an ellipsis:

If the weather is . . .

hot and dry
cool and wet

Column Subheadings

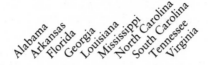

13.43 Include subheadings in column headings if needed. Typically subheadings show units of measurement. Enclose column subheadings in parentheses. Do not put units or other information in a column subhead unless it applies to all items in the column:

Thickness (mm)	Dielectric Strength (Kv/mm)	Volume (ohm/cm)	Surface Resistivity (ohms)

Column Numbers

13.44 To number columns for reference from the text, number them with consecutive arabic numerals in parentheses. Such numbers appear as column subheadings. Number only the lowest levels of column headings, not column spanner headings:

Thickness (1)	Dielectric Strength (2)	Volume (3)	Surface Resistivity (4)

Sometimes column heads contain combinations of information. In such cases, always place the column number as the last element:

Thickness (mm) (1)	Dielectric Strength[1] (Kv/mm) (2)	Volume (ohm/cm) (3)	Surface Resistivity (ohms) (4)
NOTE: Explain the content of note 1 here.			

Column Spanners

13.45 Column spanners label two or more column heads. They are also called *decked heads:*

Table 7-1: Tests of Propellant Vent Ports

Propellant Vent Port	Flow		Pressure	
	Specified (psi)	Actual (psi)	Specified (psi)	Actual (psi)
	Model 635			
External	100.0	102.7	10.2	10.4
Internal				
Aft	28.0	28.0	3.0	2.9
Mid	14.0	14.7	1.5	1.6
Stern	28.0	28.1	2.5	2.5
	Model 635-AX			
External	120.0	122.7	12.0	11.4
Internal				
Aft	30.0	28.9	3.0	2.9
Mid	15.0	14.0	1.5	1.5
Stern	32.0	29.3	2.5	2.4

NOTE: Tests performed at 72°F and 65% relative humidity.

13.46 Use column spanners to show the hierarchy of columns. In the preceding example, the column spanners include *Flow* and *Pressure,* which each span a pair of columns headed *Specified* and *Actual.* In the following example, *B* is a column spanner over columns *B.1, B.2,* and *B.3. B.3* is itself a column spanner over columns *B.3.a* and *B.3.b:*

Stub Head	A	B			
		B.1	B.2	B.3	
				B.3.a	B.3.b

13.47 Include column spanners only in horizontally ruled tables. Extend the rule beneath the column spanner over all columns to which it applies. If column spanners are used, use horizontal rules to show which columns are spanned:

	Thermal Conductivity			
	Wet		Dry	
Material	Rated	Tested	Rated	Tested

In such unboxed headings, align column heads with the bottom. Place rules about half a line below the heading; otherwise the rule may be mistaken for an underline.

13.48 Place an ellipsis (. . .) after any column spanner continued in lower column headings:

If the test solution is . . .			
red and . . .		blue and . . .	
cloudy	clear	cloudy	clear

Field Spanners

13.49 Field spanners cut across all columns of the field and apply to all items below them. Field spanners identify a third variable. Row headings are repeated for each field spanner. Field spanners are also called *cut-in* heads:

Table 7-1: Tests of Propellant Vent Ports

Propellant Vent Port	Flow		Pressure	
	Specified (psi)	Actual (psi)	Specified (psi)	Actual (psi)
	Model 635			
External	100.0	102.7	10.2	10.4
Internal				
Aft	28.0	28.0	3.0	2.9
Mid	14.0	14.7	1.5	1.6
Stern	28.0	28.1	2.5	2.5
	Model 635-AX			
External	120.0	122.7	12.0	11.4
Internal				
Aft	30.0	28.9	3.0	2.9
Mid	15.0	14.0	1.5	1.5
Stern	32.0	29.3	2.5	2.4

NOTE: Tests performed at 72°F and 65% relative humidity.

13.50 Place the top field spanner just below the column heads. Never place the first field spanner above the column heads. Never use a single field spanner in a table. Do not extend the field spanner into the stub column.

13.51 Field spanners occur only in the field of a table, never in the headings. They essentially start the table all over again for another category of data. Hence, all of the stub headings are repeated beneath each field spanner. For this reason, it is never correct to have a single field spanner.

13.52 Using field spanners is an alternative to having separate tables, each with the same column headings and stub heads.

Stub

13.53 The stub is the leftmost column of the table, like the stub of a checkbook. It lists the items about which information is provided in the columns to the right.

Stub Head

13.54 Title the stub with a column heading like other columns. The stub head describes the row headings listed in the stub:

Table 7-1: Tests of Propellant Vent Ports

Propellant Vent Port	Flow		Pressure	
	Specified (psi)	Actual (psi)	Specified (psi)	Actual (psi)
	Model 635			
External	100.0	102.7	10.2	10.4
Internal				
Aft	28.0	28.0	3.0	2.9
Mid	14.0	14.7	1.5	1.6
Stern	28.0	28.1	2.5	2.5
	Model 635-AX			
External	120.0	122.7	12.0	11.4
Internal				
Aft	30.0	28.9	3.0	2.9
Mid	15.0	14.0	1.5	1.5
Stern	32.0	29.3	2.5	2.4

NOTE: Tests performed at 72°F and 65% relative humidity.

13.55 Avoid diagonally split stub heads. If information applies to all column headings, put it in the title, in a column spanner, or in table notes:

Unacceptable

Resistance / Material	Wet		Dry	
	Rated	Tested	Rated	Tested

Acceptable

Material	Resistance			
	Wet		Dry	
	Rated	Tested	Rated	Tested

13.56 Omit the stub head only if the table's title fully and clearly identifies the row headings in the stub.

Row Headings

13.57 Row headings generally represent independent variables or categories. Row headings should

- Clearly and concisely label the rows of the table,
- Uniquely identify each row, and
- Apply to all items in the row.

Table 7-1: Tests of Propellant Vent Ports

Propellant Vent Port	Flow		Pressure	
	Specified (psi)	Actual (psi)	Specified (psi)	Actual (psi)
	Model 635			
External	100.0	102.7	10.2	10.4
Internal				
Aft	28.0	28.0	3.0	2.9
Mid	14.0	14.7	1.5	1.6
Stern	28.0	28.1	2.5	2.5
	Model 635-AX			
External	120.0	122.7	12.0	11.4
Internal				
Aft	30.0	28.9	3.0	2.9
Mid	15.0	14.0	1.5	1.5
Stern	32.0	29.3	2.5	2.4

NOTE: Tests performed at 72°F and 65% relative humidity.

13.58 Keep row headings short. Omit secondary information or put it into a table note and cite it in the row heading.

13.59 Left-justify row headings. One or two levels of indentation may be used in row headings to show the hierarchical organization of the stub. Indent subordinate row headings about four spaces. Also indent row headings for totals, subtotals, grand totals, and averages:

Power Generation

Hydroelectric
    Public
    Private
    Subtotal
Geothermal
Fossil fuel
    Coal
    Oil
    Wood
    Refuse
        Subtotal
Nuclear
        Total

13.60 If rows are clearly spaced, indent runover lines, but indent them less than subordinate heads, about two spaces:

Characteristic

Force of
    gravity at
    surface
Distance from
    earth
    (average)

If rows are not closely spaced, left-justify runover lines.

13.61 If the row headings are uneven in length, use leaders (.........) to direct the eye to the next column:

Material

Asbestos ....
Asphalt..................................
Bakelite, wood mixture ....
Jute, impregnated with asphalt
Pressboard.............................

13.62 Some useful alternatives to leaders include

- Using horizontal rules to separate rows in the table. This can cause some confusion because of the number of rules.
- Printing every other row on a background tone or color. This method has the disadvantage of implying a difference between even- and odd-numbered rows. It also reduces legibility of half of the rows.
- Right-justifying the row headings. This results in a jagged left edge that is impossible to scan rapidly.

13.63 Capitalize row headings as sentences, not headlines. Capitalize the first word and proper nouns and adjectives in English. For other languages, follow the capitalization style common in that language for sentences. Omit periods at the end of row headings. Do not number items in the stub:

Unacceptable	**Acceptable**
1. Tensile Strength (psi)	Tensile strength (psi)
2. Yield Strength. (psi)	Yield strength (psi)
3. Melting Range (°C).	Melting range (°C)

Field

13.64 The field or body of the table is the array of cells below the column headings and to the right of the row headings:

Table 7-1: Tests of Propellant Vent Ports

Propellant Vent Port	Flow		Pressure	
	Specified (psi)	Actual (psi)	Specified (psi)	Actual (psi)
	Model 635			
External	100.0	102.7	10.2	10.4
Internal				
Aft	28.0	28.0	3.0	2.9
Mid	14.0	14.7	1.5	1.6
Stern	28.0	28.1	2.5	2.5
	Model 635-AX			
External	120.0	122.7	12.0	11.4
Internal				
Aft	30.0	28.9	3.0	2.9
Mid	15.0	14.0	1.5	1.5
Stern	32.0	29.3	2.5	2.4

NOTE: Tests performed at 72°F and 65% relative humidity.

13.65 Arrange rows and columns to make finding an individual item in the table simple. Order rows and columns with a logical scheme that the reader will readily grasp and can use to locate the appropriate column and row headings quickly:

- Use time order for information recorded in a sequence,
- Arrange major and minor items in order from whole to parts,
- Group together items of the same type or size,
- Place items that will be compared close together, and
- Put numbers that will be summed or averaged in the same column.

13.66 Avoid mixing too many types of information in a single column. Information may vary by column or by row—but not both at the same time. Especially avoid unnecessary changes of units—for example, from inches to centimeters.

13.67 Align cell entries vertically with the row heading for that row. If the row heading is more than one line, align one-line cell entries to the baseline of the last line of the row heading. If both row heading and column entries are more than one line, align the first lines of both:

Amino Acid	**Isoelectric Point**	**Isolated by**
Alanine	..............6.0....................	Schutzenberg
Aspartic	..............2.8....................	Ritthaus and
acid	....	Prentis
Glutamic acid	..............3.2....................	Ritthaus
Threnonine	..............6.2....................	Gartner and
	....	Hoffman

Rules

13.68 Rules are horizontal and vertical lines that divide the parts of a table. Tables can be drawn with few or many rules. The choice depends on

- The table's complexity,
- How it is scanned, and
- The amount of material presented horizontally and vertically.

Do not vary the use of rules from table to table. Use the same style of rules for all tables in a single work.

13.69 Draw table rules bold enough so that they stand out. Keep in mind the quality of the printing process.

13.70 Avoid overuse of rules in tables. Rules separate closely spaced columns, show relationships among column heads, and guide the eye, but they should not be used as decoration.

13.71 Space columns horizontally and spread out rows vertically to make the table easier to scan horizontally. Align items within a column to make it easier to scan vertically.

13.72 For simple, informal tables—no more than three columns and five rows—rules can be omitted altogether:

Abrasive	Knoob Hardness (Kv/mm)	Melting Point (° C)
Corundum	2000	2050
Diamond	6500	1000
Emery	2000	1900
Garnet	1360	1200
Quartz	820	1700

Abrasive	Knoob Hardness (Kv/mm)	Melting Point (° C)
Corundum	2000	2050
Diamond	6500	1000
Emery	2000	1900
Garnet	1360	1200
Quartz	820	1700

13.73 The weight of rules depends on what other rules are present. Use as thin a rule as possible to visually separate adjacent areas. Start with the thinnest rule visible—for example, a hairline—and increase width as more thickness is needed. For printed works, use rules in this order

hairline
1/2 point
1 point
2 point

Side-by-Side Table Separators

13.74 If a table is too long and narrow to fit the available space, divide it into two halves presented side by side. A rule or extra space is needed to separate the two halves of the table:

Abrasive	Knoob Hardness (Kv/mm)	Melting Point (° C)	Abrasive	Knoob Hardness (Kv/mm)	Melting Point (° C)
Alumina	2000	2050	Diatomite	830	1700
Borazon	4700	2000	Emery	2000	1900
Carbide	2400	2500	Garnet	1360	1200
Corundum	2000	2050	Quartz	820	1700
Diamond	6500	1000	Tripoli	820	1700

The side-by-side table separator should be the most emphatic rule in the table.

Head-Stub-Field Separators

13.75 Head-stub-field separators are the heavy or double vertical and horizontal rules used to separate column and row headings from the body of the table. Make separator rules more emphatic than border, column, or row rules but not as emphatic as side-by-side table separators.

Table Border

13.76 The table border is a box that extends around the outside of the table. If the table appears on a single-column page with margins to left and right, the left and right vertical borders may be omitted. Make the table border the same weight as column rules. These should be lighter than head-stud-field separators but heavier than row rules.

Column Rules

13.77 Column rules separate closely spaced columns. Use column rules only if white space alone is not sufficient to separate closely spaced columns. Make column rules the same weight as border rules. These should be lighter than head-stud-field separators but heavier than row rules.

Row Rules

13.78 Row rules separate rows in the field. Row rules are seldom required. Use row rules only if vertical space is not sufficient to separate rows. In long tables use row rules to divide major sections of the table. Extend row rules the full width of the table, including the stub column. If used, row rules should be the lightest rules in the table.

Table Notes

13.79 Tables can have three types of notes:

- Source—documents, much like bibliographic citations for texts, the source of the information presented in the table,
- General—explains, clarifies, expands, or elaborates on information applicable to the entire table, and
- Specific—supplies critical information applicable to a particular feature.

13.80 Place table notes in three positions:

- Enclosed in a box at the bottom of boxed tables,
- At least two rows or two picas below all entries in an unboxed table, or
- Collected in a note page immediately after a multipage table.

Table 7-1: Tests of Propellant Vent Ports

Propellant Vent Port	Flow		Pressure	
	Specified (psi)	Actual (psi)	Specified (psi)	Actual (psi)
	Model 635			
External	100.0	102.7	10.2	10.4
Internal				
Aft	28.0	28.0	3.0	2.9
Mid	14.0	14.7	1.5	1.6
Stern	28.0	28.1	2.5	2.5
	Model 635-AX			
External	120.0	122.7	12.0	11.4
Internal				
Aft	30.0	28.9	3.0	2.9
Mid	15.0	14.0	1.5	1.5
Stern	32.0	29.3	2.5	2.4

NOTE: Tests performed at 72°F and 65% relative humidity.

13.81 If all table notes appear on a note page at the end of a multipage table, cross-reference this page with this note on the bottom of each page that contains a note reference: *See notes at the end of table* (*display, exhibit,* or whatever term describes table).

Source Notes

13.82 Source notes go at the foot of the table (or the first page of a multipage table) before any other notes.

13.83 Introduce source notes with SOURCE: or SOURCES: in emphatic type (conventionally caps and small caps). Use source notes to acknowledge the source of the data, if other than the author's original research. Use the same form as for footnotes in text.

13.84 If the table is reproduced directly from another publication, follow the citation with the words *reprinted with permission of the publisher* in parentheses:

> SOURCE: John Paulson, "Tensile Strength Loss Due to Radiation," *Metallurgy Review* (September 1986): 55 (reprinted with permission of the publisher).

If the publisher requires that the citation give credit with some other wording or format, honor the publisher's request.

General Notes

13.85 General notes include information about the entire table or large parts of it. Information appropriate for general notes includes

- How the data were gathered.

  NOTE: Study included 143 undergraduate psychology students at the University of Michigan during the years 1973 through 1979.

- How accurate or reliable are the data.

  NOTE: Equipment calibration limits accuracy to two decimal places.

- How data have been calculated, manipulated, rounded off, or approximated.

  NOTE: Data from the three studies were converted to common units of measurement and merged. No adjustments were made for inconsistencies in reporting procedures.

- Other related information or cross-references.

NOTE: For a cost estimate for this project, see Exhibit 14.

Place general notes below source notes at the foot of the first page of the table. Introduce general notes with NOTE: or NOTES: set in emphatic type (conventionally capitals and small capitals).

SPECIFIC NOTES

13.86 Specific notes give more information about a specific column, row, or cell. They are referenced by superscripted symbols in a heading or field of the table. Cite specific notes only in column heads, row headings, and cells. Do not cite specific notes in the table title or stub head. Use a general note instead.

13.87 Do not mix text footnotes and table notes. Keep them entirely separate. Number them separately. Display table notes only with the table.

13.88 Mark specific notes with raised (superscripted) symbols above and to the right of the item they refer to. Separate note references that occur together with spaces, not commas. If a note occurs in a blank cell, put the note reference in parentheses:

Single note^{1}	Double note$^{2\ 3}$	(4)

13.89 Choose symbols that stand out from the data presented in the table:

Data Type	Use	Avoid
Numeric	Letters	Numbers
Text	Numbers	Letters
Formula	Special	Numbers, Letters
Statistics	Letters	* and **

Special symbols include the asterisk, dagger, double dagger, section, parallel, and number mark. To create even more special symbols, double or even triple them (**, ***, ##, ###). Do not mix special symbols (*#, #*@). In statistics * and ** are abbreviations for probabilities of .05 and .01 respectively.

Use the same type of symbols throughout the table.

13.90 Put specific notes at the bottom of the table on the page on which they occur. If the same notes occur on multiple pages, gather all such notes at the end of the table.

13.91 Place specific notes in the order they occur in the table: starting in the upper-left corner and down the table row by row.

13.92 Extend general notes across the full width of the table. Several short notes, however, may be put on a single line:

[1] First general note could be a long one. [2] Second note [3] Third note.

13.93 Set table notes in smaller or less emphatic type than table entries.

Table Contents

13.94 Table cells can contain text, numbers, or graphics. Align and format data within cells consistently and logically.

Text in Tables

13.95 Table cells can contain a word, a phrase, or an entire paragraph. Punctuate all text in table cells using the same conventions that apply to these grammatical units—phrase, sentence, paragraph—in text. Center single-line text entries in the column. Left-justify multiple-line entries. Align all items in the column the same. If a column contains more than a few multiline entries or if centering the multiline entries makes them hard to read, then left-justify all entries in the column:

Short entry	Long entry that extends to multiple lines	Short entry
Another short entry	Another long entry that extends to multiple lines	Long entry that extends to multiple lines
Yet another short one		Another short entry

13.96 If space between rows is not adequate to separate multiline entries, indent runover lines:

Unacceptable	**Acceptable**

Long entry that extends to multiple lines
Another long entry that extends to multiple lines
Still another line that extends to multiple lines
	Long entry that extends to
	  multiple lines
	
	Another long entry that
	  extends to multiple lines
	
	Still another line that
	  extends to multiple lines
	
	Better still, add space
	  between rows

Numbers as Table Content

13.97 The most common use of tables is to organize many numbers into an easily scanned whole.

13.98 Put units of measurement in a column subheading. If different units must be shown in the same column, omit the units from the column subheading and put them in each cell. Do not put units in column subheads unless they apply to all numbers in the column. (See also Section 7.31.)

Unacceptable	**Acceptable**	**Acceptable**
Thickness	**Thickness (mm)**	**Thickness (mm)**
620 mm	620	6.2×10^2
24 cm	240	2.4×10^3
2 cm	20	2.0×10^1

13.99 Use only arabic numerals in numbers. Do not spell out numbers. Do not use roman numerals:

Unacceptable	Acceptable
IX	9
Six	6
Zero	0

13.100 Right-justify whole numbers. Align decimal numbers on the decimal point and on commas, if present. Align numbers on the real or implied decimal point. Align plus (+) and minus (–) signs preceding numbers and percent signs following numbers:

3,468.62	+47%
27,891.74	– 6%
497.65	+29%
1,078.53	–12%
28.35	+21%

13.101 Omit + or – signs, percent signs, and money symbols ($, ¥, £) if the table title or column heading (or common usage) makes the units and sign obvious.

Dates in Tables

13.102 Express all dates in the same format. Center dates, except military dates (2 Jun 88). For military dates, align the three components:

3/11/88	March 11, 1988	11 Mar 88
12/16/89	December 16, 1989	16 Dec 89
4/1/92	April 1, 1992	1 Apr 92

Times in Tables

13.103 Express all times in the same format. Align hours, minutes, and seconds. Include A.M. or P.M. after the time if not using 24-hour format. Include the time-zone reference (GMT, UTC, EST) in the column subheading to avoid confusion. (See also Chapters 3 and 7.)

Unacceptable	Acceptable
Time	Time (GMT)
10:42	10:42:00
1:24 PM	13:24:00
9:14:28	21:14:28

13.104 For durations, make clear the units of measurement. The experimenter may know that the units are hours and minutes, but the reader may interpret them as minutes and seconds:

Unacceptable	**Acceptable**
Duration	**Duration (hours:minutes)**
1:35	1:35
2:02	2:02
1:45	1:45

Ranges in Tables

13.105 Use an en dash (–) surrounded by spaces to link the ends of ranges. Do not use a hyphen or the word *to:*

Unacceptable	**Acceptable**
1972-1982	1972–1982
1929 to 1934	1929–1934

Align ranges on the dash linking the ends of the range.

Formulas in Tables

13.106 Align formulas on the equals signs (=). Follow the same rules for breaking and indenting multiline formulas as in text. (See also Chapter 7.)

$$\begin{aligned} F &= ma \\ PV &= nRT \\ E &= mc^2 \end{aligned}$$

Graphics in Tables

13.107 Graphics in tables can range from realistic pictures, such as the photograph of an employee, to abstract symbols, such as the product rating dots used by *Consumer Reports.* This chart uses dots to show participation by the teams listed in the column heads:

Management Teams

Project Teams	Departments							
	Engineering	Facilities	Finance	Marketing	Production	Purchasing	Research	Sales
Amethyst	●	●	●		●			●
Diamond		●	●	●	●	●	●	
Garnet	●	●		●		●		●
Quartz	●	●				●		
Topaz	●	●	●	●		●	●	●

13.108 If symbols are not obvious, supply a key or legend.

Missing or Negligible Values in Tables

13.109 If a column heading does not apply to a particular cell:

- Insert a leader (. . .), if the cell has no data, to direct the eye to the next column. Make the leader as wide as the widest entry in the column.
- Put a zero in the cell if it has no value.
- Do not put "N/A" or "Not applicable" in the cell. Explain the absence of data, with a table footnote in parentheses:

Unacceptable	Acceptable
N/A	. . .
	0
	(7)

Repeated Values in Tables

13.110 Show repeated values. Do not use "ditto," "do," or the quotation mark ("):

Unacceptable	Acceptable
356	356
"	356
"	356

Special Tabular Problems

13.111 This section examines more complex tables and special situations and needs.

Continued Tables

13.112 If a table is too long to fit on a single page, continue it on subsequent pages. If the table is numbered, repeat the table number. Follow it with a phrase that tells where the current page fits in the sequence of pages:

Figure 4–2 (Page 3 of 6)

13.113 If the table is not numbered, repeat the short title followed by the same phrase:

Resistivity of Carbon Filaments (Page 3 of 6)

Or:

Resistivity (Page 3 of 6)

13.114 Repeat column headings on continuation pages. On continued turned (broadside) tables, repeat the column headings on the left-hand pages only; omit them from the facing right-hand page.

13.115 In horizontally ruled tables, omit the bottom rule, except on the last page of the table. Do not number the separate pages of a multipage table separately, as in 3a, 3b, 3c, and so forth.

13.116 Do not break up the text with many multipage tables. Collect such tables at the end in an appendix. If necessary, summarize the table(s) in the text.

13.117 In works that are primarily tables or other displays, organize and sequence the displays in the same manner as paragraphs and subsections.

Referring to Tables

13.118 If tables are referred to from the text and cannot appear directly where they are referenced, number each table and refer to it by number in the text. Number tables in the order they appear and are cited.

Cite figures and tables directly:

Our study confirms Smith's results, as shown in Table 6–1.

Or parenthetically:

> Our study confirms Smith's results (see Table 6–1).

13.119 If the parenthetical reference occurs within a sentence, capitalize and punctuate it as a phrase. If it occurs between sentences, treat it as a sentence:

> Our study confirms Smith's results. (See Table 6–1.) Later studies add further support.

13.120 Do not refer to a table by its position—for example, *the above table* or *the following table.*

13.121 Avoid unnecessary cross-references. Repeat short, simple tables if necessary. In the general notes, explain that this table is a repeat:

> Note: This table is the same as Table 6–2.

Oversize Tables

13.122 If a table is too long or tall to fit on a single page, continue it on the next or subsequent pages. Or divide the table into multiple tables.

13.123 If a table is too wide, there is no ideal solution. Try remedies in this order:

1. Use a moderately condensed typeface. The typeface should not be so condensed as to severely reduce legibility.
2. Turn the table so it runs horizontally (broadside) on the page with the top on the left. Omit column and stub heads from right-hand pages if the turned table is continued.
3. Reduce the type size and spacing of columns. Never reduce size below 6 or 8 points in printed tables.
4. Extend the table horizontally across two facing pages. Proportion the table evenly between pages. Align the rows so the reader can move back and forth between pages without getting lost.
5. Print the table on oversize paper and bind it as a foldout.

Abbreviations in Tables

13.124 Spell out all titles, labels, and notes. Abbreviate only where necessary. Use only common abbreviations, for instance, for units of measurement. (See also Chapter 5.)

Unacceptable	**Acceptable**
Strut Dsgn Crit (pounds per square inch)	Structural Design Criteria (lb/in^2)

Simplifying Comparisons in Tables

13.125 Convert numbers to common units and round off to the same accuracy. Express such numbers in the same format and with the same number of decimal points (see also Chapter 7):

Unacceptable	**Acceptable**
12 mm	12.000 mm
.023 mm	0.023 mm
2.34 mm	23.400 mm
1.03456 mm	1.035 mm

This rule does not apply to original observations, which should represent the actual units and accuracy measured.

Grouping and Summarizing Data in Tables

13.126 In complex tables, group or summarize data. Avoid braces in the field or columns to show groupings. Use multilevel row headings or field spanners to group data:

Unacceptable

Propellant Vent Port	Pressure			
		Specified (psi)	Actual (psi)	
External		100.0	102.7	
Aft		28.0	28.0	Model
Mid	Internal	14.0	14.7	635
Stern		28.0	28.1	
External		120.0	122.7	
Aft		30.0	28.9	Model
Mid	Internal	15.0	14.0	635-AX
Stern		32.0	29.3	

Acceptable

Propellant Vent Port	Pressure	
	Specified (psi)	Actual (psi)
	Model 635	
External	100.0	102.7
Internal		
Aft	28.0	28.0
Mid	14.0	14.7
Stern	28.0	28.1
	Model 635-AX	
External	120.0	122.7
Internal		
Aft	30.0	28.9
Mid	15.0	14.0
Stern	32.0	29.3

13.127 In long columns of numbers, insert a blank line every five to ten rows:

Squares, Cubes, and Roots

Number	Square	Cube	Square Root	Cube Root
1	1	1	1.00000	1.00000
2	4	8	1.41421	1.25992
3	9	27	1.73205	1.44225
4	16	64	2.00000	1.58740
5	25	125	2.23607	1.70998
6	36	216	2.44949	1.81712
7	49	343	2.64575	1.91293
8	64	512	2.82843	2.00000
9	81	729	3.00000	2.08008
10	100	1000	3.16228	2.15443
11	121	1331	3.31662	2.22398
12	144	1728	3.46410	2.28943

13.128 Other methods of grouping rows include underprinting alternate groups with a light screen or a dot pattern. These methods are not recommended because they display equivalent groups of data differently. Never use a dot pattern (screen) if numbers in the table contain decimal points.

Common Types of Tables

13.129 The remainder of this chapter considers the design of specific types of tables.

Look-Up-a-Value Table

13.130 Most tables provide easy access to specific values. In such tables the reader scans the stub and column headings for ones that describe the target item. The target item is then located at the intersection of the row and column bearing the appropriate headings.

How we read a table

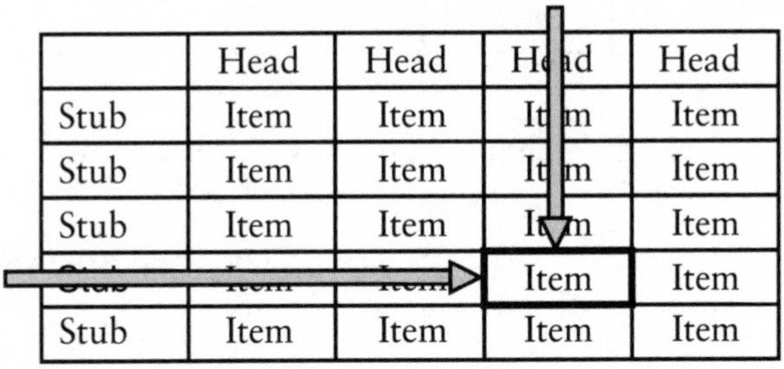
Head Head Head Head
Stub Item Item Item Item
Stub Item Item Item Item
Stub Item Item Item Item
Item Item
Stub Item Item Item Item

13.131 Design column and row headings for scanning:

- Make them distinctive,
- Keep them short,
- Use white space or rules to guide the eye, and
- Use column and row headings and subheadings to express a hierarchy of choices.

Decision Tables

13.132 Decision tables help the reader make complex decisions by simplifying alternatives, and summarizing if-then conditions.

13.133 Divide the decision table horizontally into conditions on the left and actions on the right. The reader scans down the left columns to find a row that matches the current conditions and then performs the actions to the right in that row:

Check Acceptance Policy			
Regular customer?	**Check is for . . . ?**	**ID required**	**Approval required**
Yes	Purchased items	None, unless over $50	None
	Cash	Driver's license	Supervisor
No	Purchased items	Driver's license	Supervisor
	Cash	Driver's license	Manager

13.134 Conditions may be listed as yes/no questions in the column headings with the individual cells containing symbols for the answers. Or if the conditions and responses are few and simple, they may appear as text in the cells.

Distance Tables

13.135 Distance tables show distance or other values relating to pairs of categories. Construct distance tables by listing in a triangular grid the values related to a corresponding column and row head. One of the categories is omitted from the rows and another from the columns to prevent empty boxes.

	Tokyo	Rome	Rio	Beijing	New York	Moscow	Los Angeles	London
Bombay	4,190	3,845	8,620	2,964	7,795	3,130	8,700	4,465
London	5,940	890	5,770	5,055	3,460	1,550	5,439	
Los Angeles	5,470	6,325	6,295	6,250	2,450	6,070		
Moscow	4,650	1,475	7,180	3,600	4,660			
New York	6,735	4,275	4,830	6,825				
Beijing	1,310	5,050	10,768					
Rio	11,535	5,685						
Rome	6,124							

13.136 Distance tables are not limited to presenting distance. They can present any mutual relationship among a list of items.

Matrix Charts

13.137 Matrix charts are a hybrid of charts and tables. Matrix charts are grids in which the stub and column headers are uniformly spaced scales or categories, and each cell presents information about the combination of values represented by the cell.

Modern Periodic Table

H 1																	He 2
L 3	Be 4											B 5	C 6	N 7	O 8	F 9	Ne 10
Na 11	Mg 12											Al 13	Si 14	P 15	S 16	Cl 17	Ar 18
K 19	Ca 20	Sc 21	Ti 22	V 23	Cr 24	Mn 25	Fe 26	Co 27	Ni 28	Cu 29	Zn 30	Ca 31	Ge 32	As 33	Se 34	Br 35	Kr 36
Rb 37	Sr 38	Y 39	Zr 40	Nb 41	Mo 42	Tc 43	Ru 44	Rh 45	Pd 46	Ag 47	Cd 48	In 49	Sn 50	Sb 51	Te 52	I 53	Xe 54
Cs 55	Ba 56	La 57	Hf 72	Ta 73	W 74	Re 75	Os 76	Ir 77	Pt 78	Au 79	Hg 80	Tl 81	Pb 82	Bi 83	Po 84	At 85	Rn 86
Fr 87	Ra 88	Ac 89															

Lanthanides

Ce 58	Pr 59	Nd 60	Pm 61	Sm 62	Eu 63	Gd 64	Tb 65	Dy 66	Ho 67	Er 68	Tm 69	Yb 70	Lu 71

Actinides

Th 90	Pa 91	U 92	Np 93	Pu 94	Am 95	Cm 96	Bk 97	Cf 98	Es 99	Fm 100	Mv 101	No 102	Lw 103

CHARTS

13.138 This section will enable the reader to prepare technically correct and aesthetically pleasing charts.

When to Use Charts

13.139 Charts are tools for thinking about quantitative information. Charts can show the values of numerical data, changes or trends over time, patterns in data, and relationships among various factors.

13.140 Use charts to

- Interest the reader in the data. Well-prepared charts can convince the casual browser to read the text.
- Forecast and predict future values. Charts are superior to tables or text when the reader must extrapolate from given values.
- Present large amounts of complex data without overwhelming the reader. Charts express complex ideas precisely and efficiently.
- Emphasize trends, relationships, and patterns in the data to make a convincing point.
- Add credibility. People trust pictures and distrust words.

When Not to Use Charts

13.141 Do not use charts to

- Make boring numerical data more palatable. Find an interesting aspect of the data, put it in an appendix, or eliminate it.
- Break up text. Charts are too expensive to use as decoration. Use charts to emphasize ideas in the text.
- Show precise numbers. Use a table to allow detailed comparisons.
- Communicate to an audience unfamiliar with charts. Engineers and scientists sometimes forget that a lay audience may be confused by charts.
- Interpret, explain, evaluate, and review data. Use text instead.

Characteristics of an Effective Chart

Coherent

13.142 Each chart should demonstrate coherent relationships among the data. Design charts to focus on the values and variations in the data. The chart's design must be subdued; it must not vie with the data for attention.

Simple

13.143 The chart's main idea should be simple and visually obvious. The reader should immediately grasp the chart's main idea. Each chart should express one main idea. Effective charts direct attention to the data's significance, not the chart's form or style.

Unacceptable

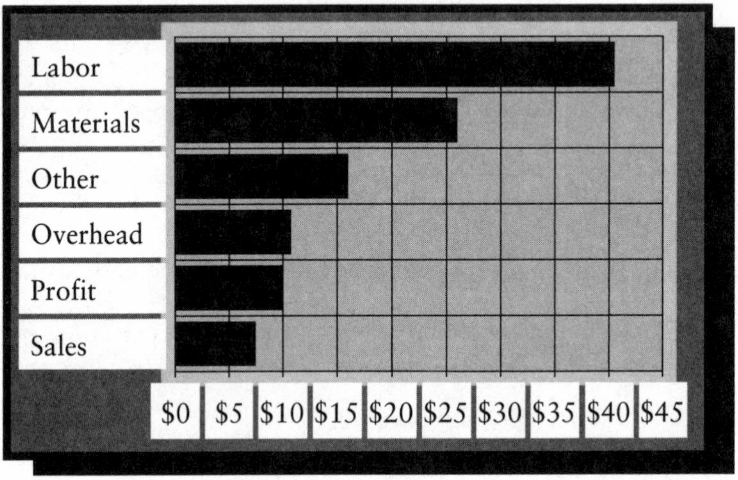

Acceptable

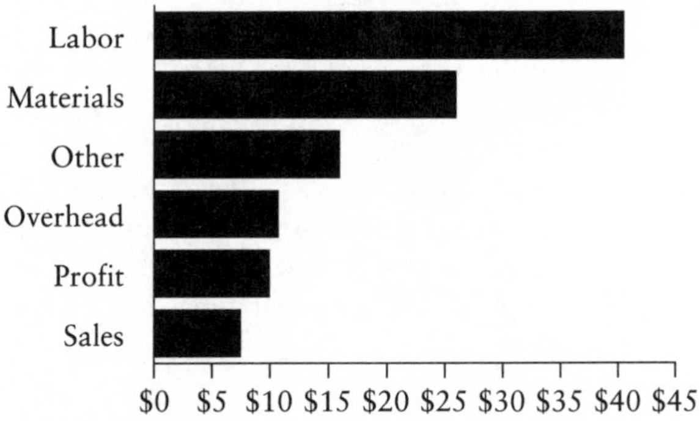

13.144 Avoid complex coding schemes of color, shading, or symbols. Likewise, avoid using chart design as decoration.

Concise

13.145 Conciseness means communicating the maximum amount of information to the greatest number of people in the shortest time with the smallest amount of print.

13.146 Eliminate redundant information, within reason. Allow redundancy that provides a context and organization for the data or simplifies comparison across the data, but do not repeat information without good reason.

13.147 If you cannot chart the full set of data, simplify it first by clustering, averaging, or smoothing.

Honest

13.148 Graphs should never lie, mislead, or confuse. Distortion happens when the data representation is inconsistent with its numeric values. Express all data in common units. For instance, dollars are not dollars, unless first compensated for inflation. Similar factors that can change the basis of measurement include population growth, changes in foreign exchange rates, and shifts in the value of assets.

Graphical lies frequently occur when the size of the graphic is not a simple, visually obvious multiple of the number it represents: For instance, a bar chart may be composed of cubes whose edges represent the value of a variable. A tripling of value results in a 27-fold increase in the volume of the cube:

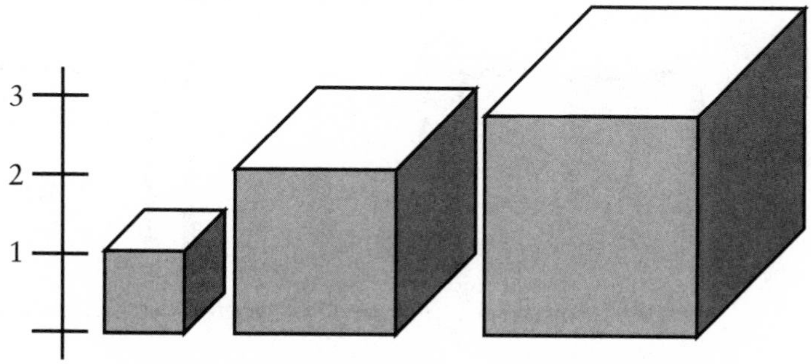

Such distortions are common in political advertisements showing the effects of inflation or growth in jobs.

Labels, annotations, and posted values can compensate for graphical distortion, but they cannot eliminate it.

Visually Attractive

13.149 Visually balance charts. Each should have a center of attention with visual elements balanced about this center. In charts this center will be near the bottom, above the horizontal scale. In all cases it should be in the field near the data.

13.150 Provide sufficient contrast to separate adjacent elements so that they are distinct and primary information stands out from secondary details:

Unacceptable

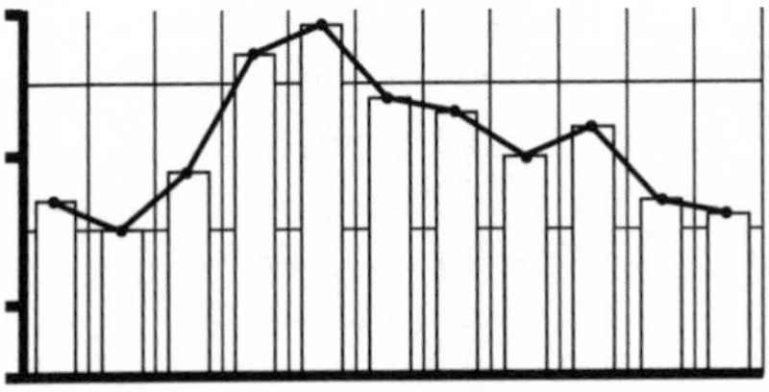

Acceptable

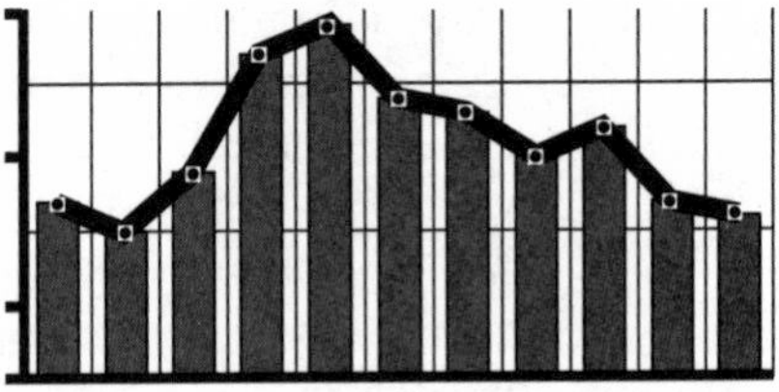

13.151 Use landscape (horizontal), as opposed to portrait (vertical), orientation for charts because they can show longer data series and text labels fit better:

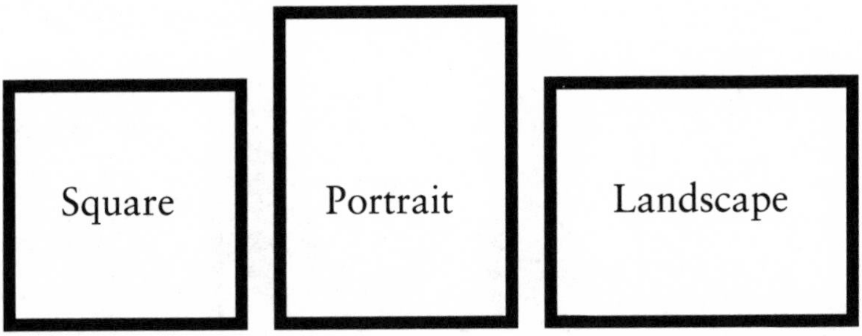

13.152 The chart's proportions and shape must fit the data. Within that restriction, recommended proportions range from 1:1.2 to 1:2.0. Right in the middle of this range is the golden rectangle, with proportions of 1:1.618:

Fit Charts to the Medium

13.153 Design charts to take advantage of the features of the medium in which they are displayed and to respect the limitations of that medium:

Medium	Characteristics	Special Consideration
TV	Color Low resolution Lay audience 3 × 4 aspect ratio	Use color Only very simple graphs Only common types Size accordingly Minimum text size = 1/20 chart height
Book	Read at leisure Color is expensive High resolution	Much detail possible Avoid color Minimum text size 6–8 points (10 points if reproduction is poor or lighting is dim)
Slide	Shown in dark room Color common Paced by presenter Medium resolution 2 × 3 aspect ratio	Light subject on dark background Use color Keep simple, trends only Minimum text size = 1/50 chart height Size accordingly
Viewgraph	Shown in lighted room Medium resolution Paced by presenter 3 × 5 or 5 × 3 aspect ratio	Dark subject on light background Minimum text size = 1/50 chart height Keep simple, trends only Size accordingly
Computer	Color Low resolution Limited text sizes	Use if available Only simple graphs Simple typefaces

Parts of Charts

13.154 Despite the diversity of charts, most share common design elements:

Exhibit 13–2:
Parts of a Chart

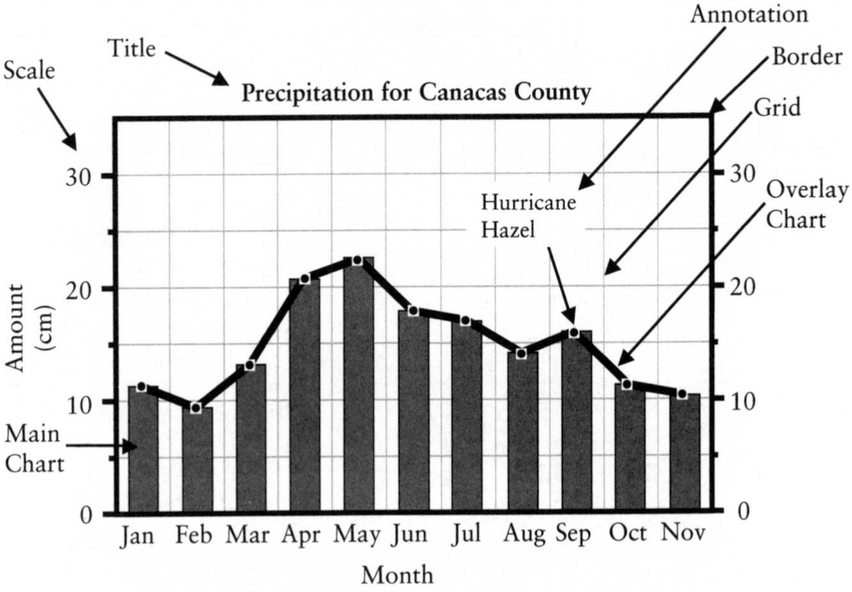

13.155 A chart should contain the following information (from most to least important):

1. Data values,
2. Title,
3. Trends in data,
4. Scales,
5. Grids, and
6. Annotations, notes.

Data Values on Charts

13.156 The most important information on a chart is the dot, line, curve, area, or other symbol that represents the data values.

13.157 Treat each item in a single data series the same. Use the same size and symbol for points. Use the same width and tone for bars:

Unacceptable

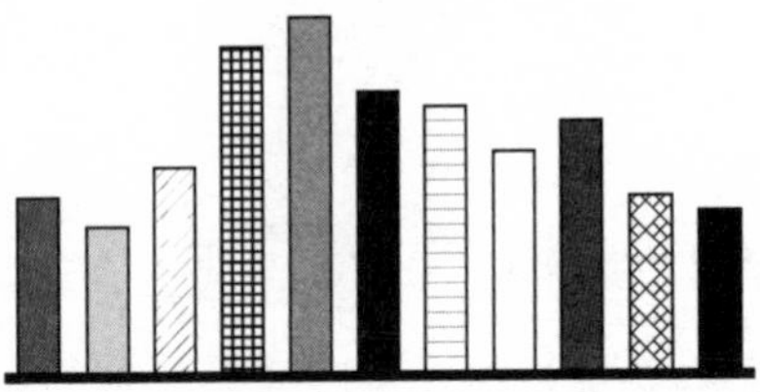

Acceptable

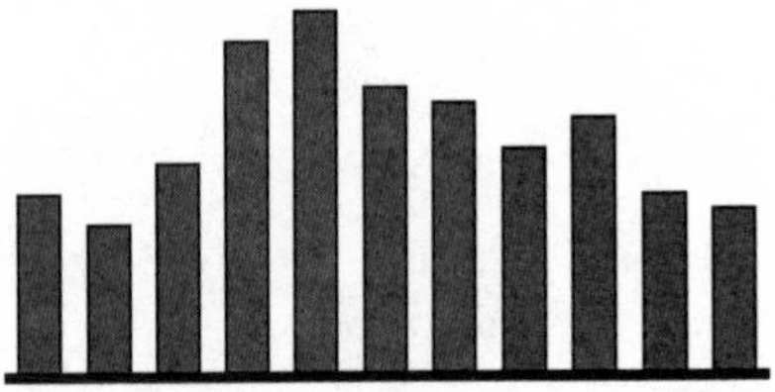

Data Point Symbols for Charts

13.158 Symbols for data points include crosses, Xs, circles, triangles, squares, and diamonds. Open shapes can be filled and shapes can be superimposed to provide more symbols. For data points, use different shapes rather than various sizes, styles, and weights of the same shape:

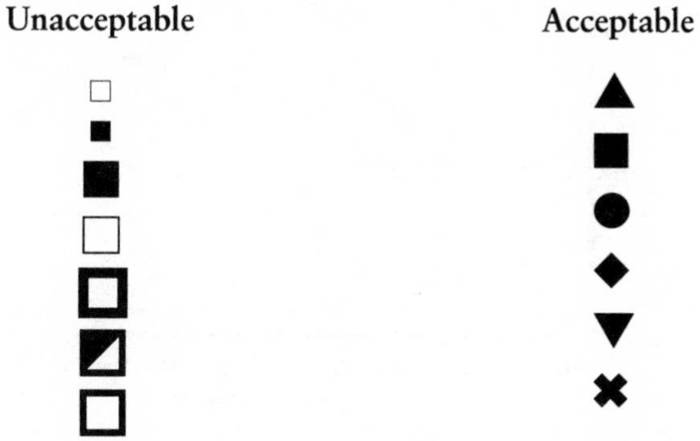

If the chart will be reduced or photocopied, avoid open or half-filled symbols, especially those with thin borders.

13.159 Plot data representing specific points in time (point data) on the grid line for that point in time. Plot data representing spans of time (period data) between the grid lines for the beginning and ending points of the time span:

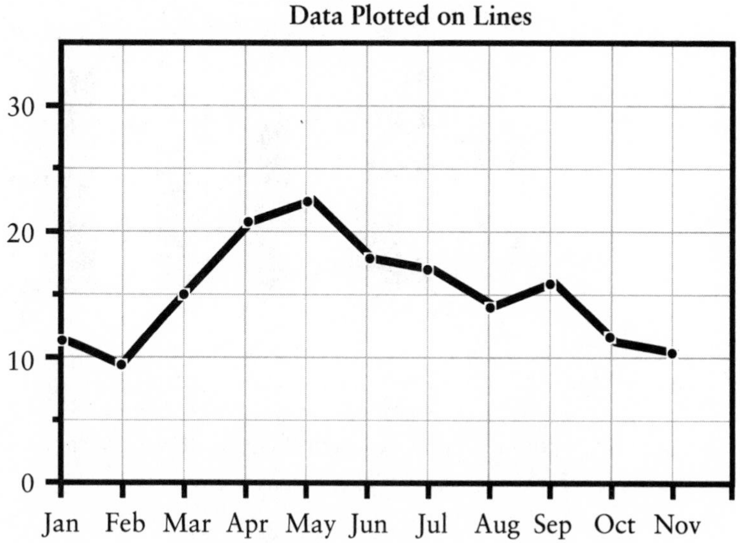

Data Plotted in Spaces

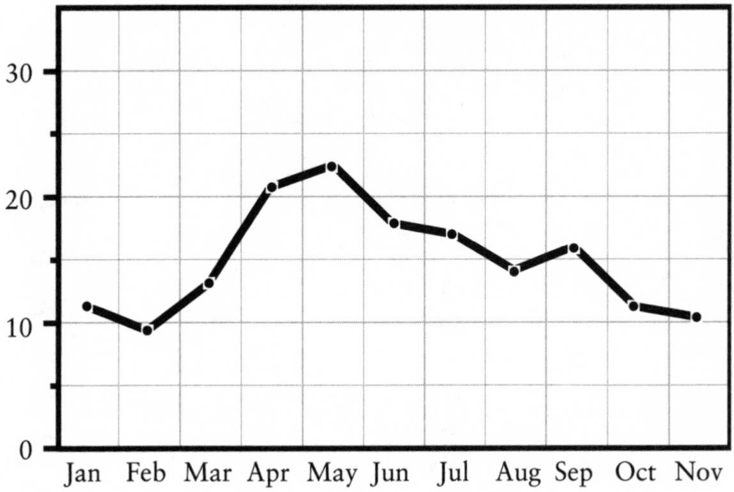

AREA OR VOLUME DESIGNATION FOR CHARTS

13.160 Readers cannot quickly compare areas and volumes, especially of irregular shapes. Avoid circles and spheres; comparing their sizes accurately is too difficult.

13.161 Comparing overlapped areas is difficult. If areas are used for comparison, separate them:

Unacceptable **Acceptable**

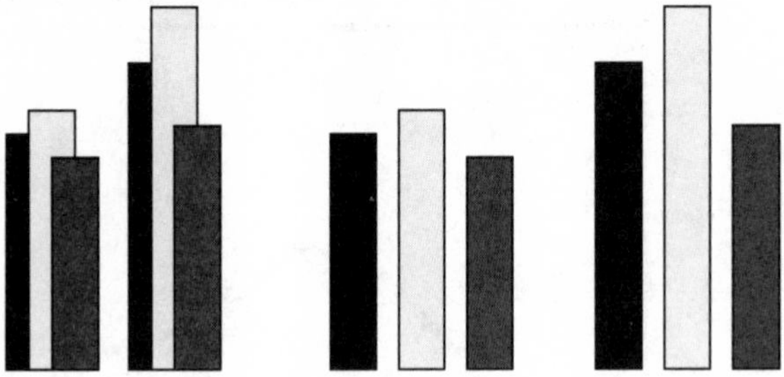

13.162 If using area to express values, show the area composed of countable blocks. The same technique works for volume as well. Make the blocks simple geometric shapes—all of the same size:

Unacceptable

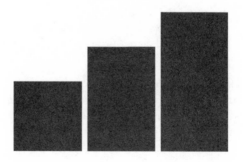

Acceptable

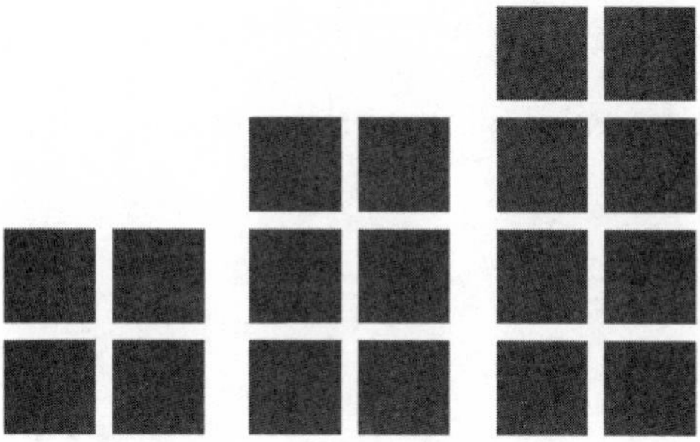

SYMBOLS OR PICTURES FOR CHARTS

13.163 Pictorial symbols add little to charts for a technical audience but may add interest for a lay audience. Use pictorial symbols as counting units. Do not vary two or three dimensions of a pictorial object to show changes in quantity.

13.164 Symbols used in a series should all have the same value. Make symbols:

- Self-explanatory;
- Simple and distinct in form and outline;
- Represent a definite unit of value;
- Represent a general concept, never a specific instance;
- Approximately the same size; and
- Easily divisible. Irregular shapes do not divide well. Never subdivide a human form. Round off numbers instead.

Avoid:

- Dated or cultural symbols. Almost any hand sign is an obscene gesture somewhere in the world.
- Symbols that require labels to be recognized.

13.165 Unit symbols do not require special graphic talent. Typewriter unit symbols are possible for many fields: $$$$$ for money, ##### for amount, %%%%% for percentages, or 00000 for other units.

Curves and Lines in Charts

13.166 Curves and lines show trends and connect data values. Curves may be straight, jagged, stepped, or straight.

13.167 Use straight lines to show linear correlations and simple trends.

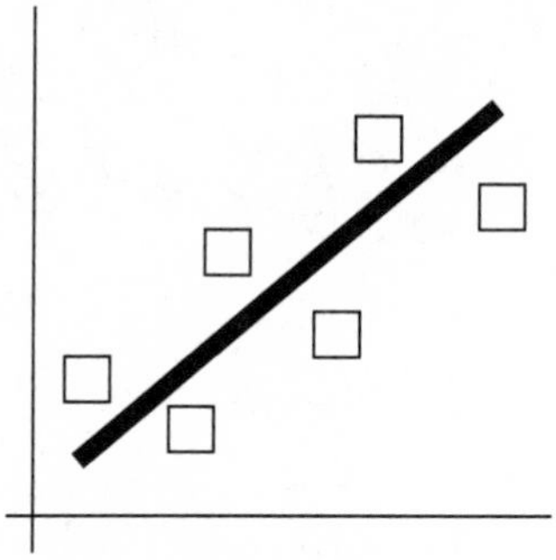

13.168 Use jagged lines to connect actual data points.

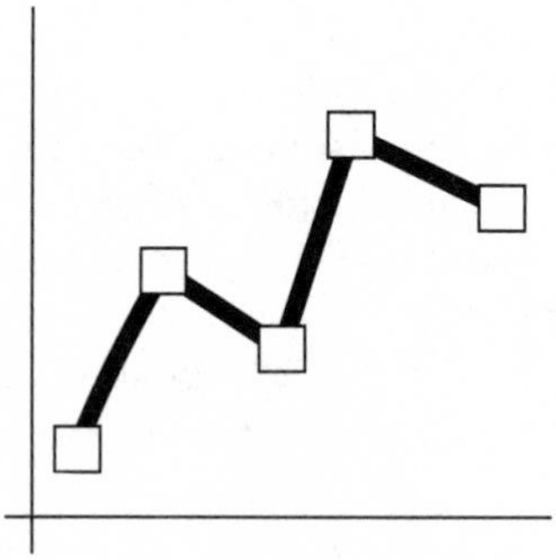

13.169 Use stepped lines to show changes in a variable at regular intervals.

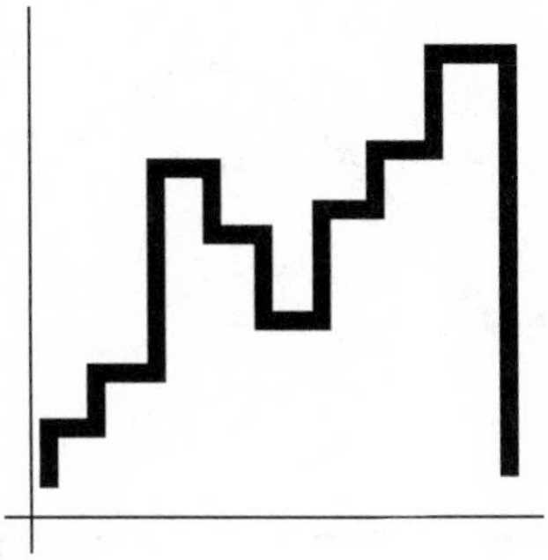

13.170 Use smooth curves to show more sophisticated trends and averaged data.

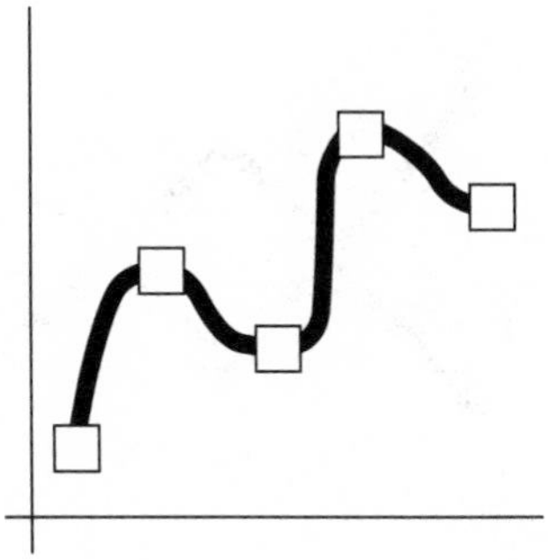

Line Weight for Chart Curves

13.171 Draw all curves the same width, about two times as thick as scales.

Line Style in Charts

13.172 Use different styles (solid, dotted, or dashed) lines to distinguish close or crossing curves. If some variables lack data, do not omit this range from the scale. Either leave this zone blank or use a dotted line to bridge zones of no data. Clearly indicate that the dotted line is not based on actual data:

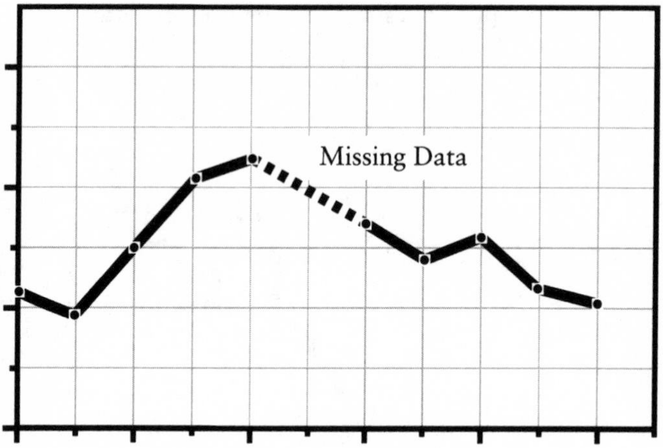

13.173 When lines cross, interrupt one of them, as if it passes under the other curve:

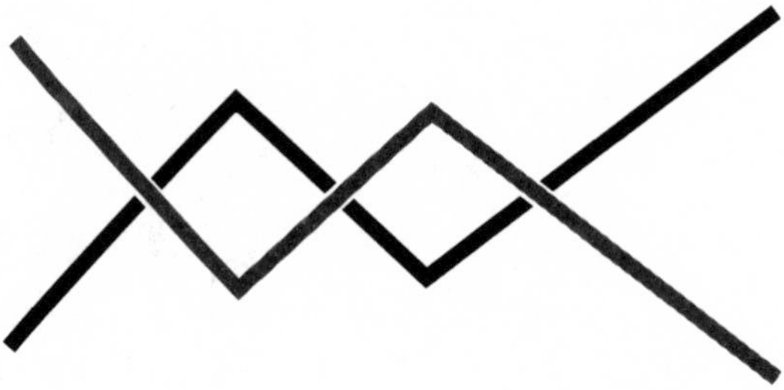

CURVE LABELS IN CHARTS

13.174 Place curve labels horizontally. Do not align curve labels along curves. Do not use arrows to connect labels to curves unless absolutely necessary. Do not box labels.

Scales for Charts

13.175 Vertical and horizontal scales indicate the value of data plotted in charts. The vertical scale is commonly called the *y-axis* or *abscissa;* the horizontal scale, the *x-axis* or *ordinate.* Place independent variables along the horizontal axis and dependent variables along the vertical axis. Time is shown on the horizontal axis. Values increase from bottom to top and from left to right.

13.176 Scales should start from zero. The horizontal axis corresponds to zero. However, if values dip below zero, extend the vertical axis below the minimum value and show the zero value as a horizontal reference line.

13.177 Place the vertical scale on the left side of the chart. However,

- If the data rise to the right or if the data on the right require special attention, put the vertical scale on the right side of the chart; and
- If the data vary throughout their range, put a vertical scale on both left and right.

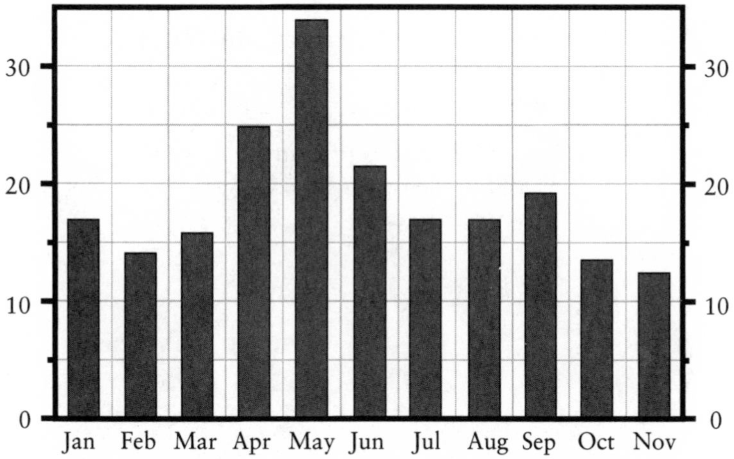

13.178 If the chart is used to look up accurate values, repeat the scale on left and right and on top and bottom. If data are plotted to two different scales, place one scale on the left and the other on the right. Make clear which scale applies to which data. Start both scales from a common baseline of zero and use the same distance between major divisions.

PLACING SCALE LABELS IN CHARTS

13.179 Scale labels show the value, unit, or count of the variable plotted along that axis. The scale may be in absolute values, percentages, rates, or named categories:

- Label the vertical scale horizontally above or to the left of the scale, or

- Label the horizontal scale horizontally below the right end of the scale.

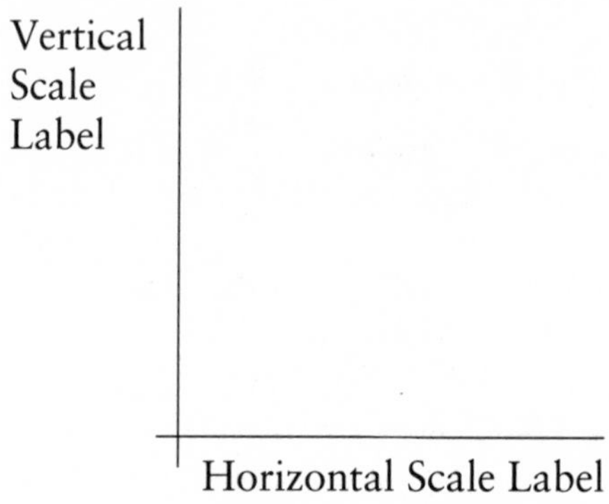

13.180 To save horizontal space the vertical scale label may be

- Turned clockwise to read from bottom to top, or
- Placed to the right of the vertical scale if this positioning does not cover the data.

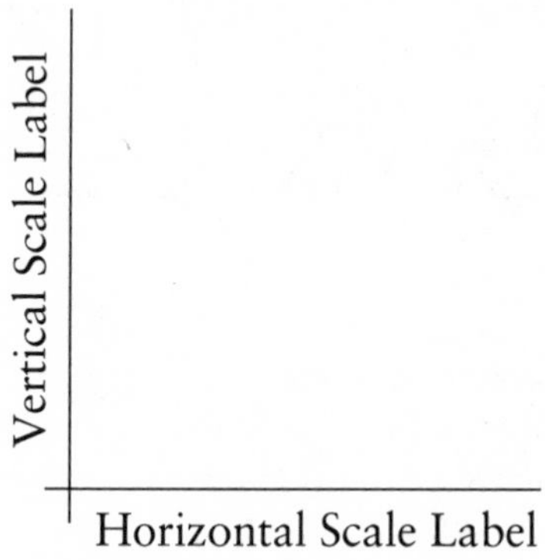

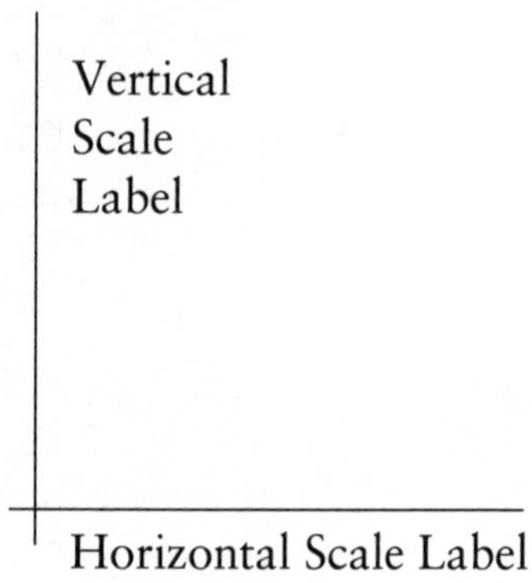

13.181 Label each scale with the variable and units of measurement. In some simple charts one or both of the scale titles may be omitted:

- If all variables are plotted against a single vertical scale and that scale is implied by the title, do not repeat the title in the scale label. Omit the vertical scale label.
- If the horizontal scale is obvious from the intervals plotted, the horizontal scale label may be omitted. For instance, if the horizontal scale is divided into intervals labeled *January, February, March,* and so on, the scale for *Months* need not be labeled.

Scale Divisions

13.182 Indicate major scale divisions with tick marks and minor divisions with smaller tick marks. On a scale, place figures at every major scale division. Major scale divisions should be every second, fourth, or fifth minor division.

13.183 Tick marks can go inside, outside, or across scale lines.

13.184 Label the scale starting with 0 at the left or bottom and proceed in even steps along the scale. Label scales with multiples of 2, 5, and 10; for instance, 0, 5, 10, 15, 20; not 1, 4, 7, 11, 14. Minimize the number of major scale divisions; five or six are all right but 10 or 12 are too many.

13.185 If scale figures are units of 1000s or higher, drop the trailing zeroes and express the size of the units as a phrase in the scale label ("in $millions") (see also Sections 7.23 and 7.31).

Unacceptable **Acceptable**

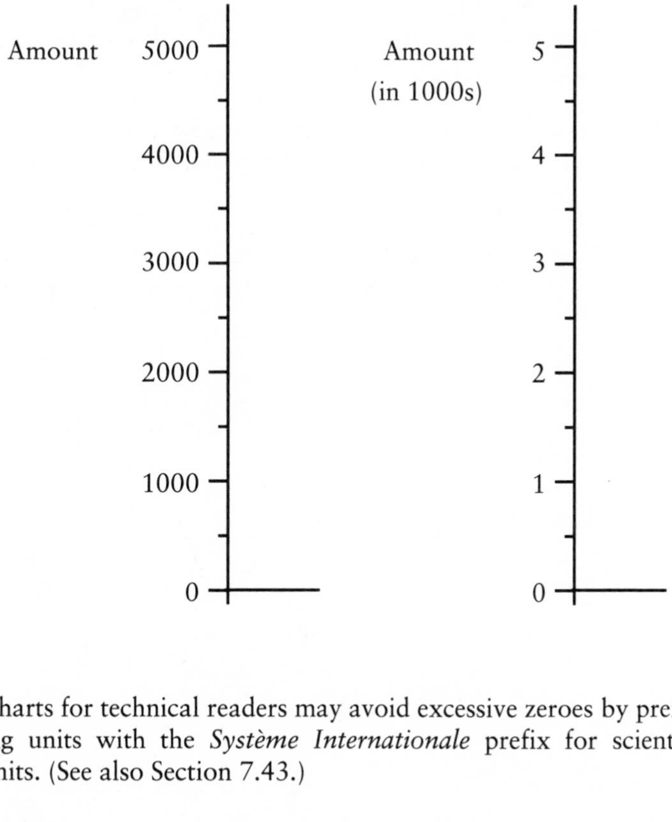

13.186 Charts for technical readers may avoid excessive zeroes by prefixing units with the *Système Internationale* prefix for scientific units. (See also Section 7.43.)

Multiplier	Prefix	Symbol
10^{-18}	atto	a
10^{-15}	femto	f
10^{-12}	pico	p
10^{-9}	nano	n
10^{-6}	micro	μ
10^{-3}	milli	m
10^{-2}	centi	c
10^{-1}	deci	d
10^{1}	deka	da
10^{2}	hecto	h
10^{3}	kilo	k
10^{6}	mega	M
10^{9}	giga	G
10^{12}	tera	T
10^{15}	peta	P
10^{18}	exa	E

Scale Ranges for Data Plotting

13.187 The scale range should be slightly larger than the data range. Sometimes the plotting intervals extend beyond the end of the data to help predict future data.

13.188 Always show a zero baseline except when the data represent index numbers relative to some other reference value, for instance, 100.

13.189 When the data values are such that much of the chart area is not used, break the grid and scale to make better use of available space. Keep the zero baseline and clearly indicate the break in the grid and scales:

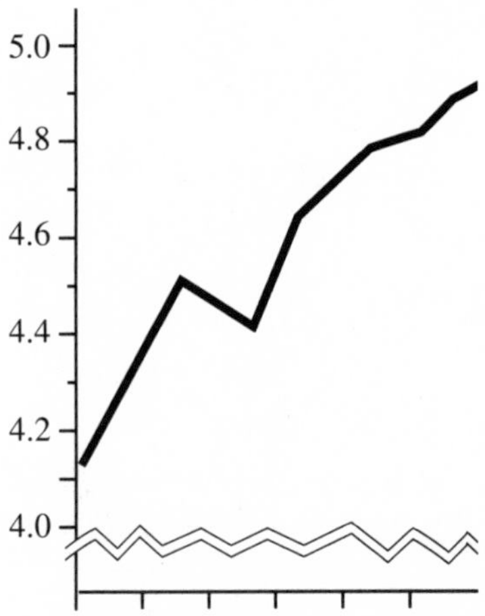

Scaling Function

13.190 Charts represent numeric values as distances on paper. The scaling function determines the relationship between the numeric value and the distance on the chart. Several scaling functions are used: linear, logarithmic, and geometric.

Linear Function

13.191 In linear scales the chart's distance or length is a constant multiple of the value it represents. If a 1-inch bar represents 100, a 2-inch bar equals 200. This practice is simple:

Distance = value × scale factor

Use only linear scales in charts for a lay audience.

Logarithmic Function

13.192 With a logarithmic scale the distance or size is proportional to the logarithm of the value:

Distance = scale factor × log value

13.193 Use a logarithmic scale to

- Show proportional or percentage relationships.
- Emphasize change rate instead of absolute values or absolute amounts of change. Use a logarithmic scale to show proportional rates of change.
- Compare relative changes for data series that differ greatly in absolute values, for instance, when comparing the growth rate of a city to that of a state. Use a logarithmic scale when a single data series includes both very small and very large values. Use logarithmic scales when baseline quantities vary greatly.
- Compare rates of change measured in different units.
- Compare a large number of curves of widely varying scales. Stack separate logarithmically scaled charts atop one another on the same page. Logarithmic scales accurately display numerical values throughout a wide range of values.

13.194 Do not use logarithmic scales

- To compare exact amounts of widely varying values,
- If the change in values is less than about 50 percent, or
- For an unsophisticated or inattentive audience.

GEOMETRIC FUNCTION

13.195 With a geometric scale distances are proportional to some power of the value:

Distance = Constant × value$^{\text{scale power}}$

13.196 Use geometric scales to prevent distortion when values are represented by an area or volume:

Value Represented by . . .	Scale power
Length	1
Area	1/2
Volume	1/3

Grid Lines

13.197 In curve charts used for locating specific values, include a background grid. Eliminate all grid lines not necessary for reading data values.

13.198 Make grid lines the lightest in the chart, about one-third the weight of data points or curves. If data lines are 2 point and axes are 1 point, make grid lines hairlines.

Reference of Baseline

13.199 Extend reference or index lines the length of the horizontal axis. They should be lighter than data curves but heavier than the scales:

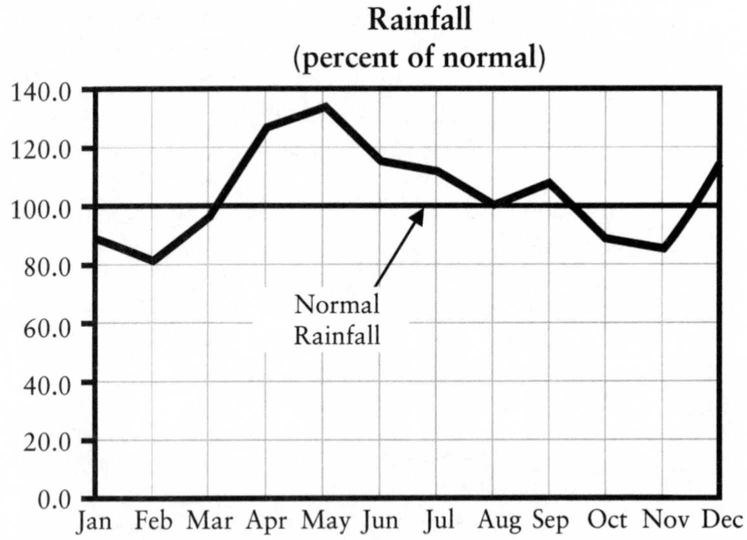

Posted Data Values

13.200 If actual data values are important, post them on the chart next to the symbol representing the value. If many data values are needed, put them in a table. Or use a curve chart with a grid from which the reader can read curve values.

If data values are posted at each point, the scale and grid may be unnecessary.

Annotations

13.201 Include annotations to explain the chart's data. Explain sudden shifts in trends, errant data points, hypothetical causes for trends, and interesting data patterns:

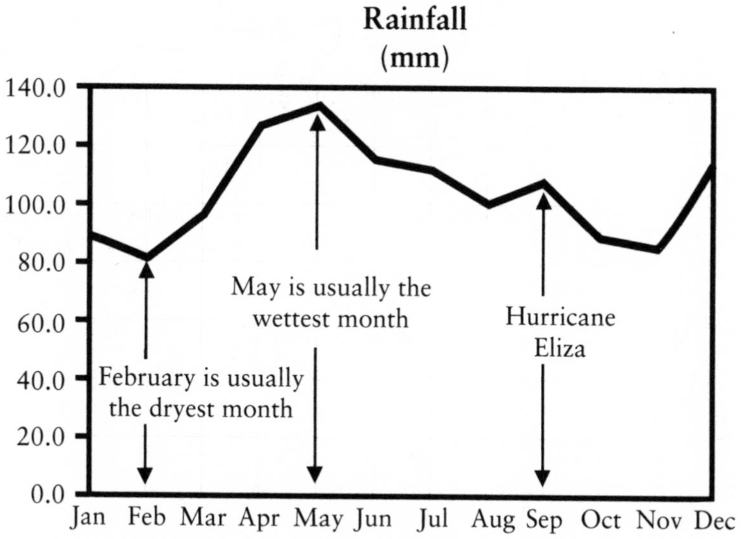

13.202 Annotation must not compete with the data. If there is insufficient room for full annotations, post key numbers or letters on the chart and provide the full annotations elsewhere. Follow the guidelines as for table notes. (See Section 13.79.)

13.203 If the chart has a background grid, blank out the grid behind annotations and labels:

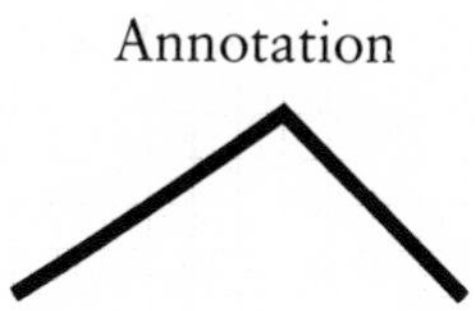

Inserts

13.204 Inserts are areas of additional information included within a chart to supplement or extend its basic elements:

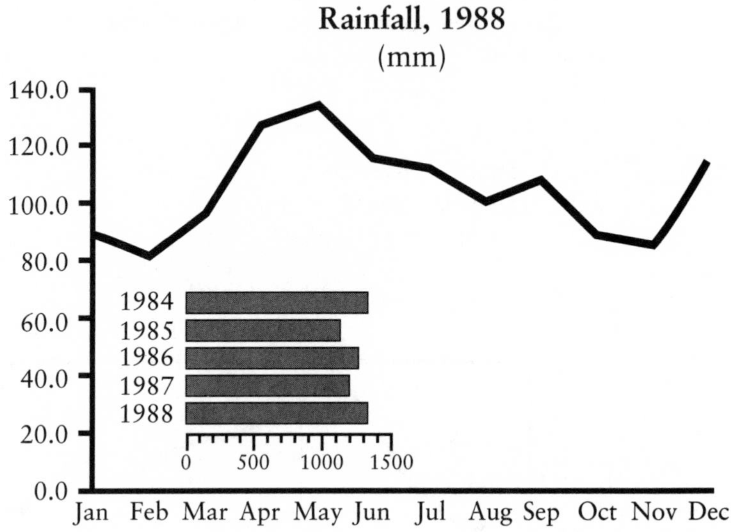

13.205 Typical uses for inserts include

- Explanatory text,
- Source of data,
- Related chart,
- Table of source data,
- Formulas,
- Key or legend of symbols,
- More recent data,
- Enlargement of a region of dense data, and
- Totals.

13.206 Box the insert only if it might otherwise be confused with the main chart. For example, box a line chart inserted on another line chart.

Notes for Charts

13.207 Use chart notes to indicate the data sources and any ways in which it has been manipulated:

SOURCE: *Cancer in Americans,* National Institute of Health Report CDR-88-4, Washington, DC.

NOTES: Data from earlier than 1945 have been excluded because of questions about their validity.

Follow the same format as for table notes. (See Section 13.79.)

Border or Frame for Charts

13.208 Borders are seldom necessary to make charts stand out. Borders are, however, useful to

- Reinforce page margins,
- Repeat reference lines,
- Establish visual horizontal and vertical directions, and
- Contain grids or scale lines.

Keys and Legends for Charts

13.209 Label parts of the chart directly. If necessary, use arrows to connect the label to its referent. If shading is used, a key or legend may be necessary:

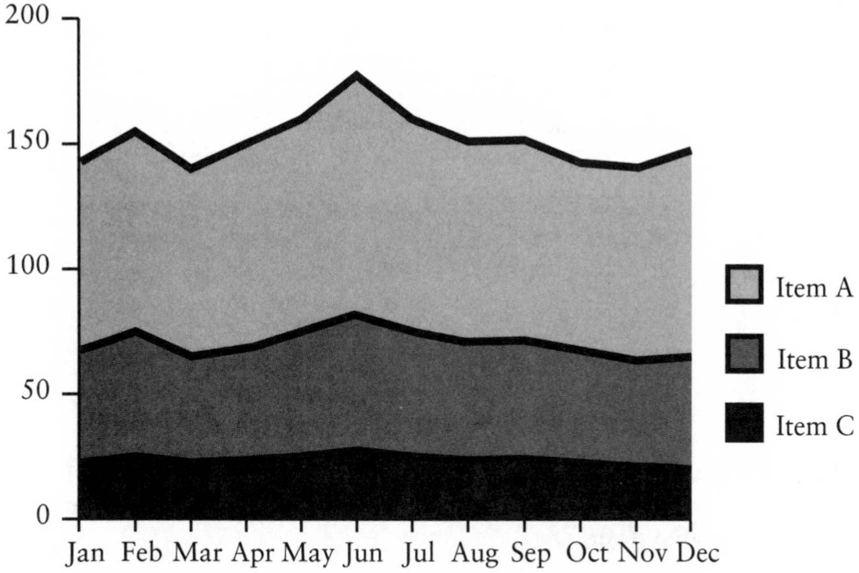

13.210 Place the key or legend, which contains small swatches of the shades, in a vacant area of the chart or to the right of the chart.

Special Topics and Effects

Size of Chart Lettering

13.211 The size of lettering in charts depends on how they are displayed and where they appear. For charts appearing in books, technical reports, and other published works, typical sizes are

Part	Typical Size (pts)	Notes
Title	12 bold	Same size as third-order heading.
Scale or axis labels	12	Same size and style as body text.
Scale figures	10	
Notes	10	Same size as text footnotes and endnotes.
Data values	10	
Annotations	8	Smallest legible size.

Weights of Chart Lines

13.212 Lines in charts should be sized to make data and trends stand out from the background of reference marks:

Component	Weight (× l width*)	Weight (points)
Curves	2–3	2–3
Scale axes	1	1
Grid, tick marks	0.5	hairline

*Units of l width are the width of the letter l in the text letter size.

Abbreviations

13.213 Spell out all titles, labels, and notes. Abbreviate only where necessary. Use only common abbreviations, for instance, for units of measurement:

Unacceptable	Acceptable
Strut Ld (pounds per square inch)	Structural Load (lb/in^2)

Patterns, Textures, and Shading

13.214 Emphasize the differences between areas by applying a different texture, tone, or color to each area. Although this technique does visually distinguish the areas, it also makes accurately comparing the size of areas harder for the reader.

13.215 In using tones and textures, follow these rules:

- Put the darkest tone (or brightest color) in the smallest area. Or when using different shadings within a bar, put the darker shading at the left. In a column, put it at the bottom.
- Alternate tones or colors to maintain adequate contrast between adjacent areas. Jacque Bertin, for instance, suggests the following values for tonal shading:

Suggested Tonal Values

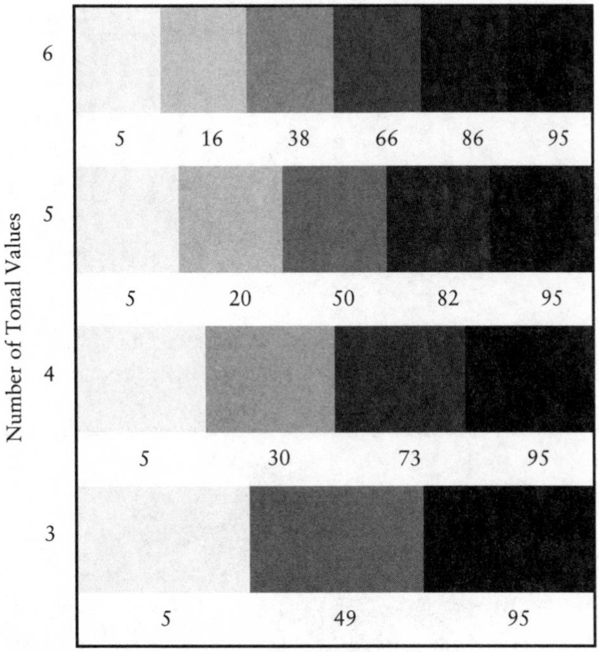

- Avoid stripes and crosshatches. Stripes make comparing areas almost impossible. They also lead to an unpleasant visual effect:

Unacceptable

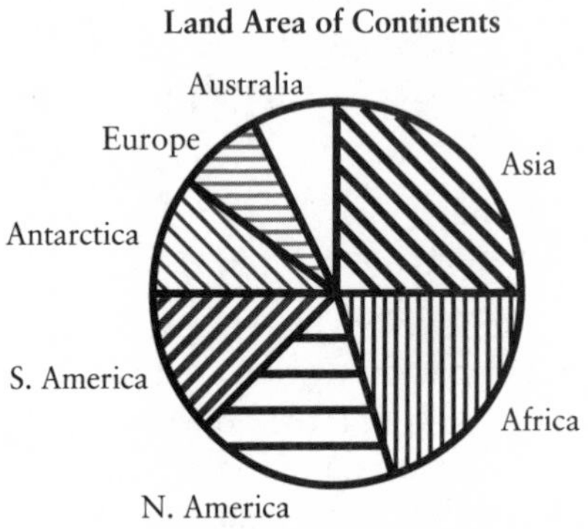

Acceptable

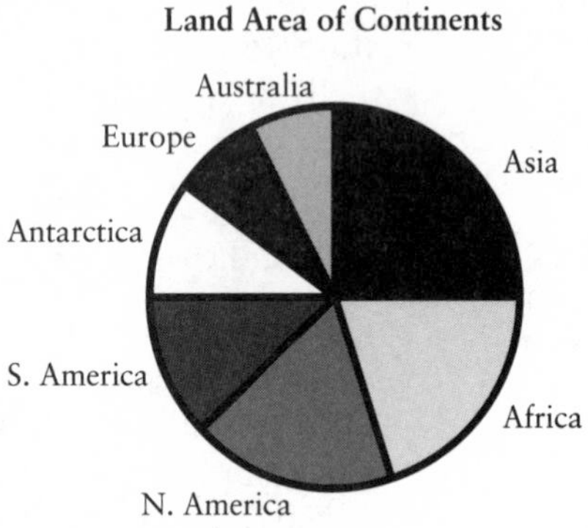

- When using stripes and patterns, establish a hierarchy similar to tonal values that makes discriminating them easy:

13.216 Use the same tone or pattern for all areas of the same data series.

13.217 Do not use black to shade large areas of any chart. Large areas of black can overwhelm the rest of the chart:

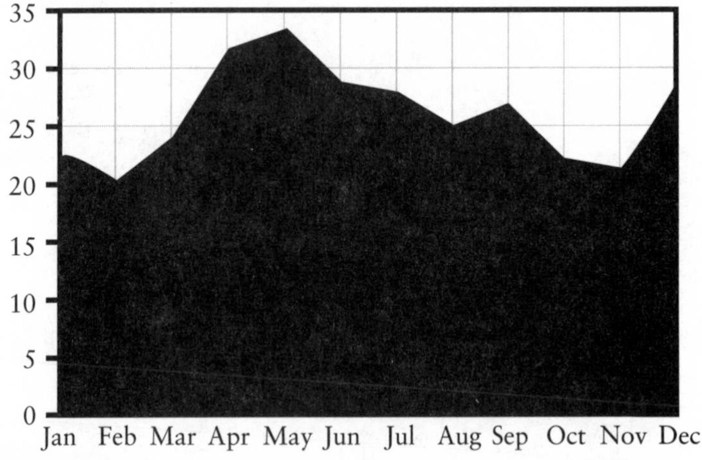

13.218 When using shading, label areas directly. Avoid a key or legend if possible. If you cannot label areas directly, provide a small key, containing small swatches of the shades, each labeled with its meaning.

Color Combinations

13.219 Never use color as the sole distinguishing trait of a point, curve, or area. Color codes will not reproduce in black-and-white photocopies.

13.220 Use color to draw attention or to emphasize. Do not use color to make objects recognizable, make critical distinctions, or add information.

13.221 Rather than a multicolor scale, use gradations in lightness for a single color or gradations between two compatible colors. Use bright warm colors to draw attention to important data. Use muted, cool colors to suppress data.

13.222 Adjacent colors should have at least 30 percent contrast in lightness. Separate colors with a thin black line.

13.223 Maintain contrast between an element and its background. The most important element in the graphic should have the brightest color and the greatest contrast with its background. Warm colors, such as yellow-green, yellow-orange, orange, red-orange, attract attention, especially on a darker color.

13.224 Do not place complementary colors of similar value or colors from opposite ends of the spectrum (red and blue, for example) next to one another. They will appear to pulse.

13.225 Avoid color choices that will make the chart incomprehensible to color-deficient or color-blind persons. Red is especially troublesome; blue can be distinguished from other colors by most people:

Some Legible Color Combinations

Subject	**Background**
Black	Yellow
Yellow	Black
Black	White
Dark Green	White
Dark Blue	White
White	Dark Blue

Three-dimensional Effects

13.226 Use three-dimensional symbols to make the chart appear to jump, slide, or float off the page. Such effects, while adding interest, can distort the representation of data and make accurate comparison impossible. Use them where only approximate comparisons are important.

OBLIQUE VIEWS

13.227 To create an illusion of depth, draw the chart as if viewed from an oblique angle. To maintain legibility, however, do not tilt text. Caution: On an oval pie chart the size of a slice depends on its position around the circle. Slices on the sides appear smaller than the same slices on front or back.

Land Area of Continents

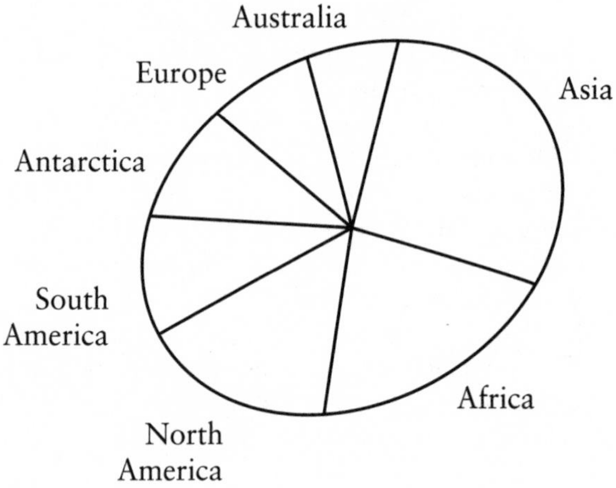

THICKNESS

13.228 Draw a pie chart as if cut from a squat cylinder.

Land Area of Continents

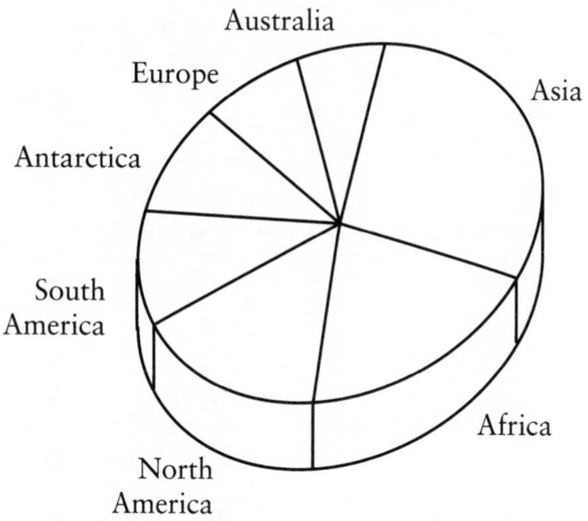

Drop Shadows

13.229 Draw a drop shadow behind the chart or its border to make the chart appear to float above the surface of the paper.

Land Area of Continents

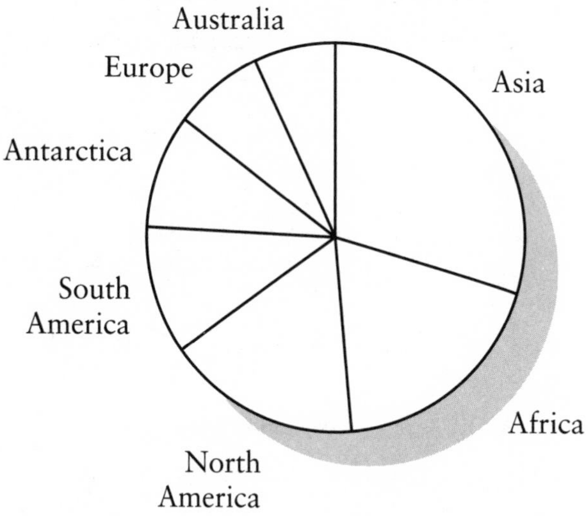

Pictorial Elements

13.230 Use pictorial elements to add interest for a lay audience. Acceptable pictorials include

- Picture as an insert or background,
- Simple picture as a border, and
- Symbols used as text labels on charts.

Overlays and Combinations

13.231 To compare dissimilar data series or to show two aspects of the same data, combine two or more types of charts into one. For example, to show cyclical data, superimpose data for corresponding time periods (days or years).

Referring to Charts

13.232 Number every chart and refer to it by number in the text. Place charts in the order they are cited in the text. Number charts in sequence with other figures. Cite figures and charts directly:

> Our study, as shown in Figure 6–4, confirms Smith's results.

Or parenthetically:

> Our study confirms Smith's results. (See Figure 6–4.)

13.233 If the parenthetical reference occurs within a sentence, capitalize and punctuate it as a phrase. If it occurs between sentences, treat it as a sentence:

> Our study confirmed Brinegar's results. (See Figure 6–4.) Later studies added further corroboration.

13.234 Do not refer to a chart by its position, for example, "the above chart" or "the following chart."

13.235 Avoid unnecessary cross-references. Repeat short, simple charts if necessary. In the general notes, explain that this chart is a repeat:

> Note: This chart is the same as Figure 6–2.

Common Types of Charts

13.236 This section examines the special requirements of specific, common types of charts.

Scale Charts

13.237 Scale charts show the ranking or value of a single variable along a scale.

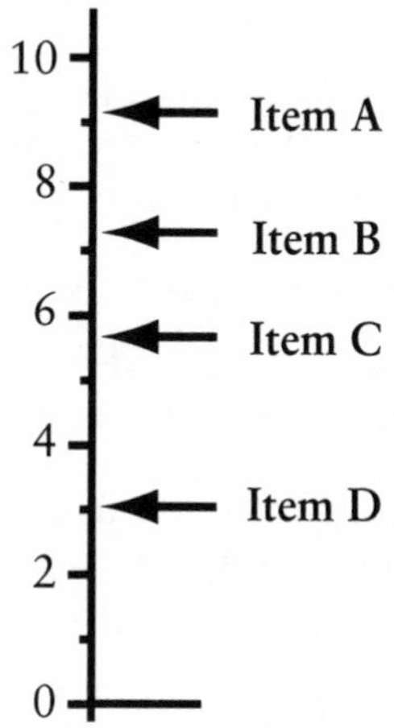

13.238 Do not use scale charts for ordered data values, such as a time series; use a line or column chart instead of a scale chart.

13.239 Arrange data items along the scale in positions corresponding to their values. Place labels next to their points on the scale. If the values are too close together, use arrows to connect labels to scale values.

Scatter Charts

13.240 Scatter charts show the correlation between two variables. Scatter charts are also known as *dot charts* or *scattergrams.*

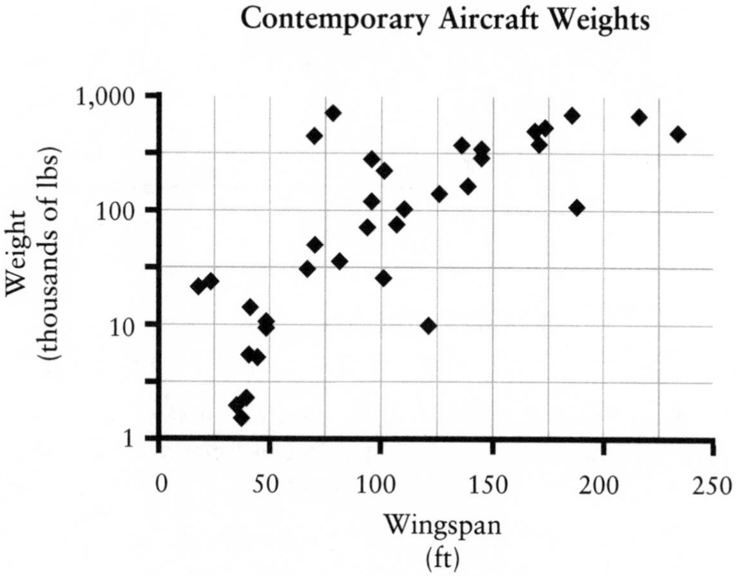

13.241 Use scatter charts to

- Present many data values (hundreds or thousands),
- Show correlations between two variables, and
- Draw conclusions about relationships in the data.

13.242 Do not use scatter charts for

- Time series data, or
- A small number of data points.

13.243 Use the same symbol for each point in a single data series. Use different symbols for different data series.

Triangular Charts

13.244 The triangular chart shows the makeup of three-component items. Each edge of the triangle is a scale of the percentage or proportion of each component present. Triangular charts are also known as *trilinear charts*.

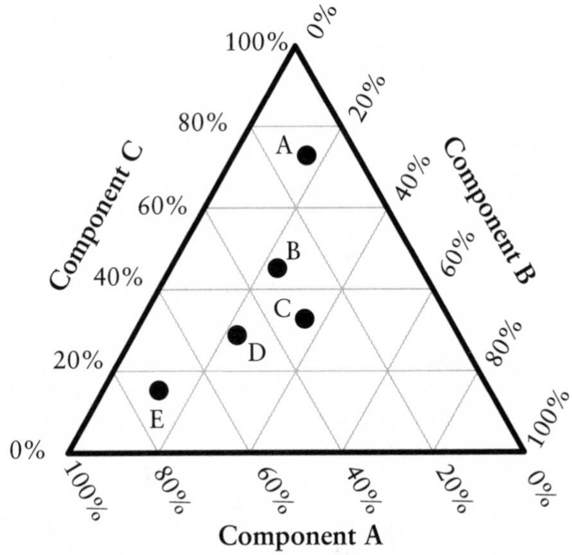

13.245 Use triangular charts to show the chemical composition of compounds and alloys with three components.

13.246 Do not use triangular charts for subjects made up of more or fewer than three components or for which the three components do not add up to 100 percent.

13.247 Use distinctive symbols for data points. Label each data point. Label scales outside the triangle.

13.248 Include a light grid inside the triangle. Make the lines of the grid parallel to each of the three edges.

Circular Charts

13.249 Circular charts use angle and distance from a center point to show how values vary over cyclical ranges. Circular charts are also known as *rose diagrams* and *polar plots.*

13.250 Use circular charts to show how a small number of variables vary over a circular range, such as the azimuth, hours of the day, or months of the year.

13.251 Do not use circular charts if the variable is not cyclical. Circular charts may prove too technical for lay audiences.

13.252 Use the angle to represent the independent variable. This variable must be an actual angle or some other variable that varies cyclically, returning to its starting value. Represent the independent variable by a distance from the center of the circle. This distance along a radius always starts at zero. Include no more than one or two variables on a circular chart.

Curve Charts

13.253 Curve charts use lines to show the relationship between two variables measured along the horizontal and vertical scales. Curves can be straight, jagged, smooth, or stepped. The term *curve* refers to trend or data lines, regardless of whether they are straight or curved. Curve charts are also known as *line graphs, line charts,* and *curve graphs.*

13.254 Use curve charts to

- Show trends over time and relationships among variables;
- Present large amounts of data in a limited space;
- Summarize data, especially if it must be understood fast;
- Show a trend over time;
- Compare several data series; or
- Interpolate known data values.

13.255 Do not use curve charts

- For lay audiences. Use column charts instead.
- To compare relative size. Use bar or column charts.
- When change amount or rate is more important than trends.

13.256 Use a log scale if ratios are important or if the data values vary over a wide range.

FREQUENCY DISTRIBUTION CHARTS

13.257 Use frequency distribution charts to plot the number of occurrences of individual values within some range of a continuous variable. Annotate the curve to explain deviations from a normal (bell shaped) or other expected distribution:

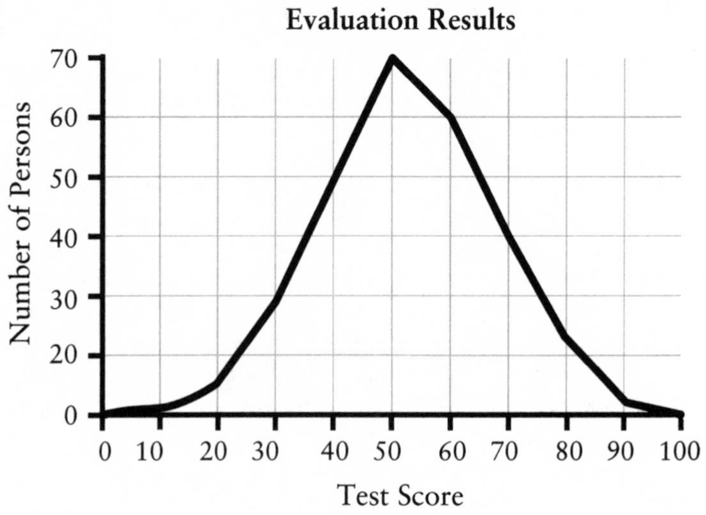

CUMULATIVE CURVE CHARTS

13.258 For each horizontal unit, cumulative curve charts plot the total of its amount and all prior units. Always clearly label cumulative charts as such:

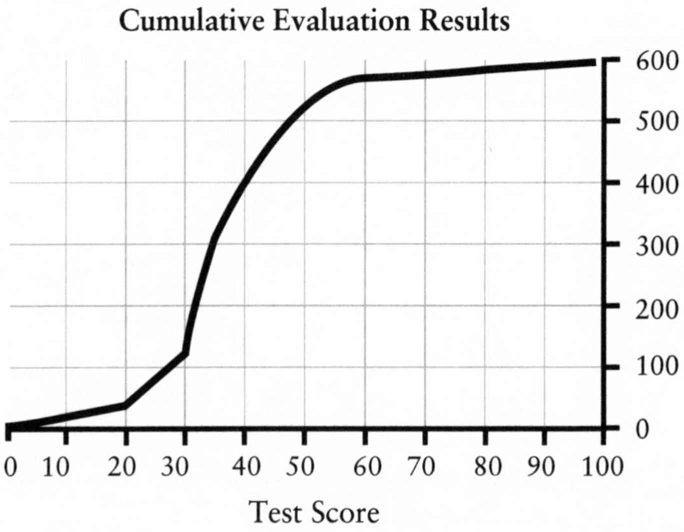

Surface Charts

13.259 Surface charts are curve charts in which the area between the curve and the baseline is shaded. Surface charts emphasize the difference between two series:

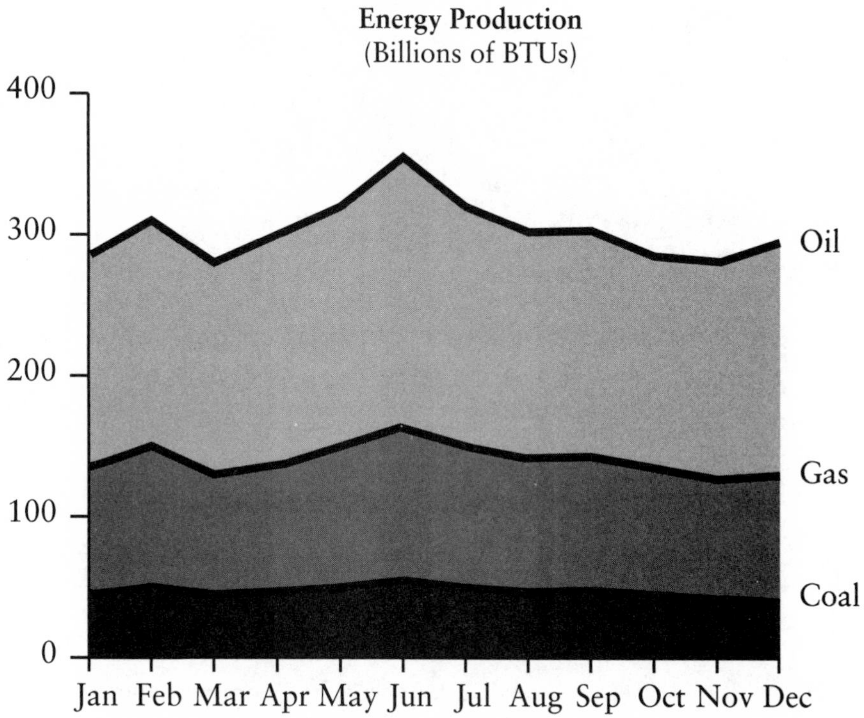

13.260 Label the curves to the right of the chart. Place labels to the right of the corresponding lines.

13.261 Do not confuse surface charts with *band surface charts.* Surface charts merely shade the area beneath the curve. They cannot be used for curves that overlap. The curve's height still represents the value.

Band Surface Charts

13.262 In band surface charts, the width of each band represents values. The widths of bands on the chart represent the values of data series. Only the tops of the first and top (total) band plot absolute values:

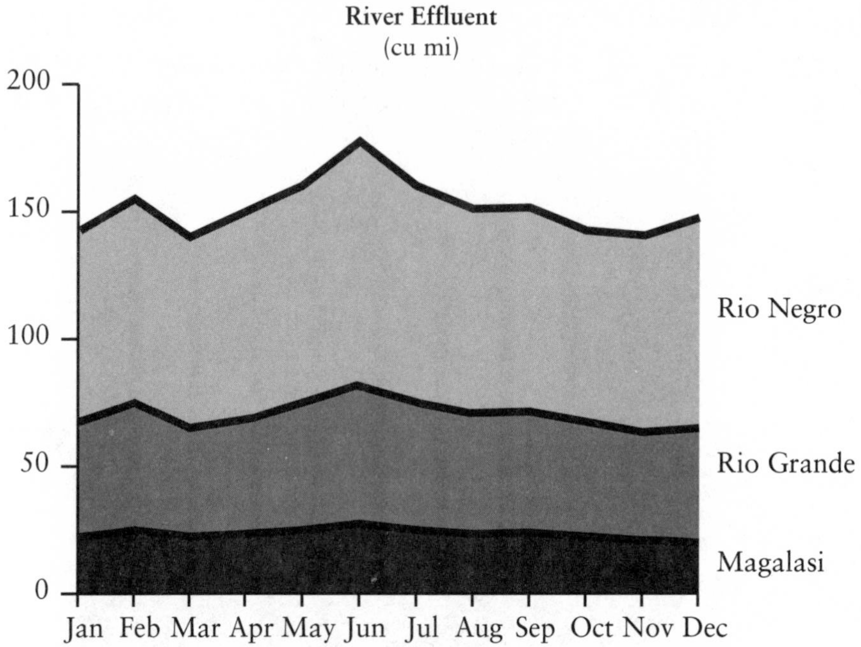

13.263 Use band surface charts to compare trends of components and a whole. Do not use a band surface chart if several of the layers vary considerably or if strong upward trends create an optical illusion of squeezed layers.

13.264 If one layer of a band surface chart has more pronounced variations, place it at the top:

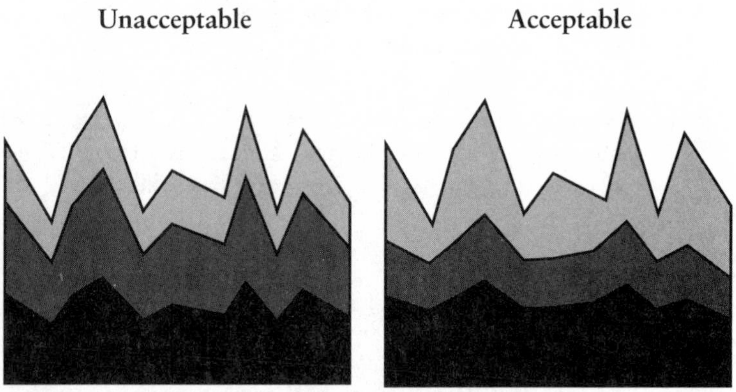

13.265 Label bands to the right of the curves, beside the corresponding bands. Avoid using a key or legend. Use arrows, if necessary, to connect labels with bands.

100 PERCENT SURFACE CHARTS

13.266 Hundred percent surface charts show trends among components of a whole. The values of the bands always total 100 percent. Label the bands to the right:

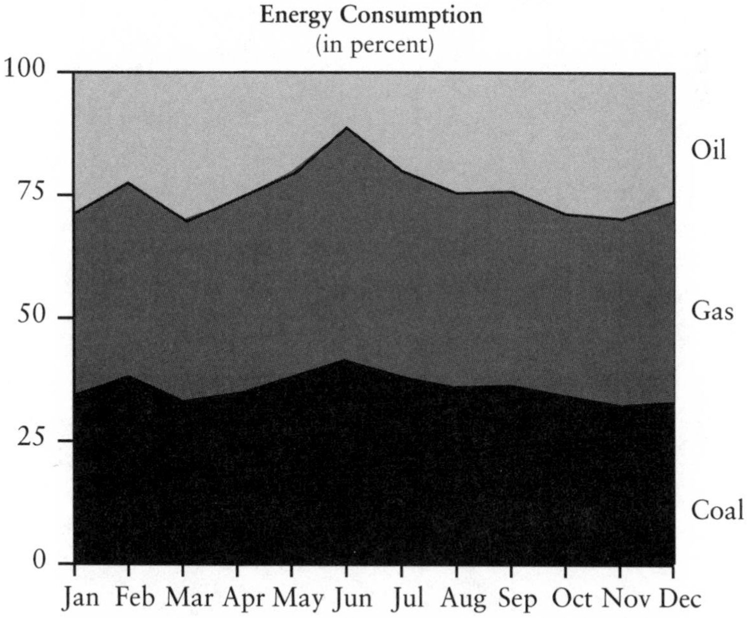

NET-DIFFERENCE SURFACE CHARTS

13.267 Net-difference surface charts shade the difference between two series, either of which may be greater. These charts show surpluses and deficits or profits and losses. Use a different tone or color for pluses and minuses:

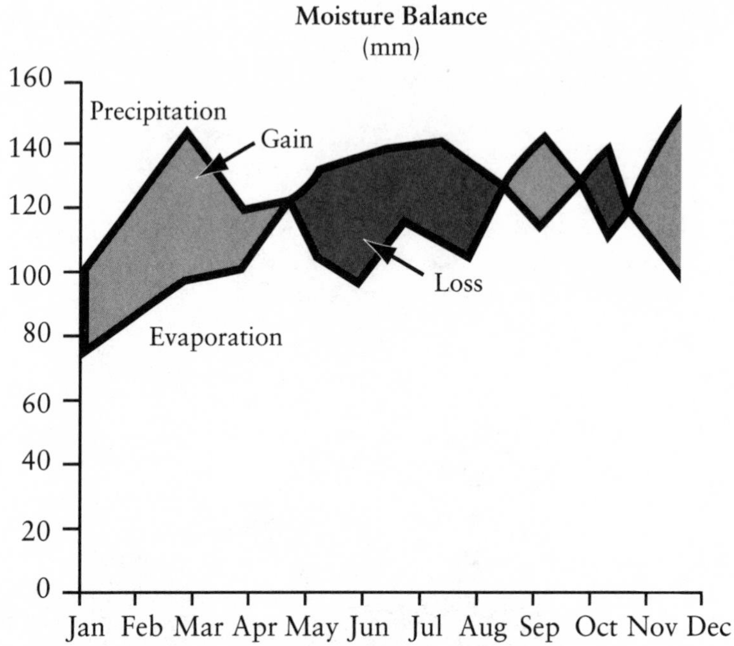

Column Charts

13.268 Column charts use vertical bars to represent quantities for a series of categories. The categories may be specific values of an independent variable or named categories:

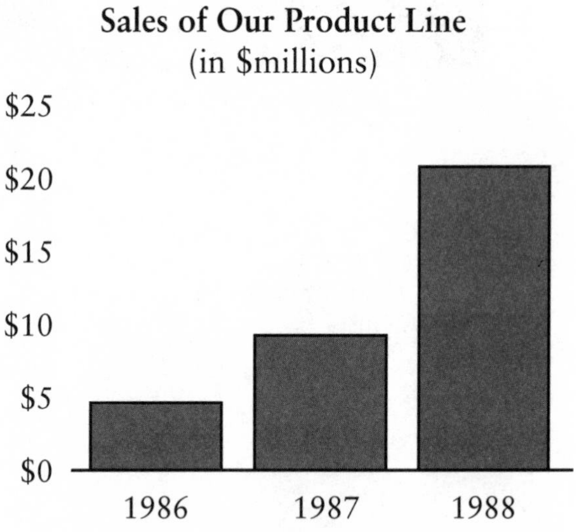

13.269 The main difference between bar and column charts is that column charts use vertical bars and bar charts horizontal.

13.270 Use column charts to compare the size or magnitude of different items or one item at different times. Use column charts to emphasize the individual amounts or differences in a simple, regular data series.

13.271 Do not use column charts to show trends. Use a curve chart instead. Do not use a column chart simply for comparing values. Use a bar chart instead. Do not use column charts if quantities do not vary significantly.

13.272 Shade columns to emphasize the data and sharpen the contrast between items being compared:

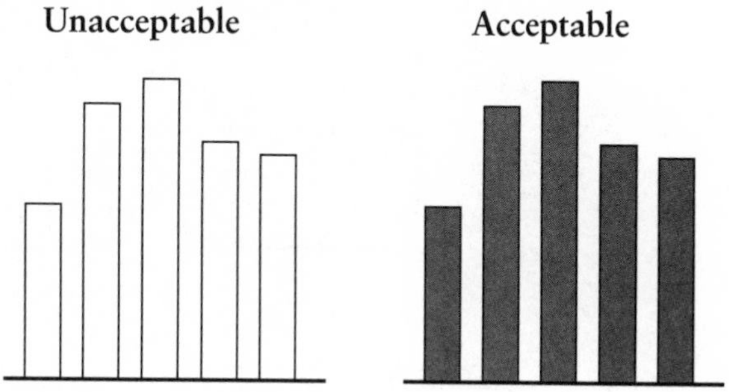

13.273 Label columns at the bottom, along the horizontal axis. Do not label columns within the columns.

13.274 Place data values based on these rules:

- If the column is shaded, post the value beyond the end of the column, not inside the column. If figures are placed at the end of the column, be sure they do not make the end of the column indistinct. Keep these figures small (but legible) and separate them a little from the end of the column.

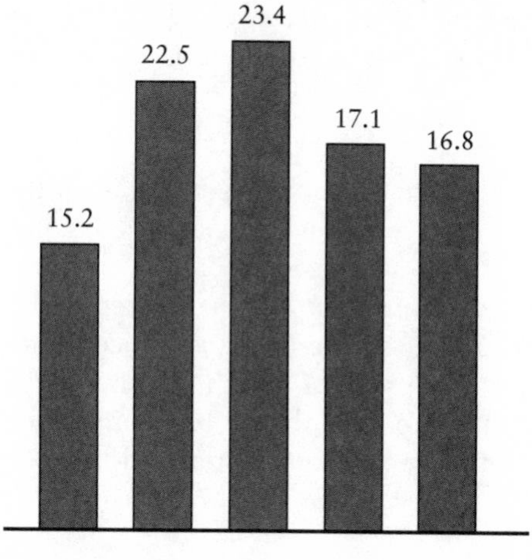

- If the column is not shaded and you can post the value within the column without making the end of the column indistinct, do so. Otherwise post the value at the base of the column.

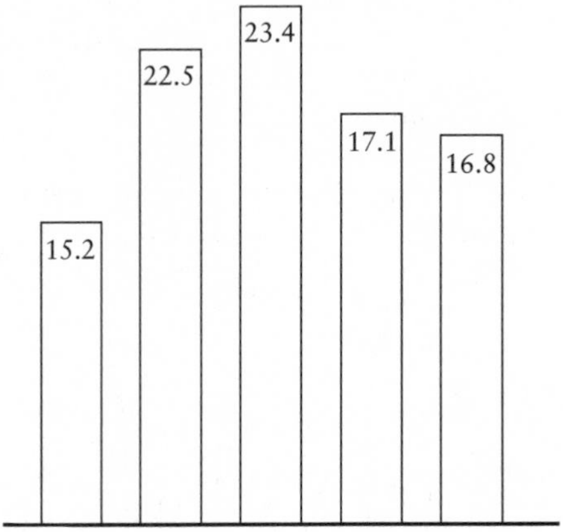

13.275 Columns should be of a uniform width and evenly spaced. The width and spacing of columns depends on the number of columns and the space available. The best spacing for columns is from 25 to 100 percent the width of a single column, with 50 percent the ideal. Connected columns are better than closely spaced, narrow columns:

Unacceptable **Acceptable**

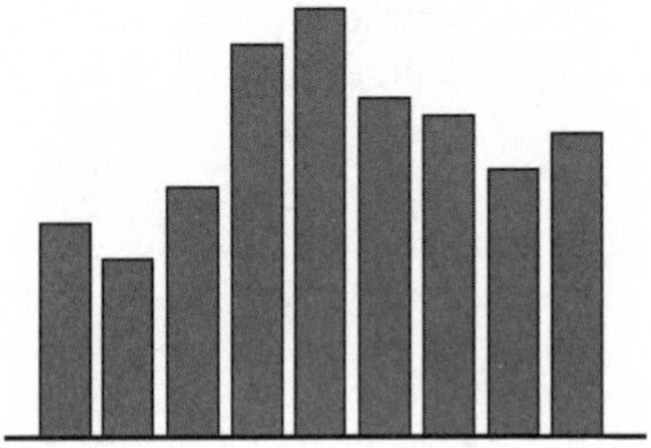

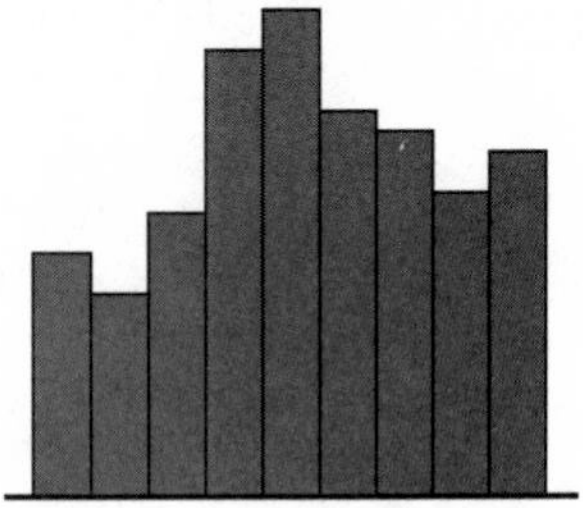

13.276 In column charts, include grid rules for the amount scale if necessary to read the values of specific columns.

13.277 Arrange columns in a visual (largest to smallest) or logical (by date) order. Columns can be compared more accurately if arranged in order of length. Begin the first column about half a column width from the scale. Columns should be taller than wide and not more than one column width apart:

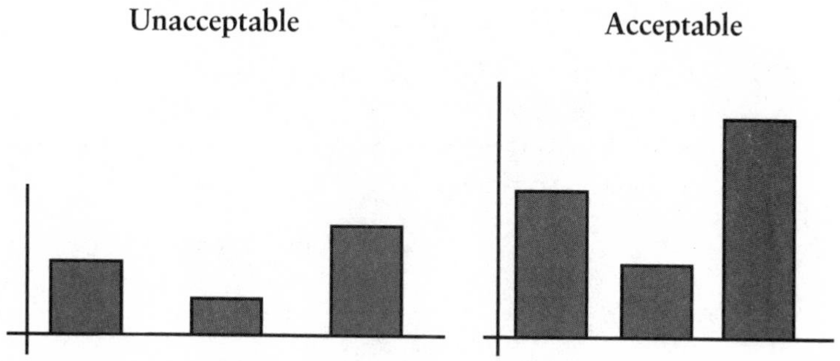

HIGH-LOW CHARTS

13.278 Use high-low charts to show daily, weekly, or yearly fluctuations in prices, temperature, or other variables:

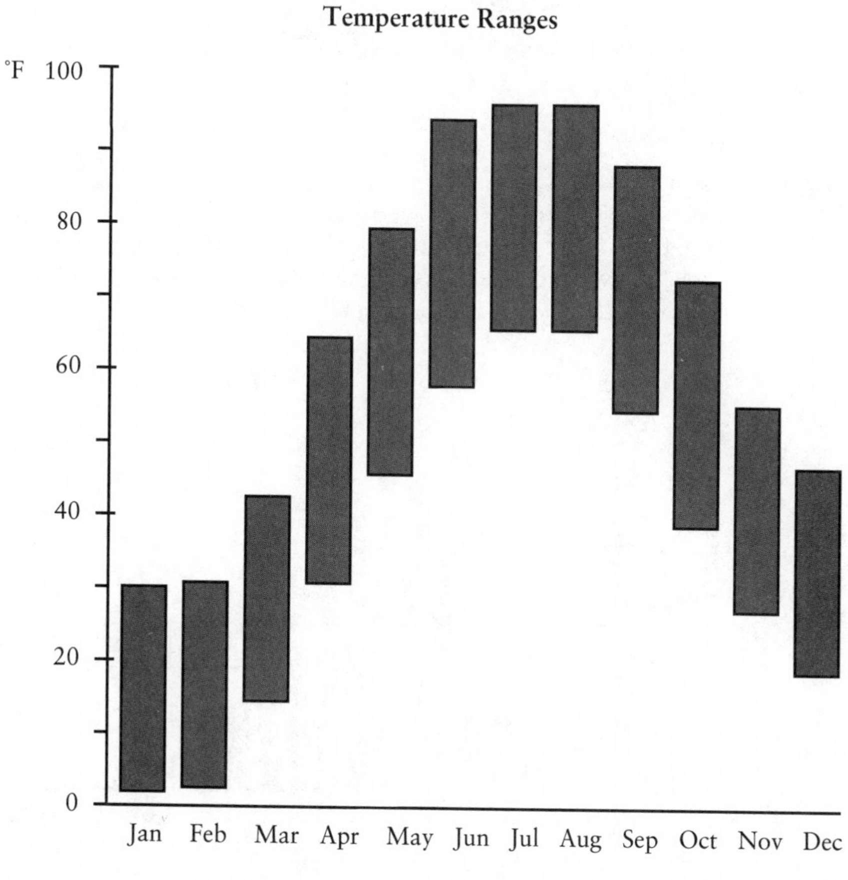

100-PERCENT COLUMN CHARTS

13.279 Use the 100-percent column chart to show the proportional composition at different times or different values of some other variables. In a 100-percent column chart, space the bars and use lines to connect sections:

Energy Consumption

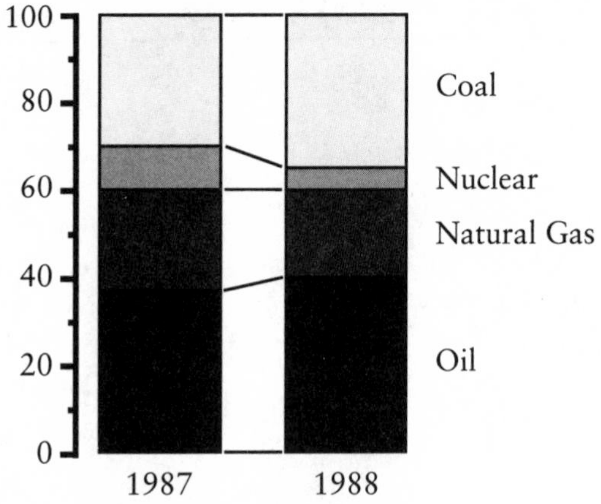

HISTOGRAMS

13.280 Use a histogram to show the relative frequency of various values within ranges. Normally columns should all be the same width. In histograms with unequal intervals, the width of the bar should be made the width of the interval and its height reduced so that the area of the bar represents the frequency of this range:

Test Results

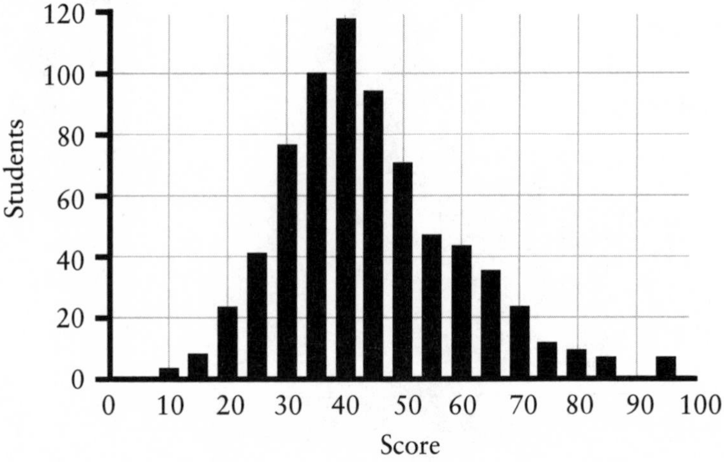

Bar Charts

13.281 Use bar charts to compare size or magnitude between different items at one time:

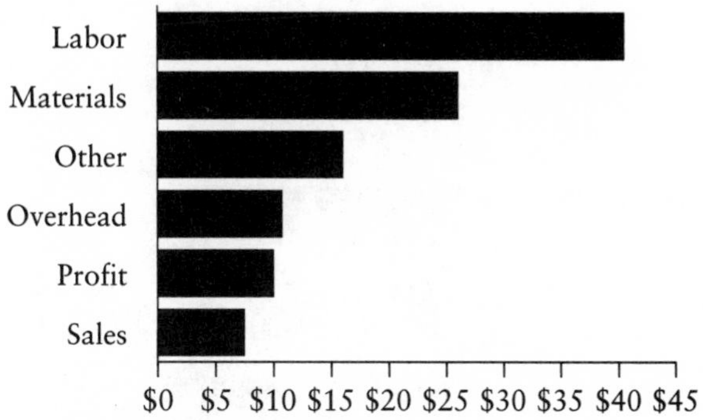

13.282 Do not use bar charts to compare size or magnitude for one item over different periods of time. Use a column chart or curve chart instead.

13.283 Shade bars to make them stand out and to simplify comparisons:

Unacceptable **Acceptable**

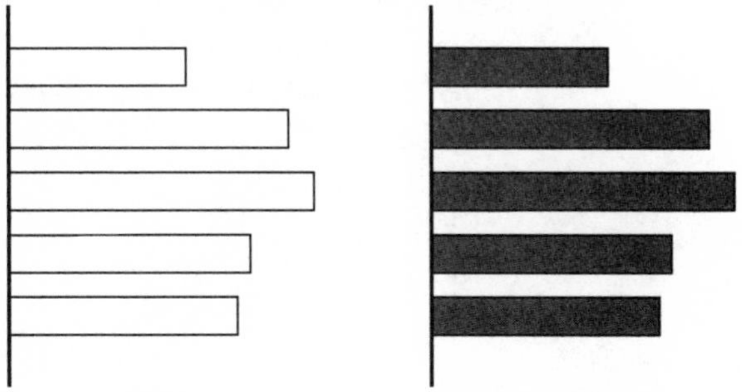

13.284 Arrange the segments of the bars in a consistent, logical order, for instance, from largest to smallest. Keep the same order for all bars of the series.

13.285 Space bars to emphasize data and leave room for labels and figures. Separated bars stand out more than touching or overlapped bars. Make bars wider than tall.

13.286 Label bars to the left of the vertical axis. Right-justify the labels. Label subdivisions of bars atop the first bar.

13.287 Place data values based on these rules:

- If the bar is shaded, post the value beyond the end of the bar, not inside the bar. If figures are placed at the end of the bar, be sure they do not make the end of the bar indistinct. Keep these figures small (but legible) and separate them a little from the end of the bar.

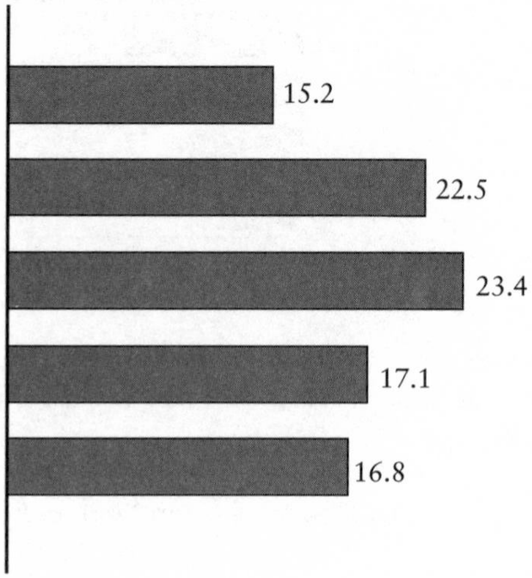

- If the bar is not shaded and you can post the value within the bar without making the end of the bar indistinct, do so. Otherwise post the value at the base of the bar.

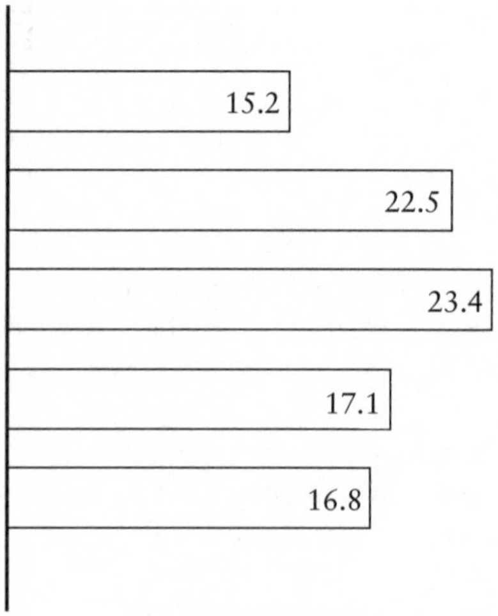

TIME LINES

13.288 Use time lines to summarize a complex schedule or procedure by showing the timetable of tasks. Time lines consist of a list of activities posted on a time grid to show the period during which each activity occurs. This grid shows the times for each activity:

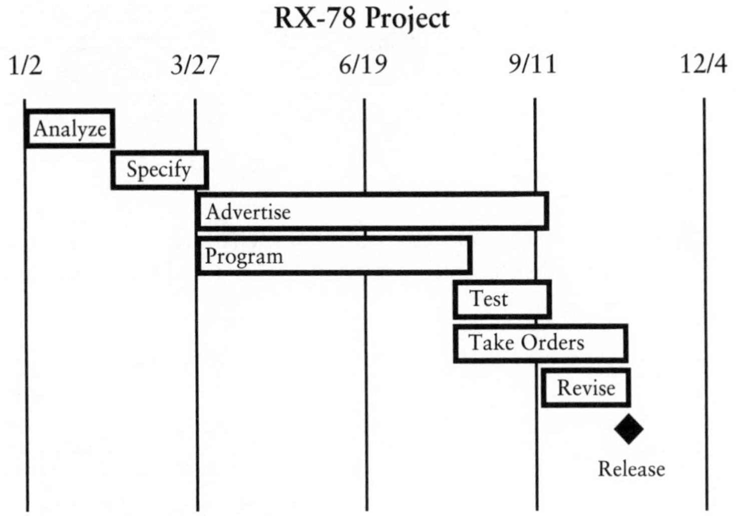

13.289 Use vertical grid lines to mark intervals. The intervals may be seconds, minutes, hours, days, weeks, months, quarters, years, or decades. List activities in chronological order. Label bars at the left edge of the chart or within bars. Do not place labels beside bars.

13.290 Use small symbols—circles, triangles, diamonds—to indicate project milestones.

PAIRED BAR CHARTS

13.291 Use paired bar charts to compare two members of a category moving in opposite directions from a common baseline. Use paired bar charts to compare two different types of data, perhaps with different units or scales:

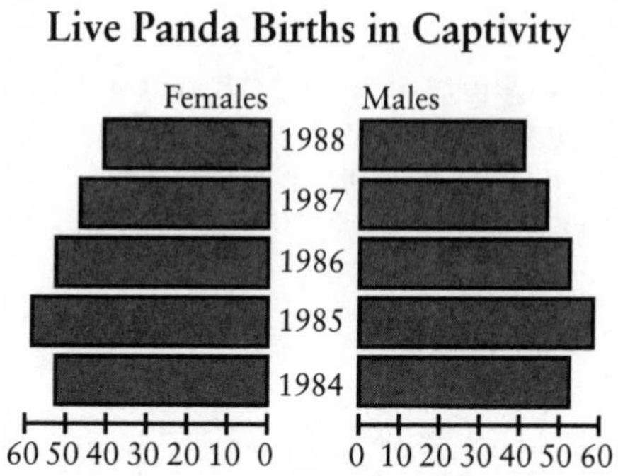

SLIDING BAR CHARTS

13.292 Use sliding bar charts to represent ranges with bars whose horizontal ends mark the beginning and ending values of the ranges:

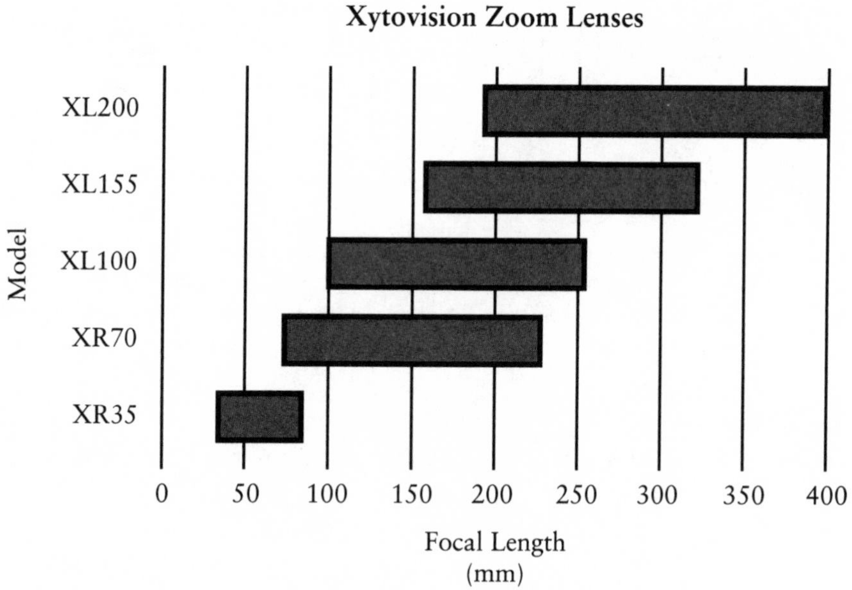

Column/Bar Chart Combinations

GROUPED COLUMN/BAR CHARTS

13.293 Use grouped column charts to present sets of columns. Each column represents a variable. Each set represents the values of the variables at a different time or under different conditions. Such charts help contrast changes among variables over time:

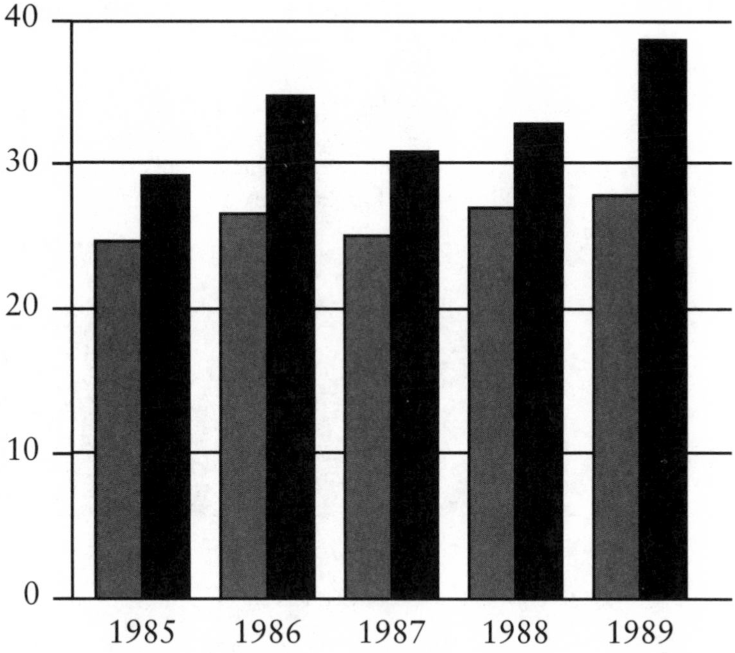

13.294 Use grouped bars to compare groups of similar information for different time periods or for some other different factor:

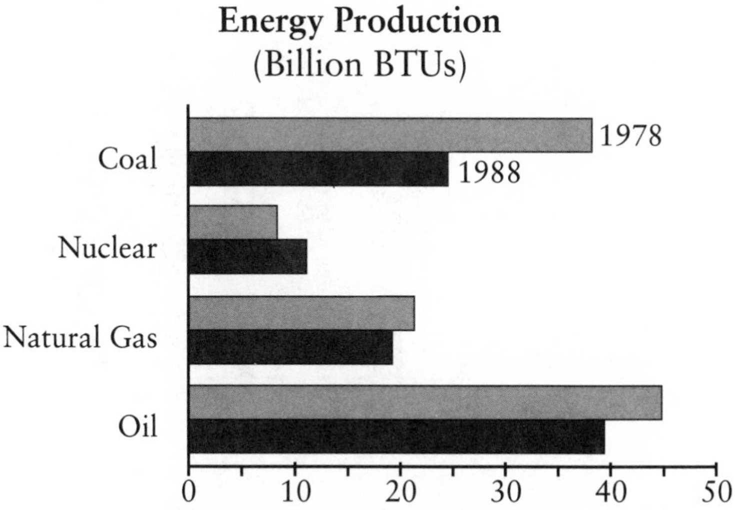

13.295 Do not use grouped bars to compare the composition of various items at one point in time; use 100-percent bar charts instead.

13.296 Separate groups of bars by at least one-quarter the width of the group or the width of a single bar.

13.297 Grouped columns may be closely spaced or touching. Don't overlap columns. Separate groups of columns by at least one-quarter the width of the group or the width of a single column. Use a different color or tone for each bar in the group:

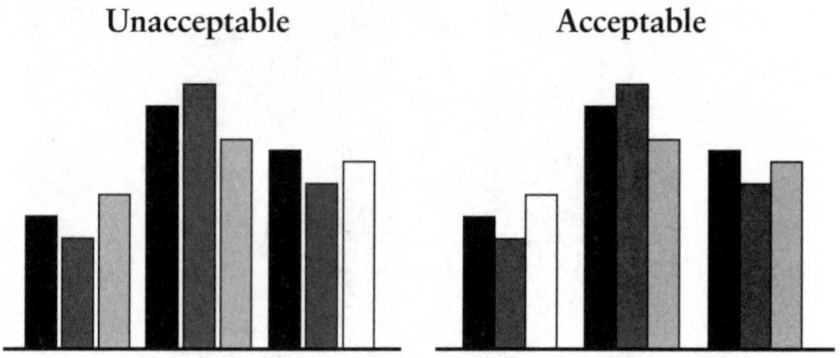

DIVIDED COLUMN/BAR CHARTS

13.298 Use divided column charts to compare components and wholes over a period of time or for changes in some other variable:

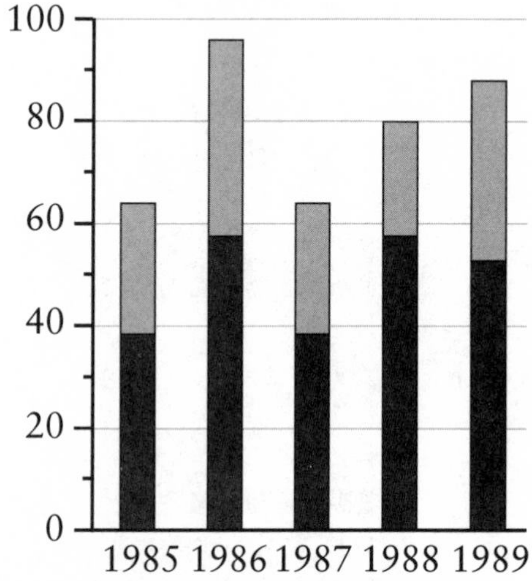

13.299 Do not use divided column charts for precise comparisons of components. Label column segments to the right of the last column, not within columns or by a separate key or legend.

13.300 Use divided bar charts to compare the component parts of a series of items, and to compare components and totals for various items at one time. Arrange the bar segments in a consistent, logical order. Keep the same order for all bars of the series:

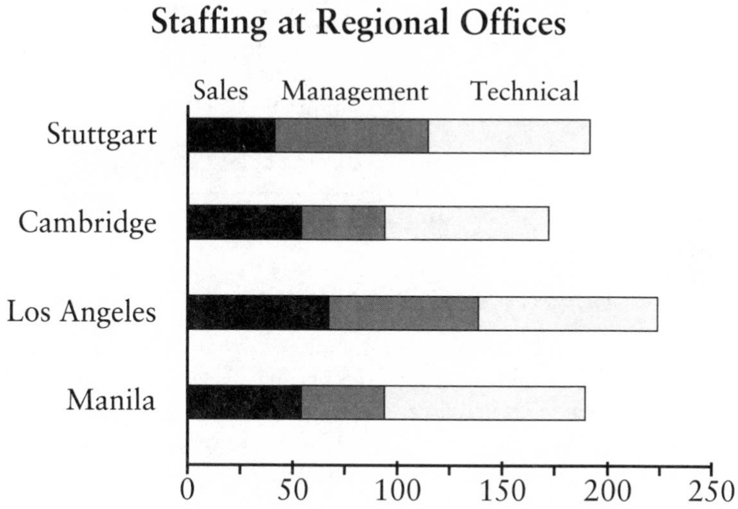

DEVIATION COLUMN/BAR CHARTS

13.301 Use deviation column charts to allow columns to extend beyond a zero baseline. They compare positive and negative values:

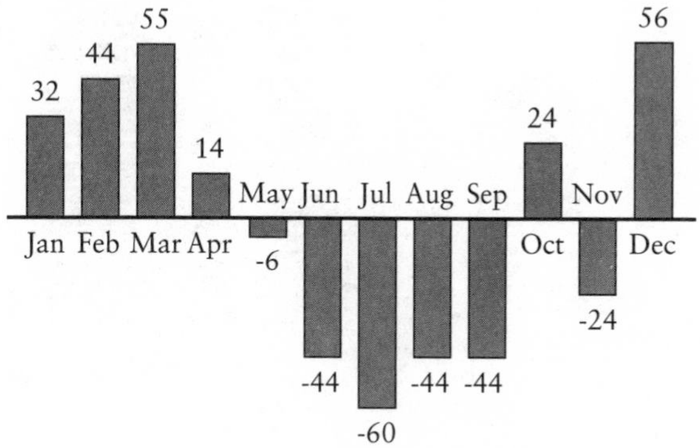

13.302 Use deviation bar charts to show positive and negative values as bars starting at a centrally located zero baseline. Positive values are to the right, negative to the left:

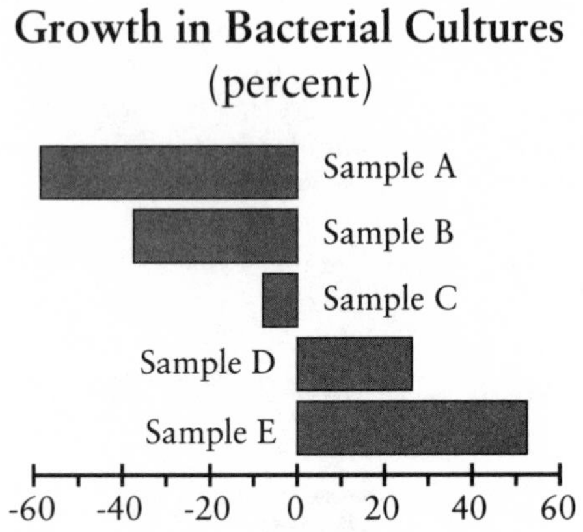

PICTOGRAPHIC COLUMN/BAR CHARTS

13.303 Use pictographic column charts when columns or stacks of pictorial symbols can represent quantities. Use pictographs only for simple comparisons, and omit the scale. Include a key explaining the meaning of each symbol, if not obvious. Also post the total at the end of the row or column of symbols:

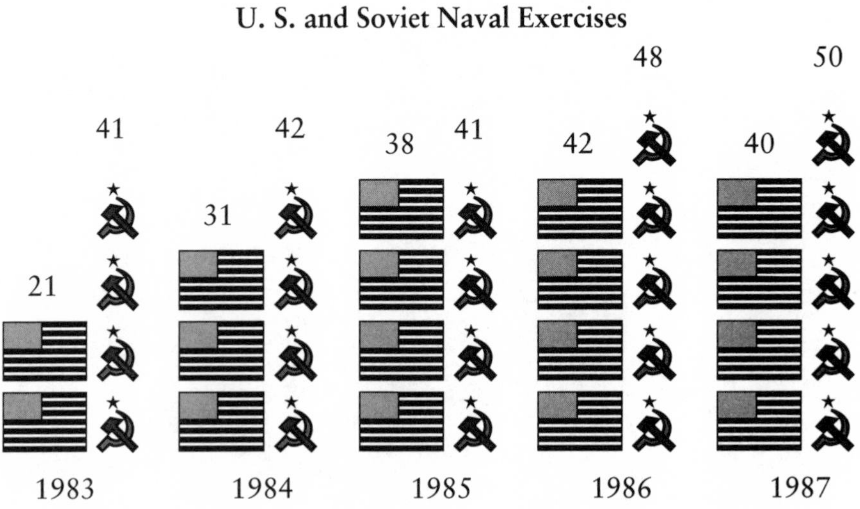

13.304 Use pictographic bar charts when pictorial symbols can represent quantities. Use pictographs for simple comparisons only. In pictographs, omit the scale. Include a key explaining the meaning of each symbol. Also post the total at the end of the line of symbols:

Photographs Taken
(millions)

Pie Charts

13.305 Use pie charts to show the relative sizes of component parts of a whole. Slices represent either percentages or fractions of the whole, which is represented by the entire circle:

Land Area of Continents

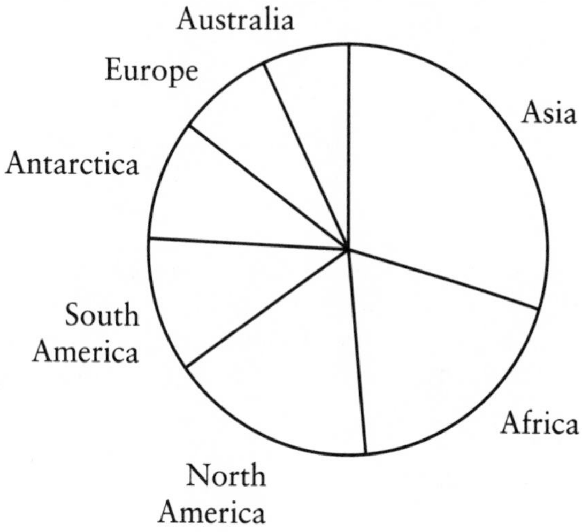

13.306 Use a pie chart when

- Showing the relative proportions of component parts,
- Simplifying complex data,
- The number of components is small, and
- General relationships are more important than precise quantities.

13.307 Do not use a pie chart when

- Comparing parts of two or more wholes,
- The number of components is large,
- Some of the components are small, and
- The exact percentages are important.

13.308 Arrange segments by size. Place the largest slice at the top extending clockwise from vertical. Continue placing progressively smaller slices clockwise. If there is an "other" or "miscellaneous" segment, place it last, even though it may be larger than other slices:

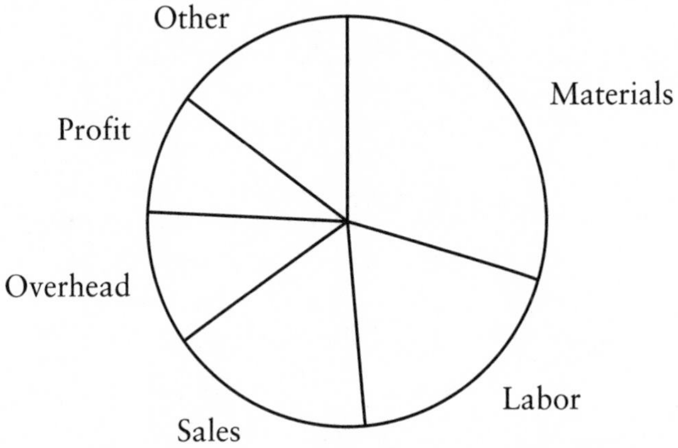

13.309 Limit the number of slices. Estimates for the upper limit range from five to eight. Include no slices smaller than about five percent (18 degrees). If you have more than a few categories of data, consolidate them into a smaller number of main categories:

Unacceptable

(too many slices)

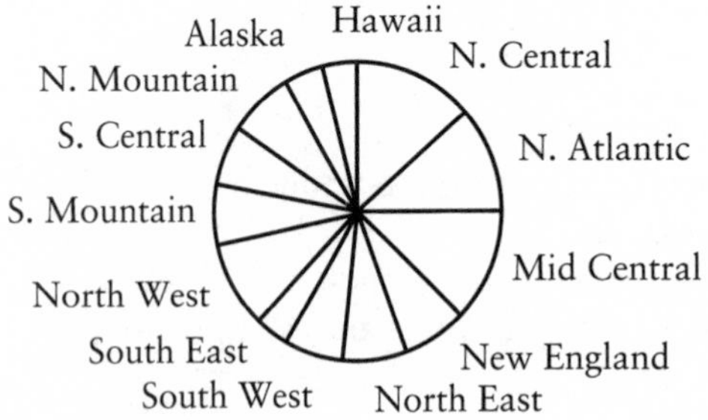

Acceptable

(slices incorporated)

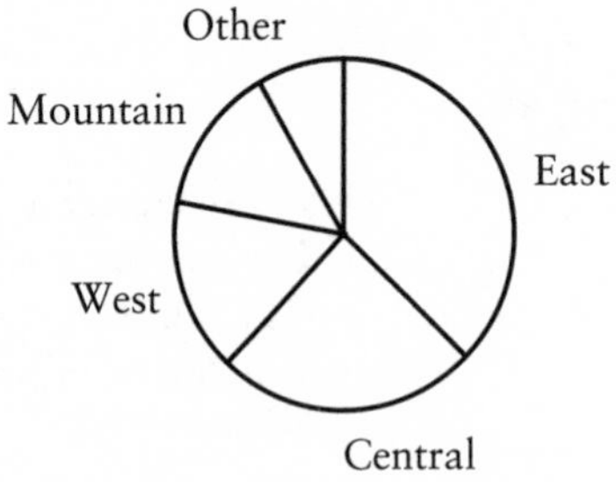

13.310 Give each slice a short, descriptive label. Place labels outside the circle. Avoid arrows, leader lines, or legends if possible. Never apply labels across the boundaries of segments.

13.311 Where space permits, include the absolute quantity, the percentage, or both in the label for each slice. Surround these quantities with parentheses: France (28%).

PICTORIAL PIE

13.312 To add graphical interest to a pie chart, cut slices from a picture of some circular, cylindrical, or spherical object:

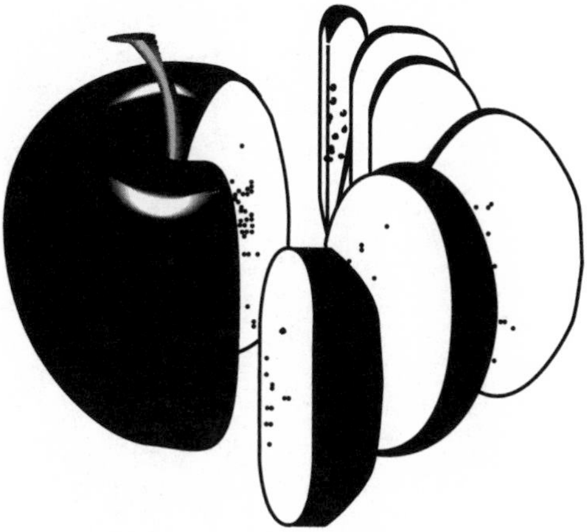

13.313 Choose an object that symbolizes the type of data shown, for example:

- Coin for money,
- Clock face for time, or
- Wheel for automotive sales.

Allocating Teachers' Time

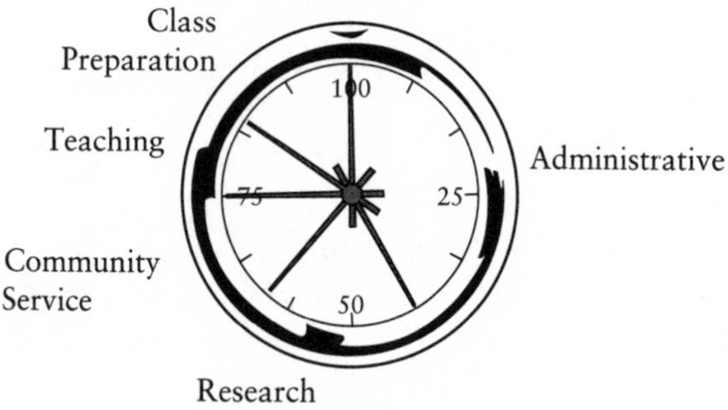

SEPARATE SEGMENT

13.314 To emphasize a single slice, detach it from the rest of the pie, sliding it outward slightly. Move the slice outward about 10 to 25 percent. Separate only one slice:

Land Area of Continents

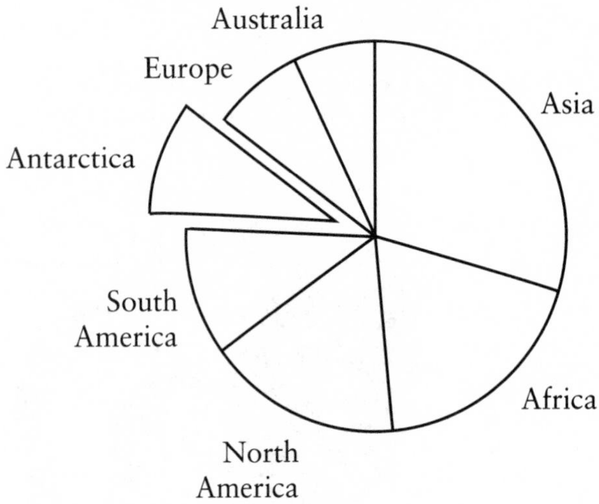

DIAGRAMS

13.315 Use diagrams to display the makeup of a system and the interrelationships among its components. Diagrams use visual means to show qualitative differences and relationships:

Relationship	Diagram
Spatial	Map
Compositional	System Organization
Responsibility	Personnel Organization
Sequence	Flowchart
Dependency	Project Tracking Chart
Connection	Topology
Influence	Causal Diagrams

13.316 Use existing diagramming techniques. Most technical disciplines, such as computer science and architecture, have formal diagramming techniques.

13.317 Diagrams are useful for

- Understanding how something operates,
- Recording interrelationships among components,
- Stimulating thinking about a complex system,
- Communicating detailed technical information concisely among technical experts,
- Documenting a system,
- Enforcing rigorous and logical organization, and
- Automating analysis by computer tools.

Components of Diagrams

13.318 Diagrams consist of symbols, which represent the objects of the system, and links, which represent the relationships among them.

Symbols

13.319 Use symbols, special labeled visual images, to represent the system components.

13.320 Symbols can be

- Solid or outlined geometric shapes: circles, squares, diamonds, or triangles;
- Silhouettes or icons of familiar objects;
- Detailed pictures of objects; or
- Words alone.

13.321 Use larger symbols for more important objects in a diagram. Do not use many symbols; more than five or six symbols is too many.

13.322 Label all symbols. Put labels next to or within the image. Put labels inside hollow shapes unless they will not fit.

13.323 If space does not permit a complete label, place a short label in or near the shape and use a footnote to provide more information. Follow the guidelines for table notes. (See Section 13.79.)

Links

13.324 Each symbol should be connected to at least one other symbol. Links express the relationships of the system. The pattern of linkages among symbols is the diagram's message. Each link shows a relationship between two system items. Links can show relationships in one direction, in both directions, or in no particular direction.

13.325 Lines, symbols, and labels define links.

13.326 Various types of connections can be shown by using different types of lines:

- Weight or width of the line: light, medium, or heavy;

- Number of lines: single or double;

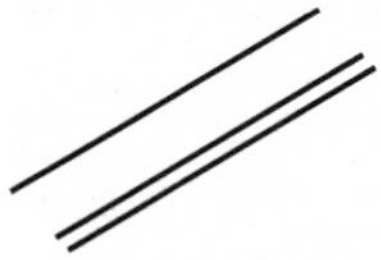

- Style of line: solid, dotted, dashed;

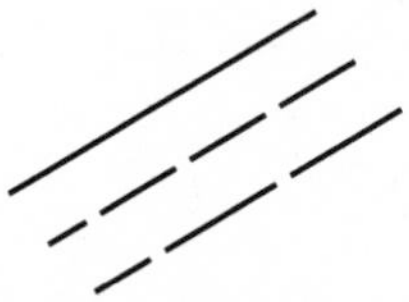

- Direction of the line: vertical, horizontal, diagonal;

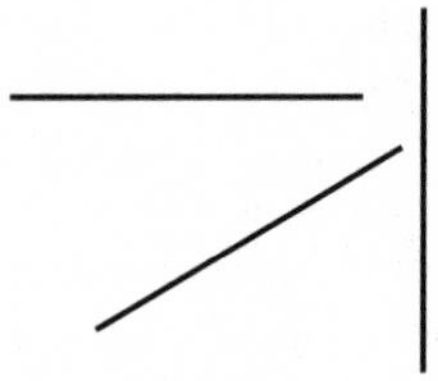

- Curvature of the line: straight, jagged, right angle, or smooth.

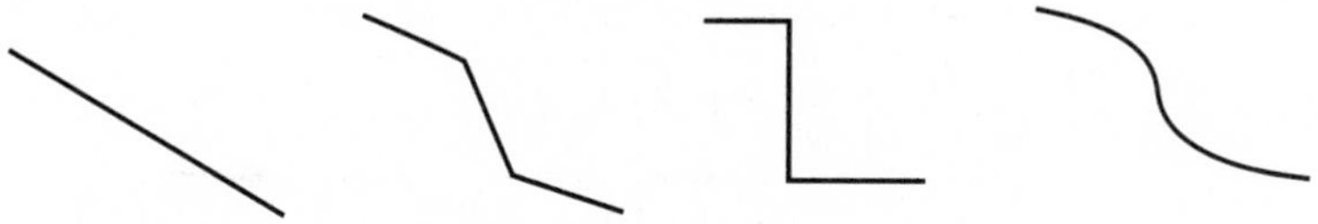

13.327 Use conventional graphic shapes such as arrowheads to show the direction of a relationship and the kind of link symbolized by the line:

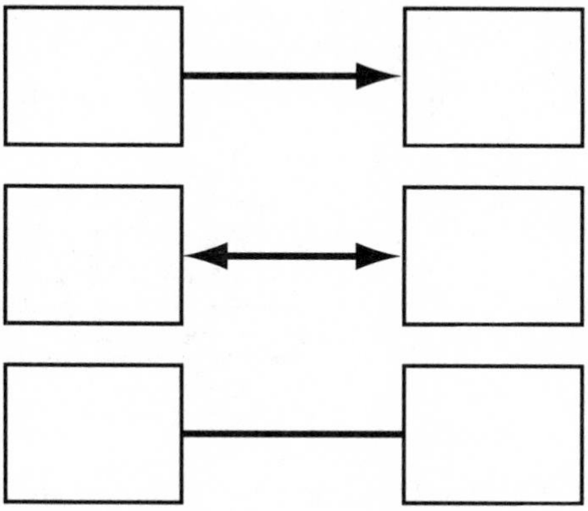

13.328 Links may be labeled. Keep the labels horizontal if possible. Place the label next to the link. Do not use arrows to connect labels to links. In some cases, however, it may be necessary to run the text along the link. Avoid curving text. Rotate the entire label, not its individual letters:

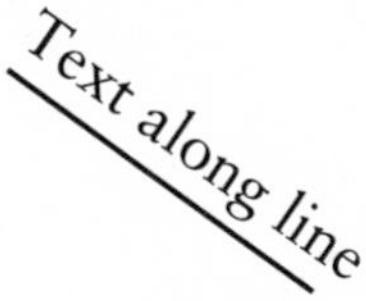

Patterns of Linkages

13.329 Linkages are commonly expressed in four ways:

- Sequence linkage—choice in two directions,
- Grid linkage—choice among a variety of logical relationships,
- Tree linkage—choice within a hierarchy, and
- Web linkage—choice within complex relationships.

SEQUENCE LINKAGE

13.330 Use sequence linkage to depict two choices: forward or backward.

GRID LINKAGE

13.331 Use grid linkage to organize and present information along two logical dimensions. For instance, the reader can read down the columns or can skim along the row to compare command syntax. Because tables have an inherent grid structure, most systems with a grid structure can be presented in a table as well as a diagram.

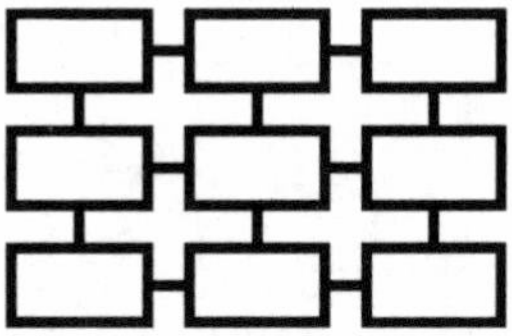

TREE LINKAGE

13.332 Use the tree structure to present simple hierarchy. The tree structure is common in organization charts.

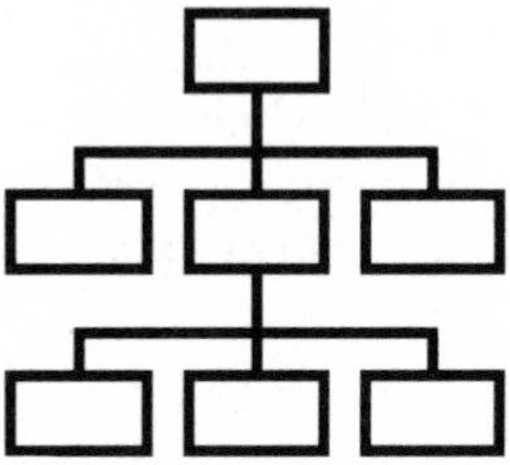

WEB LINKAGE

13.333 Use the web structure to illustrate complex relationships.

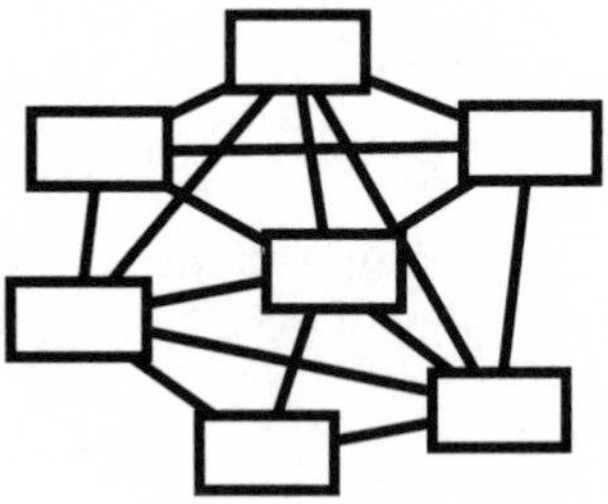

Common Types of Diagrams

Organization Charts

13.334 Use organization charts to display the interrelationships within a hierarchical organization. Organization charts show the makeup of a system. They originate from one symbol and divide into more and more symbols:

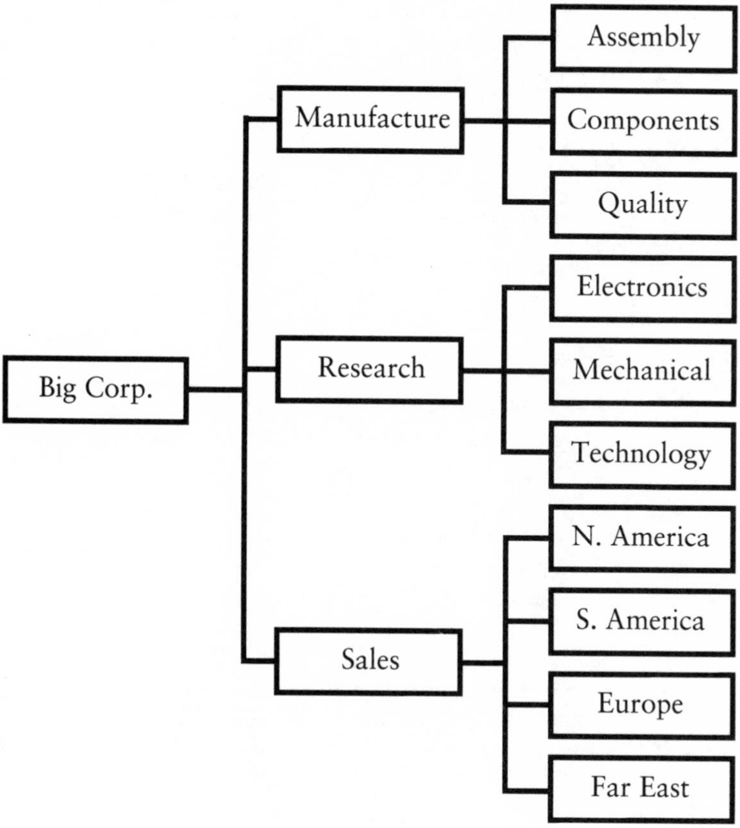

13.335 Do not mix different types of units in a single organization chart.

13.336 Organization charts can be drawn vertically (as a pyramid or tree), horizontally, or circularly. On a multilevel organization chart, the biggest problem is running out of space for boxes. Solutions include starting in the center and branching outward or turning the chart on the page.

13.337 In management organization charts, place the more important position on the top. Importance can also be shown by box size or the border weight.

13.338 In an organization chart, use solid lines to show official command and control relationships. Use dashed lines to show coordination, communication, and temporary relationships.

Flowcharts

13.339 Flowcharts

- Show the sequence of activities, operations, events, or ideas in a complex procedure or interrelated system of components;
- Summarize long, involved, detailed descriptions and instructions;
- Describe the specific series of tasks, activities, and operations necessary to perform a procedure; and
- Illustrate the movement through a process.

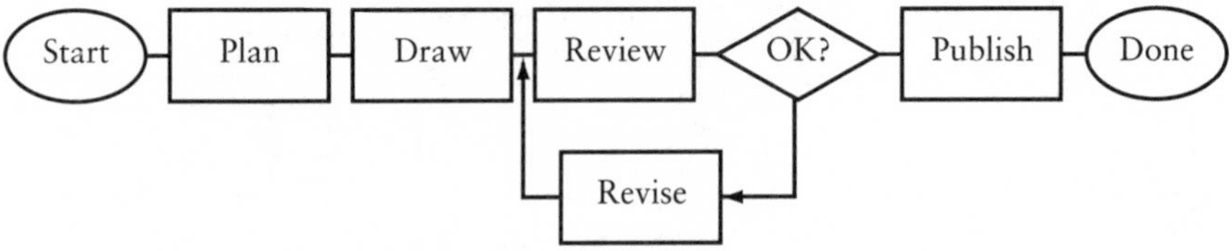

13.340 The assumed direction of flow is from top to bottom and left to right. Links that move in these directions may omit arrowheads. Links that move counter to these directions must include arrowheads.

13.341 Subdivide complex procedures into multiple simple flowcharts.

Connection Maps

13.342 Use connection maps to show permitted or existing connections. A map of a subway system is an example of a connection map.

Dependency Diagrams

13.343 Use dependency diagrams to show causal effects and influences among interrelated components. Dependency diagrams consist of labels, sometimes in circles, connected by arrows showing the direction of influence:

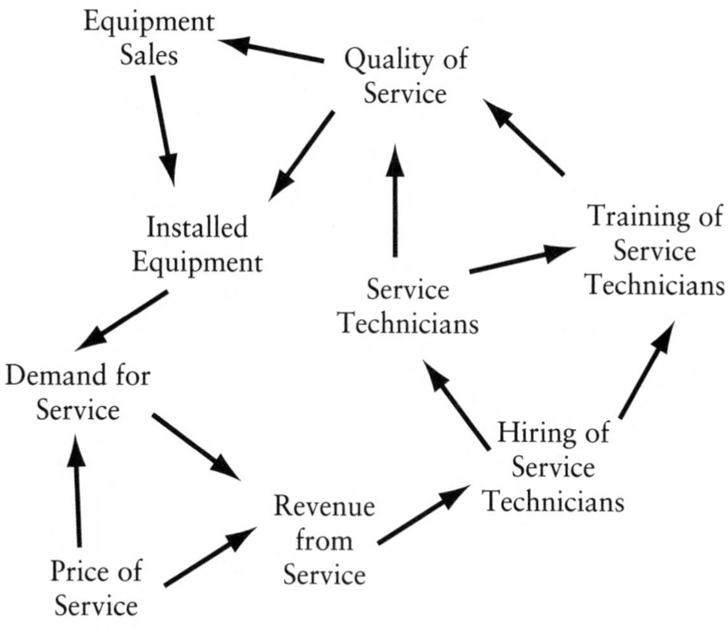

FOURTEEN

Designing Useful Documents

DESIGNING A DOCUMENT'S FORMAT

14.1 A document's graphic design includes the decisions that control its physical appearance:

- Size,
- Color,
- Layout,
- Cover illustration and design,
- Paper,
- Typography,
- Illustrative material (photographs, data displays, and the like), and
- Binding method.

14.2 Good document design

- Invites people to read the document,
- Helps readers find information easily in the document,
- Offers a good impression of the content and authoring agency, and
- Provides a sense of corporate or organizational identity.

GENERAL FORMAT AND DESIGN CONSIDERATIONS

14.3 Planning a document's graphic design requires an understanding of audience, purpose, and scope. Before making specific design decisions, define these needs by asking

- How will the document be used? Good design can promote a document's intended use. For example, if a technical manual will be used in a crowded workspace, a small size will be more effective.
- How often will the document be used? How often will it be updated? If a document must survive constant use, design it to minimize wear and tear; if it will need frequent updates, the binding must permit pages to be added easily.

- How will the document be distributed? Distribution requires a knowledge of storage, packaging, and shipping costs and methods.
- Will the document compete with other documents on a bookshelf for readers' attention? If so, it should have interesting packaging. Even if the document will be shipped directly to its audience, it will still convey an image.

14.4 This chapter presents general design rules. Although these rules are usually good ones to follow, remember that any rule can be wrong in a particular instance. Also keep in mind that good design is not simply a matter of aesthetics, it is also a matter of function and improved communication.

DESIGNING THE PAGE

(See Section 14.68.)

14.5 Many variables must be considered when creating a page format. For example, most technical documents are not read cover to cover; instead, people read only those paragraphs or sections that describe a specific procedure or concept.

14.6 Use obvious and simple formatting principles to help readers move directly to specific information without being distracted by any formatting techniques.

14.7 After developing a format, use the same format throughout the document. The single most important factor in developing a page format is consistency. Presenting information in too many ways confuses readers.

14.8 When designing the page, consider all the formal format elements:

- Page size;
- White space;
- Margins;
- Typefaces;
- Type sizes;
- Leading or line spacing;
- Line width;
- Interactions of type size, line width, and leading; and
- Highlighting techniques.

Page Size

14.9 When selecting an appropriate page size, keep in mind how the document will be used. When readers will use the document in a small space, such as on a crowded desk, select a smaller page size. However, if the text is very long, a larger page size may prevent the final document from looking thick and imposing.

14.10 Fashion can affect page size. For example, small-format manuals, typically 5″ × 8″, have become customary for office documents. Larger equipment documentation usually use the traditional 8-1/2″ × 11″ page size. Remember, too, that documents destined for foreign markets are often printed on other size pages:

Exhibit 14–1:
Typical Page Sizes and Orientation

Supporting Equipment	Size in Inches	Orientation
Office automation		
typewriters	5 × 8	landscape
computers	7 × 10	portrait
copiers	5 × 8	landscape
telecommunication	5 × 8	landscape
Quick reference		
desk reference	5 × 7	portrait
instructions	5 × 7	portrait
troubleshooting	5 × 7	portrait
Tutorial manuals	7 × 10	portrait
Brochures		
One-fold	7 × 10	portrait
Newsletters		
One-fold	8 × 10	portrait
Stitched	8 × 10	portrait
Service manuals	8 1/2 × 11	portrait
Reference manuals	8 1/2 × 11	portrait
Journals	8 × 10 5 × 8	portrait portrait
Commercial magazines	8 × 10	portrait

14.11 Also, consider that paper is cut from stock-size paper. For instance, 3″ × 6″ pages are often printed on 6-1/4″ × 12-1/2″ sheets that have been cut from 28″ × 42″ standard-size sheets:

Some Standard Page Sizes and Printer Sheets

Page Size (inches)	Paper Size (inches)	Press Sheet Size (inches)
3 × 6	28 × 42	6-1/4 × 12-1/2
5 × 7	32 × 44	10-1/4 × 14-1/2
5 × 8	35 × 45	10-1/4 × 16-1/2
7 × 10	32 × 44	16 × 22
8 × 10	35 × 45	17-1/2 × 22-1/2
8-1/2 × 11	35 × 45	17-1/2 × 22-1/2

14.12 Page size may be controlled by the available press size. The preceding table offers only a limited selection of press sheet sizes. Typically, these sheets range from 6-1/4″ × 12-1/2″ to 38″ × 50″.

14.13 Using a specific page size may not be possible because the document may be part of an existing document series for which the page size has already been determined.

White Space

14.14 Use white space, empty area around text or illustrations, to isolate and separate information. Make pages readable, visually appealing, and easily accessible by using enough white space to emphasize information.

14.15 Place white space between sections, headings, and major topics to indicate a physical break. Also use white space to bring attention to illustrations.

14.16 Establish a ratio of white space above and below elements to be emphasized in a document. Such a principle emphasizes the division and visually ties the new header or illustration title more closely with the text or illustration it introduces:

Exhibit 14–2:
Typical White Space

Level Head	Leading White Space	Trailing White Space
1	18	12
2	15	10
3	12	8

14.17 Also consider the use of a similar ratio for type size. (See Section 14.23.)

Margins

14.18 Use margins, the white space around a page's perimeter, to improve legibility.

14.19 Use ragged-right margins, uneven text in the right margin, to give a document a relaxed appearance. Readability studies indicate that people read ragged-right text faster than justified text.

14.20 Use justified margins, when all text share a common right margin, to give a technical document a more formal appearance. However, justified text may create "rivers," accidental interline spacing, that cause the reader's eye to move vertically rather than horizontally:

This text has been justified left and has a ragged-right margin. Some readability studies indicate that it is easier to read than fully justified text.

This text has been fully justified. Some readability studies indicate that it is more difficult to read than text that is only left-justified. It also has "rivers of white."

Typefaces

14.21 Typeface is the style of lettering used in a document. Typewriters and letter-quality word processing printers usually offer only a few typefaces. Laserographic printers and typesetting equipment usually offer a variety of typefaces. All typefaces fall into two categories: serif and sans serif. A serif is a short line that stems from the end of a letter. Serifs give an illusion of connecting the letters in a word. Readability studies show that, for long documents, a serif typeface is easier to read than a sans serif typeface.

14.22 Use as few different typefaces as possible in any one document or document library. Typically, very useful documents use no more than two different typefaces—one for the text and a second for any nontextual (graphic) information or for headers.

Type Sizes

14.23 Type sizes are measured in points; 72 points equal an inch. When compared with typewriter letter sizes, 10-point type is about equivalent to 12-pitch typewriter letters; 11- or 12-point type (depending on the typeface) is about equal to 10-pitch typewriter letters.

14.24 For typeface and type size considerations, documents are divided into headings and body text:

- For body text, 10- or 11-point serifed type is the most comfortable typeface and size for the average reader.
- Captions and callouts for illustrations are preferably the same size as the body text, though, in practice, they tend to be smaller. Using the same typeface for both body text and illustration callouts and text can reduce production costs. If effective discrimination methods—white space, rules, boxes, and the like—have been used, this practice should not make the text more difficult to read.
- Use larger type sizes for headings. In addition, establish a ratio among the various heading sizes to provide useful clues for the reader seeking specific heading levels. Select type sizes at least three points different in size to distinguish headings, since the typical user cannot discriminate type-size differences of less than three points:

Exhibit 14–3:
Typical Header Algorithm

Level Head	Header Type Size	Leading White Space	Trailing White Space
1	18	24	12
2	15	20	10
3	12	16	8

In creating this ratio, also consider the use of a similar ratio for white space. (See Section 14.14.)

- Footnotes and indexes can be as small as 8- or 9-point to save space.

Leading (Line Spacing)

14.25 The space from the bottom of the descenders—such as the tails of *p*s or *g*s—on one line to the top of the ascenders—usually capitals but sometimes *h*s, *l*s, or the like—on the next line is referred to as line leading. The recommended amount of leading for 10- or 11-point type is 1 or 2 points; more or less leading tends to slow reading. Note, however, that typeface as well as line width also influences leading decisions.

14.26 For documents prepared with typewriters or character printers, use single-spaced text rather than double-spaced or one-and-a-half-spaced text; it is usually easier to read. The double-spaced manuscripts required by many publications are convenient for reviewers because this provides space for annotations; the final document is usually produced with closer line spacing for readability.

Line Width

14.27 Format text with no more than 10 to 12 words per line. Most text set in 10-point type should have a line width of no more than 4 to 5 inches (24 to 30 picas), with 1 point of leading for shorter lines and 2 points for longer lines.

Interactions of Type Size, Line Width, and Leading

14.28 Type size, line width, and leading interact in a variety of ways. The best source for information on these interactions is a graphic designer. However, some general observations can be offered based on contemporary research. These observations should be incorporated into design decisions:

Exhibit 14–4:
Interaction of Type Size,
Line Width, and Leading

Type Size	Line Width	Leading
8	14 – 36	2 – 4
9	14 – 30	1 – 4
10	14 – 31	1 – 4
11	16 – 34	0 – 2
12	17 – 33	1 – 4

Highlighting Techniques

14.29 Use highlighting techniques such as boldface type, italic type, reverse type, underlining, and different type sizes to help readers find important text. Use the following guidelines for highlighting techniques:

- Keep highlighting simple; do not use all available type styles, type sizes, and paragraph formats.
- Avoid decorative uses for and combinations of type styles. Italics, boldface, underlining, and other typographic emphasis techniques work well when only about one word in every 200 is emphasized. Readers begin to ignore specialized typography when about one word in every 30 is emphasized.
- Use all capital type only for text elements that deserve unusual attention, since this type slows reading considerably. Thus, all caps may be ideal for warnings when danger to a human operator is involved.
- Boldface type, set in mixed case, is best for cueing, especially in headings. However, boldface inside paragraphs disrupts the balance of the page, creating a spotty and disorganized appearance.
- Color can also be used for highlighting. Color adds interest and improves reading proficiency when it does not overpower the text. Like any highlighting technique, color should be used sparingly and for special and specific purposes. Useful color combinations include

Exhibit 14–5:
Useful Color Combinations

Paper Color	Ink Color: White	Black	Red	Blue	Yellow
White	Not a Good Combination	Second Best Combination	Not a Good Combination	Best Combination	Not a Good Combination
Black	Second Best Combination	Not a Good Combination	Not a Good Combination	Not a Good Combination	Best Combination
Red	Best Combination	Not a Good Combination	Not a Good Combination	Not a Good Combination	Second Best Combination
Green	Not a Good Combination	Best Combination	Not a Good Combination	Second Best Combination	Not a Good Combination
Blue	Best Combination	Not a Good Combination	Not a Good Combination	Not a Good Combination	Second Best Combination
Cyan	Not a Good Combination	Second Best Combination	Not a Good Combination	Best Combination	Not a Good Combination
Magenta	Second Best Combination	Best Combination	Not a Good Combination	Not a Good Combination	Not a Good Combination
Yellow	Not a Good Combination	Not a Good Combination	Best Combination	Second Best Combination	Not a Good Combination

■ Best Combination
▒ Second Best Combination
□ Not a Good Combination

Since excessive highlighting creates distracting and hard-to-read pages, use restraint in selecting these techniques.

PAGE ELEMENTS

14.30 Page elements include all of the considerations discussed in the earlier sections:

- Page size;
- White space;
- Margins;
- Typefaces;
- Type sizes;
- Line spacing;
- Line width;
- Interactions of type size, line width, and leading; and
- Highlighting techniques.

Once each of these elements has been assigned a specific value, it can be incorporated into page grids.

Page Grids

14.31 Create page grids to establish where specific page elements will appear on a sheet of paper. Page grids provide guidance across product lines, among writers working on the same project, or across many corporate sites. They provide consistency and continuity that helps readers use documents successfully.

14.32 The following sections offer sample grid designs for four major size documents: 5 × 7, 5 × 8, 7 × 10, and 8-1/2 × 1. See Exhibit 14–1 for more information about document sizes and their use. It should be understood that these sample grids do not exhaust the variations a graphic designer could propose for a document. Instead, they are intended as guidelines to help create similar designs for both paper and electronic composition systems (such as the many electronic paste-up programs available for computers).

5 × 7 Portrait

14.33 Use the 5 × 7 portrait size to prepare such documents as desk reference, instruction, and troubleshooting manuals. The advantage to this size manual is that its footprint (the amount of space it occupies in the workspace) is very small. Unfortunately, the page is too narrow to accommodate two columns (11.5 picas wide).

5 × 7 One-Column Portrait Page Grid: Major Section Break
(dimensions in picas)

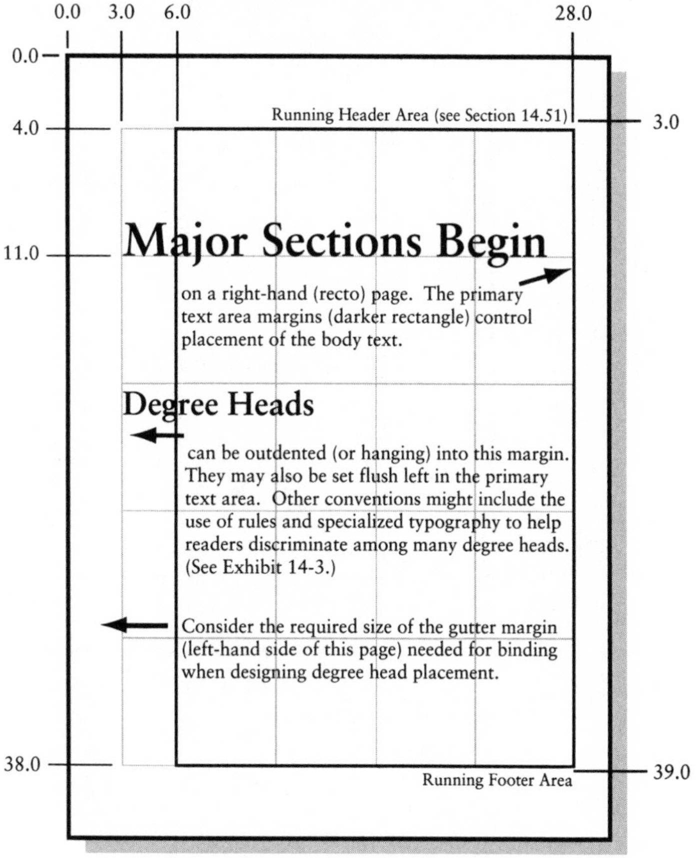

0.0
3.0
6.0
28.0
0.0
4.0
11.0
38.0
3.0
39.0
Running Header Area (see Section 14.51)
Major Sections Begin
on a right-hand (recto) page. The primary text area margins (darker rectangle) control placement of the body text.
Degree Heads
can be outdented (or hanging) into this margin. They may also be set flush left in the primary text area. Other conventions might include the use of rules and specialized typography to help readers discriminate among many degree heads. (See Exhibit 14-3.)
Consider the required size of the gutter margin (left-hand side of this page) needed for binding when designing degree head placement.
Running Footer Area

One-Column 5 × 7 Portrait

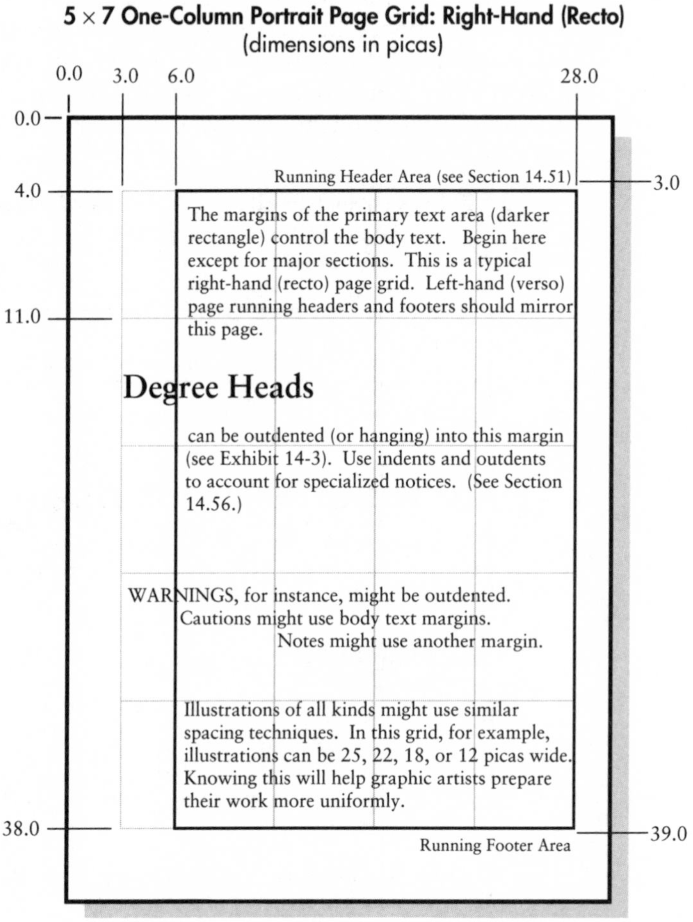
5 × 7 One-Column Portrait Page Grid: Right-Hand (Recto)
(dimensions in picas)
0.0
3.0
6.0
28.0
0.0
4.0
11.0
38.0
3.0
39.0
Running Header Area (see Section 14.51)
The margins of the primary text area (darker rectangle) control the body text. Begin here except for major sections. This is a typical right-hand (recto) page grid. Left-hand (verso) page running headers and footers should mirror this page.
Degree Heads
can be outdented (or hanging) into this margin (see Exhibit 14-3). Use indents and outdents to account for specialized notices. (See Section 14.56.)
WARNINGS, for instance, might be outdented.
Cautions might use body text margins.
Notes might use another margin.
Illustrations of all kinds might use similar spacing techniques. In this grid, for example, illustrations can be 25, 22, 18, or 12 picas wide. Knowing this will help graphic artists prepare their work more uniformly.
Running Footer Area

5 × 8 Landscape

14.34 Use the 5 × 8 landscape size to prepare such documents as typewriter, copier, and telecommunication manuals. The advantage to this size manual is that its footprint (the amount of space it occupies in the workspace) is very small. It also is wide enough so that two-column manuals can be created.

5 × 8 One-Column Landscape Page Grid: Major Section Break
(dimensions in picas)

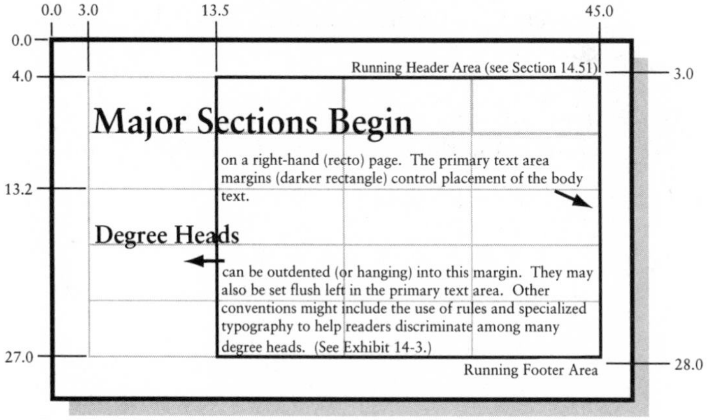
0.0
3.0
13.5
45.0
0.0
4.0
13.2
27.0
Running Header Area (see Section 14.51)
3.0
Major Sections Begin
on a right-hand (recto) page. The primary text area margins (darker rectangle) control placement of the body text.
Degree Heads
can be outdented (or hanging) into this margin. They may also be set flush left in the primary text area. Other conventions might include the use of rules and specialized typography to help readers discriminate among many degree heads. (See Exhibit 14-3.)
Running Footer Area
28.0

5 × 8 One-Column Landscape Page Grid: Right-Hand (Recto)
(dimensions in picas)

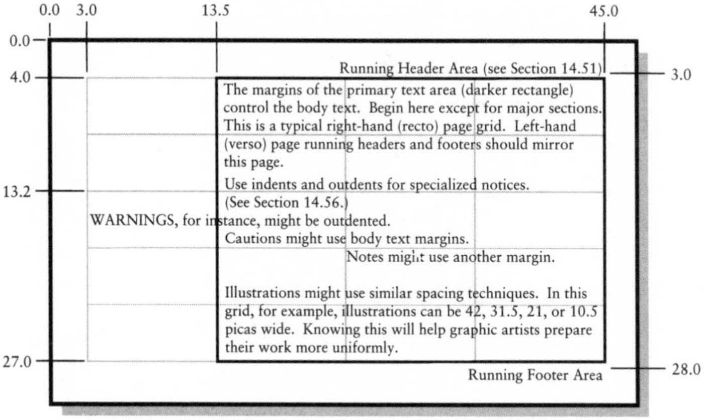
0.0
3.0
13.5
45.0
0.0
4.0
13.2
27.0
Running Header Area (see Section 14.51)
3.0
The margins of the primary text area (darker rectangle) control the body text. Begin here except for major sections. This is a typical right-hand (recto) page grid. Left-hand (verso) page running headers and footers should mirror this page.
Use indents and outdents for specialized notices. (See Section 14.56.)
WARNINGS, for instance, might be outdented.
Cautions might use body text margins.
Notes might use another margin.
Illustrations might use similar spacing techniques. In this grid, for example, illustrations can be 42, 31.5, 21, or 10.5 picas wide. Knowing this will help graphic artists prepare their work more uniformly.
Running Footer Area
28.0

Two-Column 5 × 8 Landscape

14.35 Use the two-column size to prepare such documents as typewriter, copier, and telecommunication manuals. The advantage to this manual is that its footprint (the amount of space it occupies in the workspace) is very small. It also has enough width so that two-column manuals can be created.

14.36 In some instances these two-column grids can be used to present user action-system response situations. To use this technique, place the action that the user must do in the left column and the expected (and correct) results that will be encountered in the right column.

14.37 This grid can also be used to place supporting illustrations in one column and the text explaining the illustrations in the other. This layout is called *paired pictographs.*

14.38 The disadvantages of two-column grids include limited

- Width for degree heads,
- Width for major section heads, and
- Discrimination for specialized notices.

5 × 8 Two-Column Landscape Page Grid: Major Section Break
(dimensions in picas)

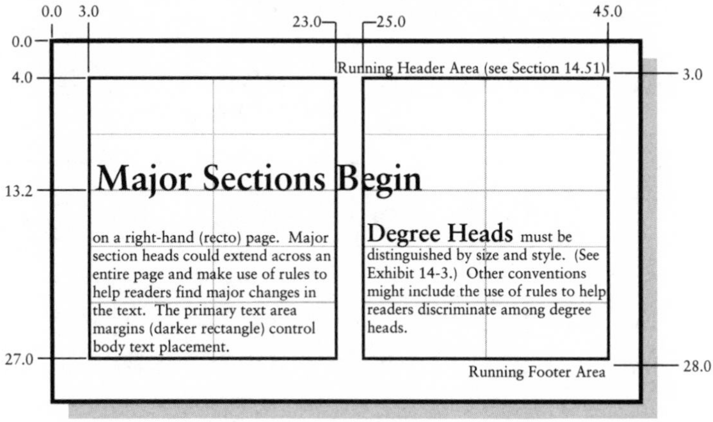
0.0
3.0
23.0
25.0
45.0
0.0
4.0
13.2
27.0
3.0
28.0
Running Header Area (see Section 14.51)
Major Sections Begin
on a right-hand (recto) page. Major section heads could extend across an entire page and make use of rules to help readers find major changes in the text. The primary text area margins (darker rectangle) control body text placement.
Degree Heads must be distinguished by size and style. (See Exhibit 14-3.) Other conventions might include the use of rules to help readers discriminate among degree heads.
Running Footer Area

5 × 8 Two-Column Landscape Page Grid: Right-Hand (Recto)
(dimensions in picas)

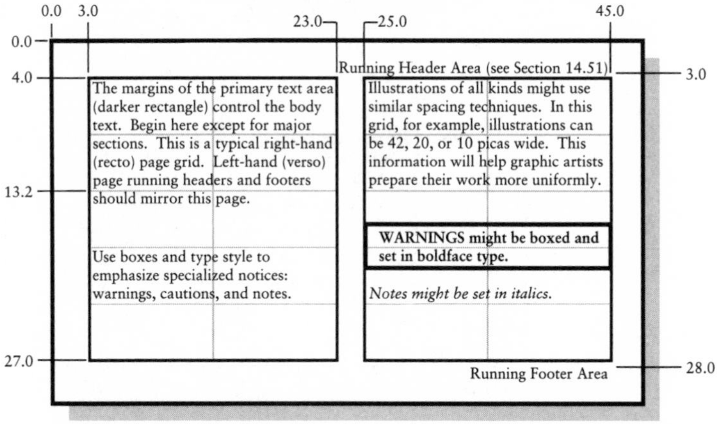
0.0
3.0
23.0
25.0
45.0
0.0
4.0
13.2
27.0
3.0
28.0
Running Header Area (see Section 14.51)
The margins of the primary text area (darker rectangle) control the body text. Begin here except for major sections. This is a typical right-hand (recto) page grid. Left-hand (verso) page running headers and footers should mirror this page.
Use boxes and type style to emphasize specialized notices: warnings, cautions, and notes.
Illustrations of all kinds might use similar spacing techniques. In this grid, for example, illustrations can be 42, 20, or 10 picas wide. This information will help graphic artists prepare their work more uniformly.
WARNINGS might be boxed and set in boldface type.
Notes might be set in italics.
Running Footer Area

7 × 10 Portrait

14.39 Use the 7 × 10 portrait size to prepare such documents as computer and tutorial manuals, and for some brochures. The advantage to this size is that its footprint (the amount of space it occupies in the workspace) is very small. It also is wide enough that two-column manuals can be created.

ONE COLUMN 7 × 10 PORTRAIT

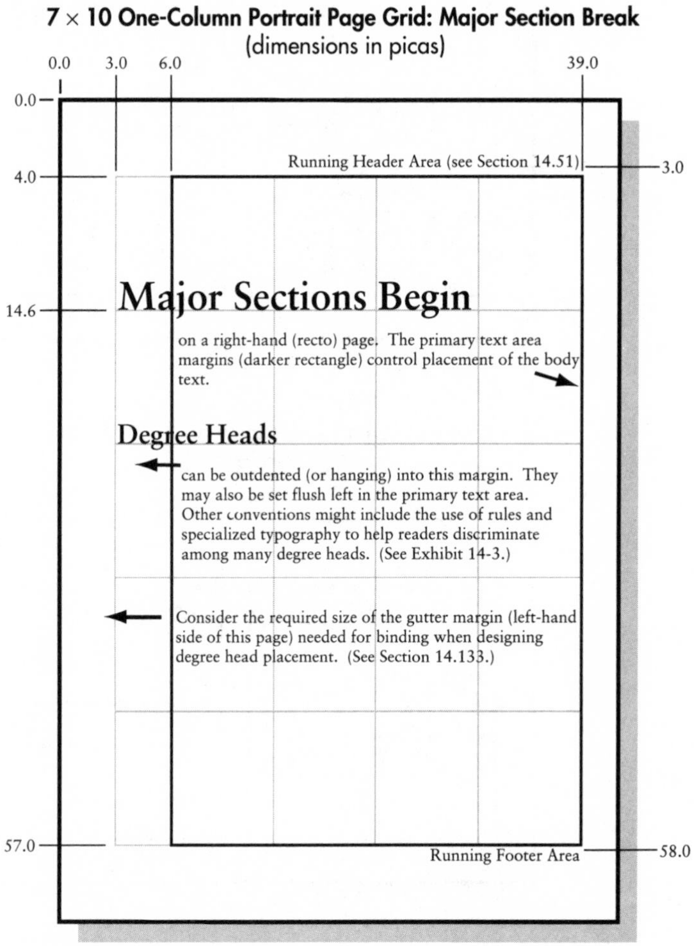
7 × 10 One-Column Portrait Page Grid: Major Section Break
(dimensions in picas)
0.0
3.0
6.0
39.0
0.0
4.0
14.6
57.0
Running Header Area (see Section 14.51)
3.0
Major Sections Begin
on a right-hand (recto) page. The primary text area margins (darker rectangle) control placement of the body text.
Degree Heads
can be outdented (or hanging) into this margin. They may also be set flush left in the primary text area. Other conventions might include the use of rules and specialized typography to help readers discriminate among many degree heads. (See Exhibit 14-3.)
Consider the required size of the gutter margin (left-hand side of this page) needed for binding when designing degree head placement. (See Section 14.133.)
Running Footer Area
58.0

7 × 10 One-Column Portrait Page Grid: Right-Hand (Recto)
(dimensions in picas)

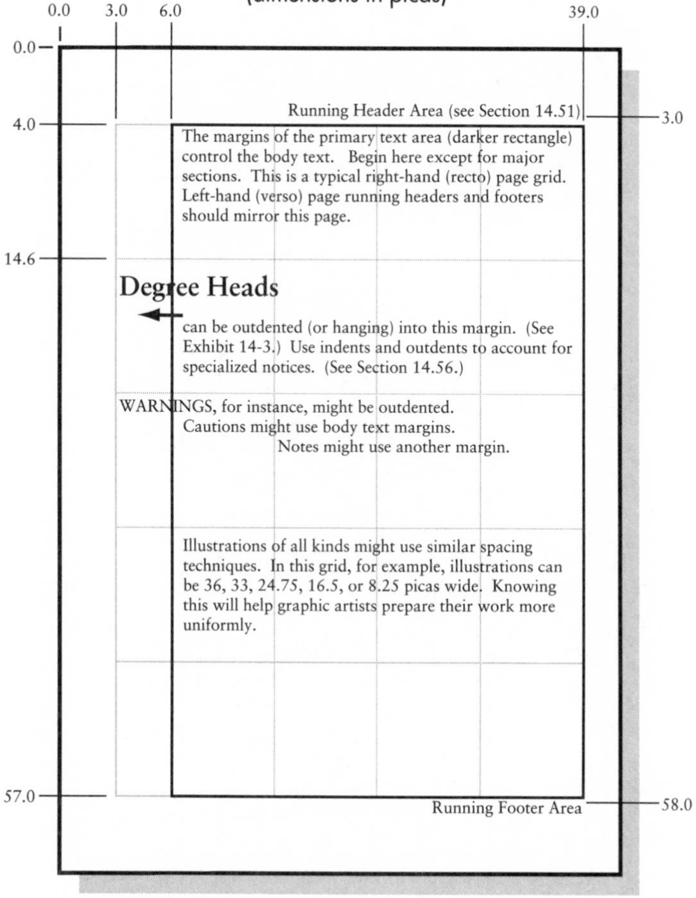

0.0
3.0
6.0
39.0
0.0
4.0
14.6
57.0
3.0
58.0
Running Header Area (see Section 14.51)
The margins of the primary text area (darker rectangle) control the body text. Begin here except for major sections. This is a typical right-hand (recto) page grid. Left-hand (verso) page running headers and footers should mirror this page.
Degree Heads
can be outdented (or hanging) into this margin. (See Exhibit 14-3.) Use indents and outdents to account for specialized notices. (See Section 14.56.)
WARNINGS, for instance, might be outdented.
Cautions might use body text margins.
Notes might use another margin.
Illustrations of all kinds might use similar spacing techniques. In this grid, for example, illustrations can be 36, 33, 24.75, 16.5, or 8.25 picas wide. Knowing this will help graphic artists prepare their work more uniformly.
Running Footer Area

Two-Column 7 × 10 Portrait

14.40 Use the two-column size to prepare such documents as computer and tutorial manuals, and for some brochures. The advantage to this size is that its footprint (the amount of space it occupies in the workspace) is very small. It is also wide enough that two-column manuals can be created.

14.41 In some instances these two-column grids can be used to present user action-system response situations. To use this technique, place the action that the user must do in the left column and the expected (and correct) results that will be encountered in the right column.

7 × 10 Two-Column Portrait Page Grid: Major Section Break
(dimensions in picas)

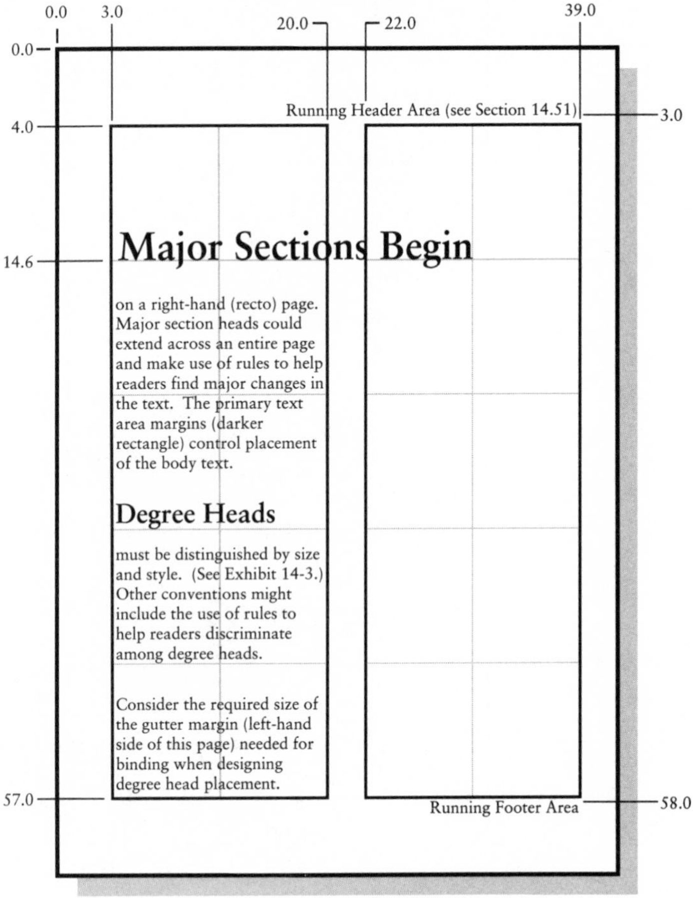

0.0
3.0
20.0
22.0
39.0
0.0
4.0
14.6
57.0
3.0
58.0
Running Header Area (see Section 14.51)
Major Sections Begin
on a right-hand (recto) page. Major section heads could extend across an entire page and make use of rules to help readers find major changes in the text. The primary text area margins (darker rectangle) control placement of the body text.
Degree Heads
must be distinguished by size and style. (See Exhibit 14-3.) Other conventions might include the use of rules to help readers discriminate among degree heads.
Consider the required size of the gutter margin (left-hand side of this page) needed for binding when designing degree head placement.
Running Footer Area

14.42 A similar use for this grid can be to place supporting illustrations in one column and the text explaining the illustrations in the other. This layout is called *paired pictographs.*

14.43 The disadvantages of two-column grids include limited

- Width for degree heads,
- Width for major section heads, and
- Discrimination for specialized notices.

7 × 10 Two-Column Portrait Page Grid: Left-Hand (Verso)
(dimensions in picas)

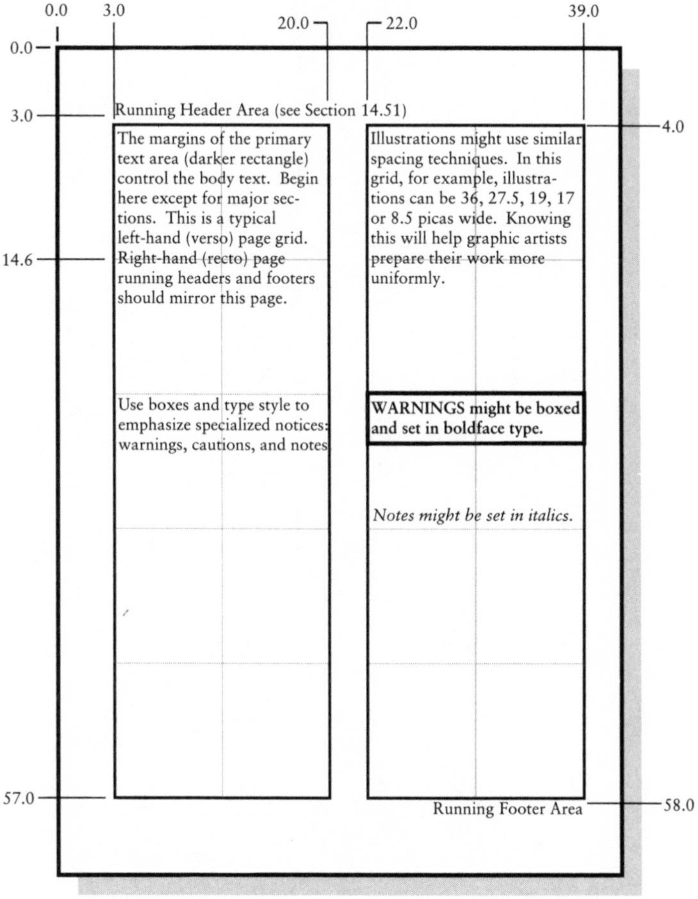
0.0
3.0
20.0
22.0
39.0
0.0
3.0
Running Header Area (see Section 14.51)
4.0
The margins of the primary text area (darker rectangle) control the body text. Begin here except for major sections. This is a typical left-hand (verso) page grid. Right-hand (recto) page running headers and footers should mirror this page.
Illustrations might use similar spacing techniques. In this grid, for example, illustrations can be 36, 27.5, 19, 17 or 8.5 picas wide. Knowing this will help graphic artists prepare their work more uniformly.
14.6
Use boxes and type style to emphasize specialized notices: warnings, cautions, and notes
WARNINGS might be boxed and set in boldface type.
Notes might be set in italics.
57.0
Running Footer Area
58.0

8-1/2 × 11 Portrait

14.44 Use the 8-1/2 × 11 portrait size to prepare such documents as service and reference manuals. The advantage to this size is that it provides the most printable area of the more common paper sizes. It is also wide enough that two-column manuals can be created.

ONE-COLUMN 8-1/2 × 11 PORTRAIT

8-1/2 × 11 One-Column Portrait Page Grid: Major Section Break
(dimensions in picas)

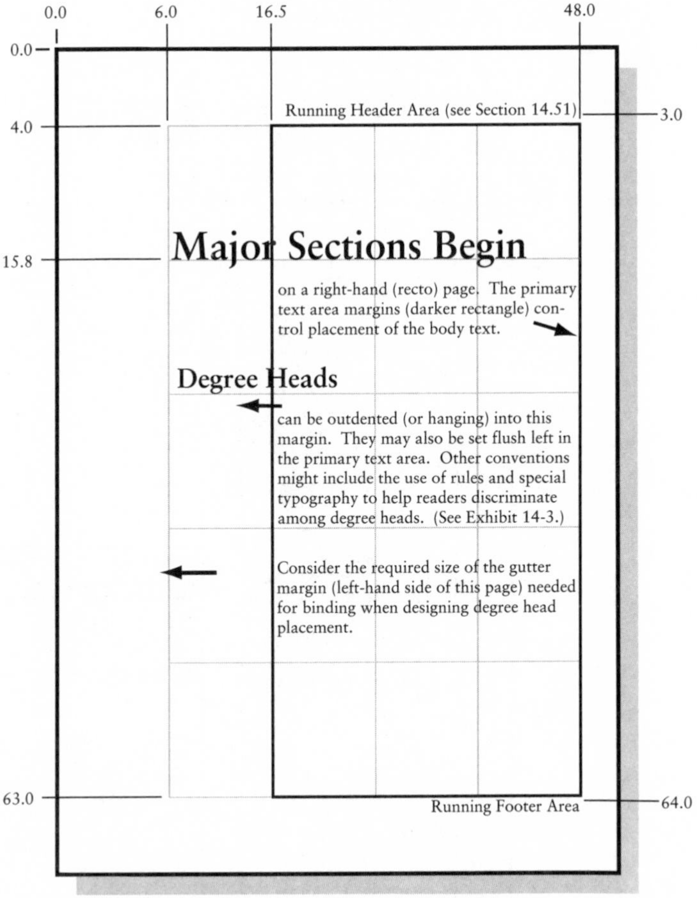

0.0
6.0
16.5
48.0
0.0
4.0
15.8
63.0
3.0
64.0
Running Header Area (see Section 14.51)
Major Sections Begin
on a right-hand (recto) page. The primary text area margins (darker rectangle) control placement of the body text.
Degree Heads
can be outdented (or hanging) into this margin. They may also be set flush left in the primary text area. Other conventions might include the use of rules and special typography to help readers discriminate among degree heads. (See Exhibit 14-3.)
Consider the required size of the gutter margin (left-hand side of this page) needed for binding when designing degree head placement.
Running Footer Area

8-1/2 × 11 One-Column Portrait Page Grid: Right-Hand (Recto)
(dimensions in picas)

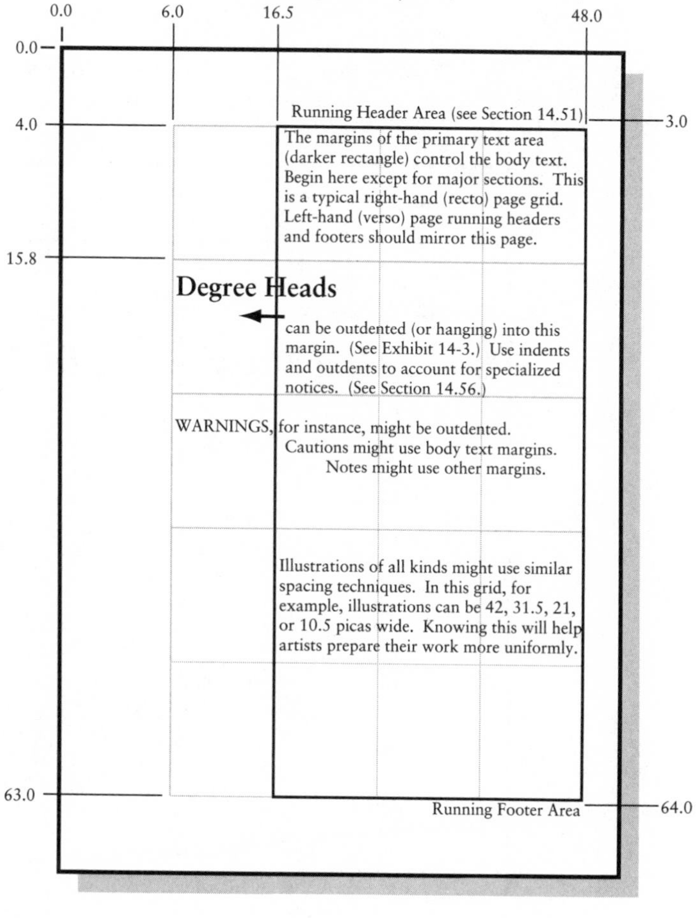

0.0
6.0
16.5
48.0
0.0
4.0
15.8
63.0
3.0
64.0
Running Header Area (see Section 14.51)
The margins of the primary text area (darker rectangle) control the body text. Begin here except for major sections. This is a typical right-hand (recto) page grid. Left-hand (verso) page running headers and footers should mirror this page.
Degree Heads
can be outdented (or hanging) into this margin. (See Exhibit 14-3.) Use indents and outdents to account for specialized notices. (See Section 14.56.)
WARNINGS, for instance, might be outdented.
Cautions might use body text margins.
Notes might use other margins.
Illustrations of all kinds might use similar spacing techniques. In this grid, for example, illustrations can be 42, 31.5, 21, or 10.5 picas wide. Knowing this will help artists prepare their work more uniformly.
Running Footer Area

Two-Column 8-1/2 × 11 Portrait

14.45 Use the two-column size to prepare such documents as service and reference manuals. The advantage to this size is that it provides the most printable area of the more common paper sizes. It also is wide enough that two-column manuals can be created.

14.46 In some instances these two-column grids can be used to present user action-system response situations. To use this technique, place the action that the user must do in the left column and the expected (and correct) results that will be encountered in the right column.

8-1/2 × 11 Two-Column Portrait Page Grid: Major Section Break
(dimensions in picas)

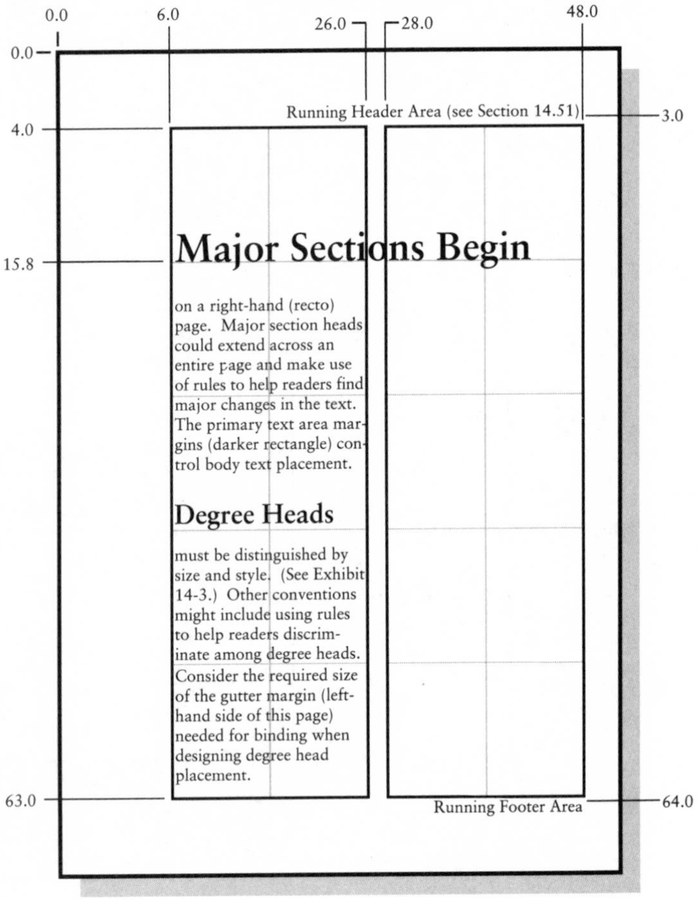

0.0
6.0
26.0
28.0
48.0
0.0
4.0
15.8
63.0
3.0
64.0
Running Header Area (see Section 14.51)
Major Sections Begin
on a right-hand (recto) page. Major section heads could extend across an entire page and make use of rules to help readers find major changes in the text. The primary text area margins (darker rectangle) control body text placement.
Degree Heads
must be distinguished by size and style. (See Exhibit 14-3.) Other conventions might include using rules to help readers discriminate among degree heads. Consider the required size of the gutter margin (left-hand side of this page) needed for binding when designing degree head placement.
Running Footer Area

14.47 A similar use for this grid can be to place supporting illustrations in one column and the text explaining the illustrations in the other. This layout is called *paired pictographs.*

14.48 The disadvantages of two-column grids include limited

- Width for degree heads,
- Width for major section heads, and
- Discrimination for specialized notices.

8-1/2 × 11 Two-Column Portrait Page Grid: Right-Hand (Recto)
(dimensions in picas)

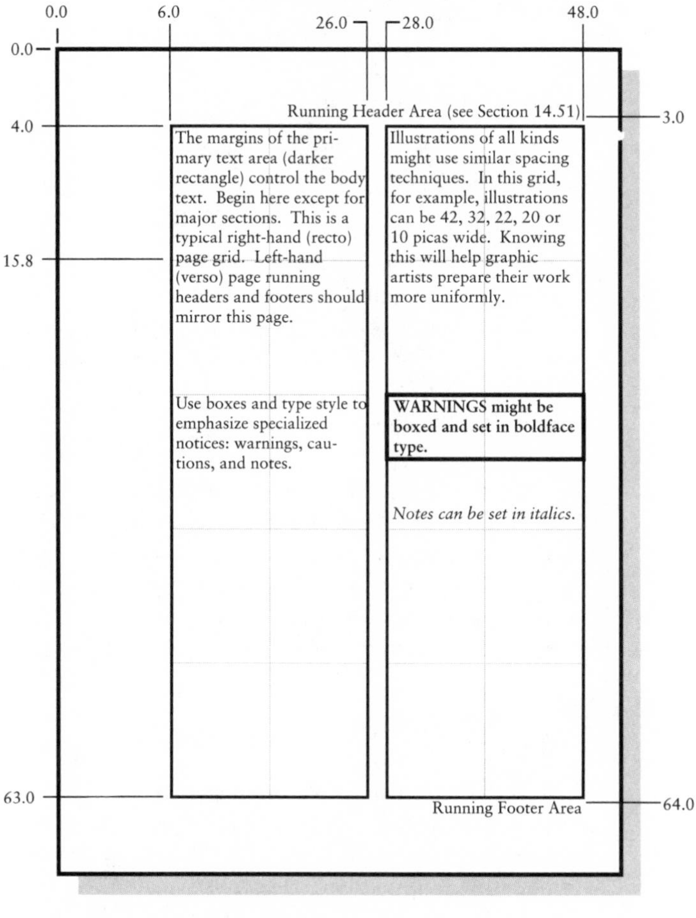

0.0
6.0
26.0
28.0
48.0
0.0
4.0
15.8
63.0
3.0
64.0
Running Header Area (see Section 14.51)
The margins of the primary text area (darker rectangle) control the body text. Begin here except for major sections. This is a typical right-hand (recto) page grid. Left-hand (verso) page running headers and footers should mirror this page.
Illustrations of all kinds might use similar spacing techniques. In this grid, for example, illustrations can be 42, 32, 22, 20 or 10 picas wide. Knowing this will help graphic artists prepare their work more uniformly.
Use boxes and type style to emphasize specialized notices: warnings, cautions, and notes.
WARNINGS might be boxed and set in boldface type.
Notes can be set in italics.
Running Footer Area

Page Numbering

14.49 Place page numbers in a visible position on the page. Page numbers usually appear either in the same place on every page (such as centered at the bottom) or in a consistent place on mirror-image pages (such as on the top left of left-hand pages and the top right of right-hand pages, with the page number always on the outside edge of the page).

14.50 Long manuals frequently use modular page numbers (1–1, 1–2, . . . , 5–1, 5–2, or 1.1, 1.2, . . . , 5.1, 5.2) rather than sequential page numbers (1, 2, 3, . . . , 19). Such a numbering system allows these manuals to be updated more easily; if information changes in only a few chapters, those chapters can be reprinted without changing the entire manual. Only the pages in the revised chapters need to be repaginated.

Running Headers and Footers

14.51 Use headers and footers to provide reference information at the top (header) and/or bottom (footer) of every page. Besides the page number, headers and footers can include

- Company logo or name,
- Document's title (or short title),
- Document's revision or printing date (or chapter or page),
- Chapter title (or short title),
- Current heading, and
- Inventory control information.

14.52 Use typeface, size, placement, and (perhaps) color in designing the header and/or footer to help the reader understand where they are in the text. Use either the same size or slightly smaller type for running headers and footers than body type. It is unnecessary to use a different type style for these page elements; their placement is a good enough separation and distinguishing feature.

Degree Heads

14.53 Use larger type sizes for headings. In addition, establish a ratio among the various heading sizes to provide useful clues for the reader seeking specific heading levels. Select type sizes at least three points different in size to distinguish headings, since the typical user cannot discriminate type size differences of less than three points:

Exhibit 14–6:
Typical Header Algorithm

Level Head	Header Type Size	Leading White Space	Trailing White Space
1	18	24	12
2	15	20	10
3	12	16	8

18 Point Level One Degree Head

15 Point Level Two Degree Head

12 Point Level Three Degree Head

18 Point Level One Degree Head

15 Point Level Two Degree Head

12 Point Level Three Degree Head

Body Text

14.54 Use 10- or 11-point serifed type for most technical documents. Avoid too many typefaces or styles in a single document. Typically, a single serifed typeface (such as Times Roman) is selected for the body text, a sans serif typeface (such as Univers or Helvetica) is used for degree heads, and one of these two types is also used for illustrations.

14.55 Include as many typographic variations as needed to emphasize specific kinds of information. Specialized notices—warnings, cautions, and notes—often need special typographic treatment.

Specialized Notices: Warnings, Cautions, and Notes

14.56 Use warnings, cautions, and notes to offer three levels of emphasis for special situations. Definitions of these components vary greatly among organizations, and are usually listed in the style standards. Some organizations use these guidelines:

- Warnings inform the reader of possible bodily injury if procedures are not followed exactly,
- Cautions alert the reader to possible equipment damage if procedures are not followed correctly, and
- Notes inform the reader either of a general rule for a procedure or of exceptions to such a rule.

14.57 Specialized notices should be designed so that readers can

- See at a glance, without any ambiguity, that these notices have a different kind of importance than either the body text or the headings.
- Distinguish among their levels of urgency. In most documents, warnings (physical danger to humans) are most urgent, then cautions (danger to the equipment), followed by notes (noncritical information of general interest).

Designers use not only typeface, type size, and color to make these distinctions, but also line width, placement, and ruled boxes.

Source Notes and References

14.58 Use a footnote or endnote when citing the authority for a statement. Because footnotes and endnotes are subordinate to the text of a manual, use smaller type sizes. See Chapters 9 and 10 for definitions, examples, and detailed information about footnotes and endnotes.

Lists

14.59 Set lists off from major text sections and use the same type conventions found in the rest of the body text. Use numbers for lists in which the reader is expected to perform an activity in a specific order. Use bullets if there is no precedence.

Illustrations, Data Displays, and Procedures

14.60 Use captions to describe the contents of an illustration or data display, and place them immediately above or below the visual. Captions are often numbered modularly according to chapter number; that is, the visuals in Chapter 2 are 2–1, 2–2, 2–3 or 2.1, 2.2, 2.3. However, some technical manuals today use captions only, not numbers.

14.61 Design decisions for these captions include

- Whether they appear above or below the visual;
- Whether the number appears on the same line as the caption text;
- How to justify the lines of the caption; and
- What typeface, type size, and placement to use.

(See also Chapters 12 and 13.)

14.62 When planning a visual, consider these guidelines

- Introduce the visual in the text that precedes it.
- Label the visual with terms used in the text.
- Leave enough white space around the visual to separate it from the text. Use a ratio similar to the heading system in the text. (See Sections 14.14 and 14.23.)
- Plan visuals so that the reader does not have to rotate the manual sideways to view them.

14.63 In designing illustrations use

- Three-dimensional line drawings to illustrate physical objects;
- Photographs to provide an accurate replica of an object and to take the place of three-dimensional drawings; and
- Two-dimensional drawings for floor plans, schematics, block diagrams, and flowcharts.

(See also Chapters 12 and 13.)

14.64 In deciding whether to use line drawings or photographs, consider

- Would a line drawing or a photograph be a more useful illustration?
- If the equipment is still under development, photographs may not be possible. The prototype may not look exactly like the final product.
- If many illustrations are needed and equipment is available, photographs may be faster and more cost-effective. Also, photographs create credibility for new products.
- Photographs require an extra step to prepare overlays with callouts of specific elements, while callouts can be drawn

directly into a line drawing. Therefore, the need for many callouts may outweigh the economy of photography.

- To call attention to a specific part of an instrument or machine (for example, to illustrate changing a diskette), a line drawing may be less distracting than a photograph that shows the entire piece of equipment.
- Similarly, if the product changes slightly, it may be possible to redraw only one or two drawings, but an entire sequence of photographs will usually need reshooting.

(See also Chapters 12 and 13.)

Revision Notices

14.65 Use revision notices for material that is updated or revised frequently. Such notices usually appear in two forms: text or revision bar. Textual revision notices often appear as part of the footer and include some corporate coding scheme that accounts for the state of each page. Revision bars are vertical rules placed in the margin that indicate which sections of a text have been changed recently.

DESIGNING A TECHNICAL MANUAL

14.66 In planning a technical manual, maintain consistency and make decisions that create a simple and straightforward format.

14.67 Consult existing style standards before designing a technical manual. Many organizations have documentation style standards for technical manuals that specify the format and design for most document components. However, most standards leave some format decisions to the author; when possible, make such decisions to coordinate with your organization's standards.

Specific Format Considerations

14.68 In designing the page, also consider these format issues:

- Navigation aids (headers, footers, and headings),
- Reader access,
- Front cover,
- Front matter,
- Chapter or section divisions,
- Appendixes,
- Glossary,

- Indexes,
- Bibliography, and
- Back cover.

Navigation Aids

14.69 Provide navigation aids (headers, footers, and headings) to help readers skim or quickly find their place in a manual.

14.70 Use headers and footers in a manual to provide reference information at the top (header) and/or bottom (footer) of every page. Besides the page number, headers and footers can include

- Company logo or name,
- Manual's title,
- Manual's revision or printing date (or chapter or page),
- Chapter title (abbreviated, if necessary),
- Current heading, and
- Inventory control information.

Design decisions for navigation aids include whether headers, footers, or both are used, and their typeface, type size, color, and placement.

14.71 Use headings (headlines) as navigation aids. Document organization determines how many heading levels each document needs. The design decisions—typeface, type size, color, and placement—for each heading level should make it easy to tell headings from text, and clarify their subordination levels. (See Section 14.29.)

Other Reader Access Decisions

14.72 Provide reader access to book sections. This access is based on such questions as

- Will the manual have separator tabs?
- Will such tabs be physical tabbed pages, or will they be blind tabs printed on the edges of the text pages?
- Will there be tabs for every chapter and appendix, or do the chapters and appendixes fall into groups, with a tab per group?
- What text will appear on each tab?
- Are the chapter titles short enough to fit? If not, how can they be abbreviated?

14.73 Print a table of contents for each chapter on the tab page that begins a new section.

Front Cover

14.74 Use the front cover to establish the company's identity, the content of the document, and the design features found in the document. For instance, if the document itself is set 5×8 landscape on a two-column grid, use those same grid dimensions to place elements on the cover. This establishes a sense of consistency and continuity from the outset.

Front Matter

14.75 Pages that appear before the first chapter or section in a technical manual are called *front matter.* Almost all organizations specify the content and format of front matter. Front matter can include

- Title page,
- Copyright page/trademark page,
- Acknowledgments,
- Abstract or Preface,
- Table of Contents,
- List of Figures, and/or
- About This Manual.

14.76 Make format decisions for the chapter pages and the cover first, then plan the front matter to be as consistent as possible with these components. Number front matter pages with Roman numerals, beginning on the back of the title page with the number ii.

Title Page

14.77 Documents often have a covering title page, often called a half title page, that offers abbreviated publication information. Typically, this page cites a short title, author's name, and publisher.

14.78 Use a title page to list the name, date, and publication number of the manual. (The publication number sometimes appears on the copyright/trademark page.) The title page of some technical documents and scientific reports often includes approval signatures granted by various administrators in charge of projects.

Copyright/Trademark Page

14.79 Provide a copyright/trademark page to display the manual's copyright notice and list of the registered trademarks used in the manual.

Acknowledgments

14.80 Use an acknowledgments page to list the contributors to the manual. (Alternatively, they may appear on the copyright/trademark page or as part of the preface.)

Abstract or Preface

14.81 Include a preface to define the manual's purpose, intended audience, organization, and prerequisites (if any) for using the manual.

Table of Contents

14.82 Use a table of contents to present the contents of the manual, according to chapters, sections, subsections, appendixes, glossary, and index. (Lists of figures and/or tables are sometimes presented separately.)

List of Figures

14.83 Use a list of figures and/or tables to catalog all the visual contents of the manual, according to chapters, sections, subsections, and appendixes.

About This Manual

14.84 Use this introduction to present an overview of its contents. (Often the introduction is the first chapter of a manual.)

14.85 Use the introduction to offer specialized instructions on how to use the manual or a summary of conventions used in the manual (such as warning, caution, and note practices).

Chapter or Section Divisions

14.86 Chapters can be considered navigation aids in a very broad sense, because they help readers locate common information.

Appendixes

14.87 Place multiple appendixes in the same order they are referred to in the body of the manual. Appendixes contain details that might break up the flow of information in a chapter or section. Long tables, error messages, and algorithms usually appear in appendixes.

14.88 Design decisions for chapters and appendixes include the layout for the opening page of each chapter or appendix, the layout for continuing pages, and the numbering method. Design the opening page of a chapter or appendix to feature its title.

14.89 Number chapters and identify appendixes with letters. Make specific design decisions for typeface, type size, and capitalization of the numbers and letters.

Bibliography

14.90 Include a bibliography for those works that rely on previous research. See Chapter 10 for information on preparing bibliographic entries.

Glossary

14.91 Design a glossary to be consistent with the rest of the text. A glossary is a special kind of chapter or appendix, with the same alphabetical access as an index.

Indexes

14.92 Design decisions for indexes should reflect the information-access nature of an index: locating index entries quickly should be the goal. Many indexes use multiple columns and smaller type than the body text. For greater simplicity, maintain as much as possible of the chapter page design. (See also Chapter 11.)

Back Cover

14.93 Continue any design principles onto the back cover if there is anything printed on that surface.

DESIGNING BROCHURES AND NEWSLETTERS

14.94 Well-designed brochures and newsletters encourage people to invest the time and effort to read them. On first glance, readers can find the document attractive and professional—or the same text can look dull, technical, or amateurish.

14.95 When preparing a brochure or newsletter, consider the audience profile, objectives, and needs:

- What purpose will the document serve? Is it a marketing tool? Will it educate customers or users?
- What is the personality or character of the product (or subject) and its market (or audience)?
- How much information must be presented? How many illustrations are needed, and what kind?
- What is the most important part of the document? What is the least important part?
- Will the document be used in-house (entirely by people within an organization), outside, or both?
- Does the organization have an established corporate identity? Are there corporate colors, logos, and styles?

(See Sections 14.3 and/or 14.66.)

Brochure (7 × 10 Portrait, one fold)

14.96 Use brochures when the shelf life of a document is expected to be short. Brochures advertise or summarize a product or service. This exhibit offers one example of the many sizes in which brochures can be prepared.

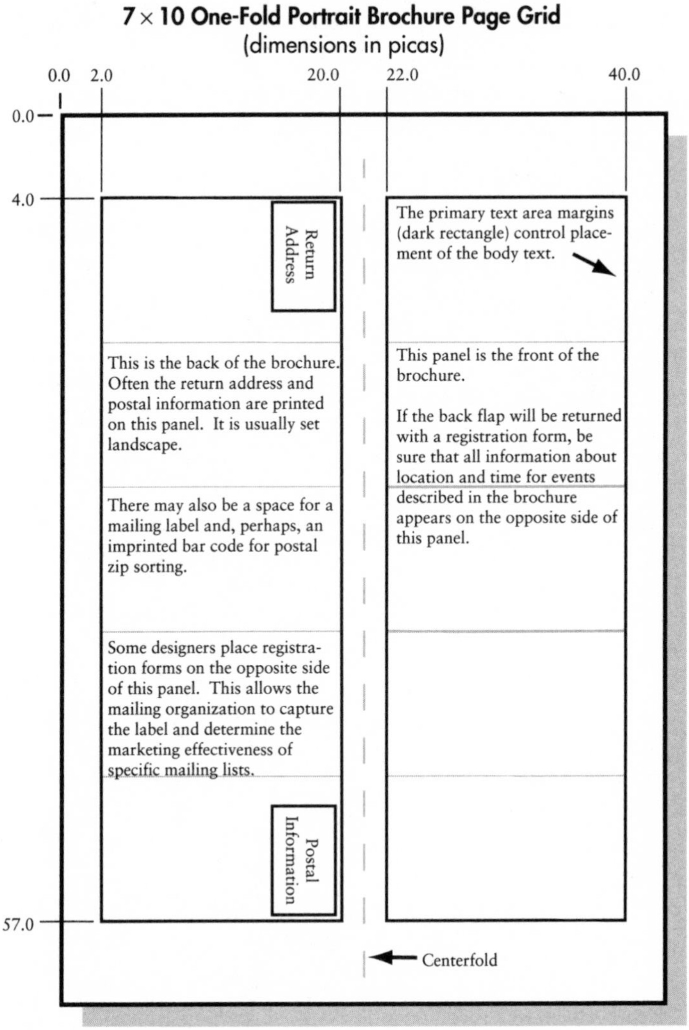

7 × 10 One-Fold Portrait Brochure Page Grid
(dimensions in picas)
0.0
2.0
20.0
22.0
40.0
0.0
4.0
57.0
Return Address
This is the back of the brochure. Often the return address and postal information are printed on this panel. It is usually set landscape.
There may also be a space for a mailing label and, perhaps, an imprinted bar code for postal zip sorting.
Some designers place registration forms on the opposite side of this panel. This allows the mailing organization to capture the label and determine the marketing effectiveness of specific mailing lists.
Postal Information
The primary text area margins (dark rectangle) control placement of the body text.
This panel is the front of the brochure.
If the back flap will be returned with a registration form, be sure that all information about location and time for events described in the brochure appears on the opposite side of this panel.
Centerfold

Newsletter (8 × 10 Portrait)

14.97 Use newsletters when the shelf life of a document is expected to be short. Newsletters convey news about an organization. This exhibit offers one example of the many sizes in which newsletters can be prepared.

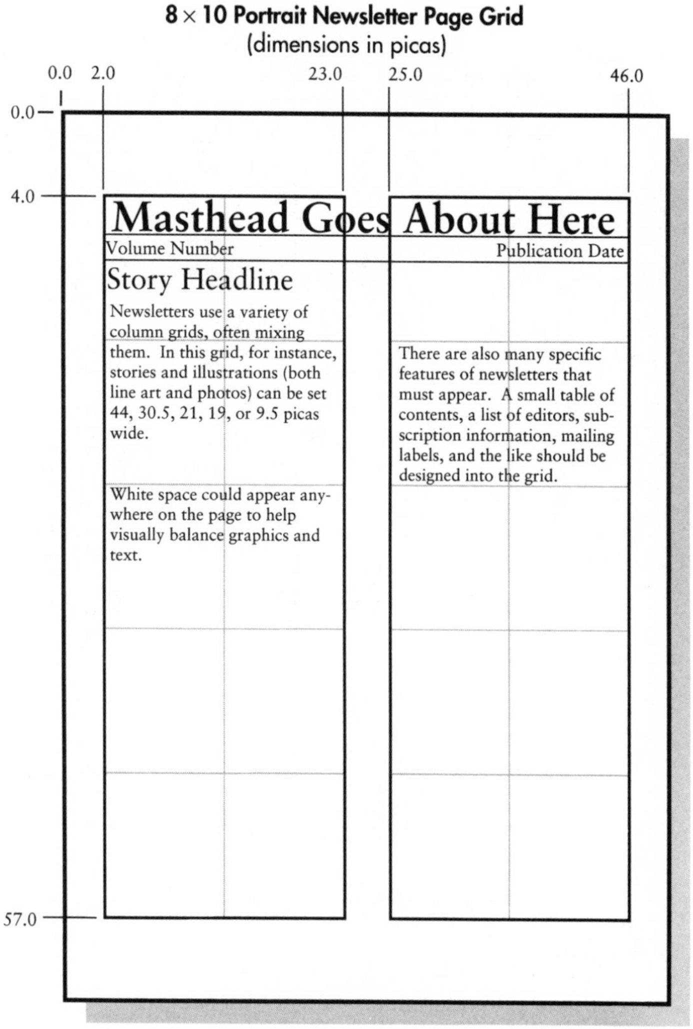

8 × 10 Portrait Newsletter Page Grid
(dimensions in picas)
0.0
2.0
23.0
25.0
46.0
4.0
57.0
Masthead Goes About Here
Volume Number
Publication Date
Story Headline
Newsletters use a variety of column grids, often mixing them. In this grid, for instance, stories and illustrations (both line art and photos) can be set 44, 30.5, 21, 19, or 9.5 picas wide.
White space could appear anywhere on the page to help visually balance graphics and text.
There are also many specific features of newsletters that must appear. A small table of contents, a list of editors, subscription information, mailing labels, and the like should be designed into the grid.

DESIGNING A4 FOREIGN PUBLICATIONS

14.98 Use A4 (8-1⁄4 × 11-3⁄4 inches or 49.5 × 70.5 picas) paper for documents intended for a foreign market. This exhibit provides a suggested grid for a typical technical document.

A4 One-Column Portrait Page Grid: Major Section Break
(dimensions in picas)

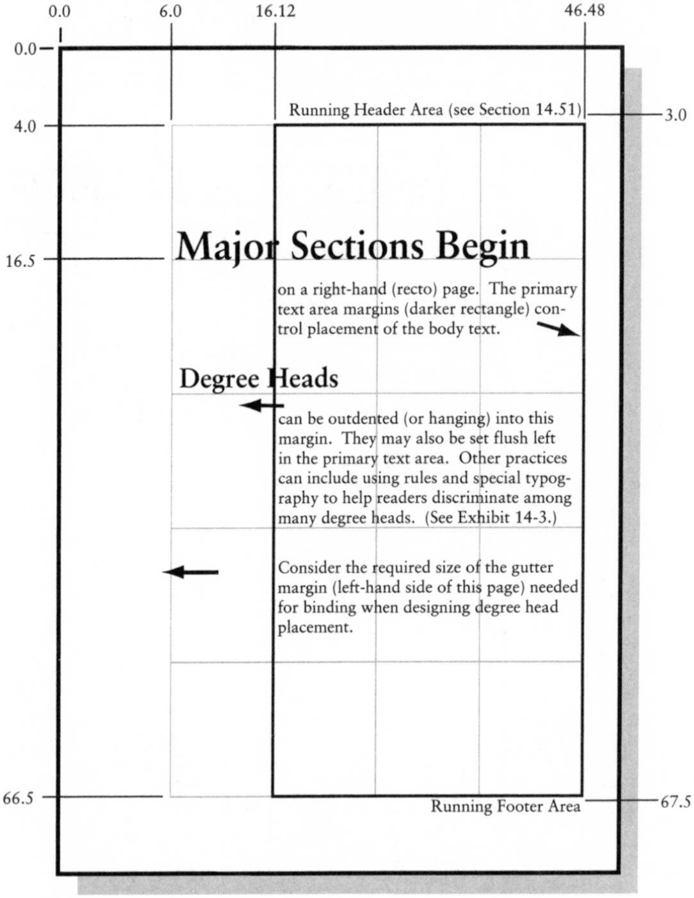
0.0
6.0
16.12
46.48
0.0
4.0
16.5
66.5
3.0
67.5
Running Header Area (see Section 14.51)
Major Sections Begin
on a right-hand (recto) page. The primary text area margins (darker rectangle) control placement of the body text.
Degree Heads
can be outdented (or hanging) into this margin. They may also be set flush left in the primary text area. Other practices can include using rules and special typography to help readers discriminate among many degree heads. (See Exhibit 14-3.)
Consider the required size of the gutter margin (left-hand side of this page) needed for binding when designing degree head placement.
Running Footer Area

PREPARING A PAPER OR ARTICLE FOR PUBLICATION

14.99 Before writing an article, select a potential journal or magazine, and prepare the article according to that publication's requirements. Because there are so many journals and magazines, and because they cover such a variety of subject matter, it is impossible to provide a single set of rules to follow for preparing a manuscript. However, the elements described next are considered to be standard parts of most papers and articles. (See also Chapters 9 and 10.)

Title

14.100 Create a specific title that describes the paper's content and emphasis. There is a fine balance between brevity and descriptive accuracy: a three-word title may be too cryptic, but a 14-word title is probably too long.

14.101 Place the byline beneath the title. Some journals require the name of the institution and city in which the author conducted the research. Use the same style as other bylines in the journal:

Applying Television Research to On-line Information:
A Case Study

Carbon-dating Evidence Based on Mesoamerican
Cooking Utensils

Abstract

14.102 Provide an abstract for academic journals that summarizes the work reported in the article; most trade journals do not require an abstract. Unless the journal must approve an abstract before considering the article, write the abstract last to insure that it reflects the article's content.

14.103 Keep the abstract succinct. Most journals generally accept a 200-word abstract; check the publication's requirements to verify this limit. A typical abstract might read:

> This article examines the genealogy of recent innovations in online information and hypermedia. In doing so it provides an overview of the more important developments and applications of these advances. While the essay celebrates those successes

that are clearly exceptional, it also assesses the more typical, and sometimes equally useful, techniques found in less dramatic and innovative environments. Finally, the essay asks the reader to consider some of the more optimistic scenarios for the future of on-line information and hypermedia in relation to the concept of literacy.

Introduction

14.104 In the introduction, clearly state the problem or theory. Indicate the significance of the work in the context of what is already known. If appropriate, provide a brief review of existing literature that is relevant to the topic.

Body

14.105 Whatever the journal type, use headings and subheadings to help readers understand the information's structure and content. The body's organization and length depend on the subject matter and the journal. Many academic journals, for instance, have strict organizational rules.

When appropriate, include figures and tables to improve clarity. (See Chapter 13.)

Conclusion

14.106 Most academic journals require a conclusion or summary. Make sure such conclusions are interpretive and not repetitious. If the research or problem was not solved, suggesting further study is appropriate. Most trade journals do not require formal conclusions; however, they can be used to stress what your article demonstrates.

References

14.107 Methods for citing references in text differ widely, so check recent issues of a journal for preferences. (See also Chapters 9 and 10.)

PREPARING BI- AND MULTILINGUAL DOCUMENTS

14.108 Prepare documents that include more than one language in one of the following formats:

- Separate publications,
- Separate joined publications,
- Two-column side-by-side publications, and
- Multicolumn side-by-side publications.

14.109 In deciding which technique to use, be sure to consider the difference in length for texts written in various languages. Typically, most foreign languages require more pages than English for the same text.

Separate Publications

14.110 Use separate publications, two separate documents, when many copies will be printed in the different languages. Use this technique only when frequent revisions are not expected.

Separate Joined Publications

14.111 Use separate joined publications when two languages must appear in a single publication. Typically, these documents are printed in normal orientation (front to back) or tumbled. A tumbled document has two front covers: one in the first language, the other in the second. Where the text in one language ends, the text for the second language appears upside down. To read this section of the document, the book must be "tumbled," or turned. Tumbled orientation is often impractical for multilanguage documents, although there are some examples of its use in short documents provided with consumer products.

Two-Column Side-by-Side Publications

14.112 Use two-column side-by-side publications when the content of concurrent revisions must be seen by all readers.

A4 Two-Column Side-by-Side Bilingual Publication
(dimensions in picas)

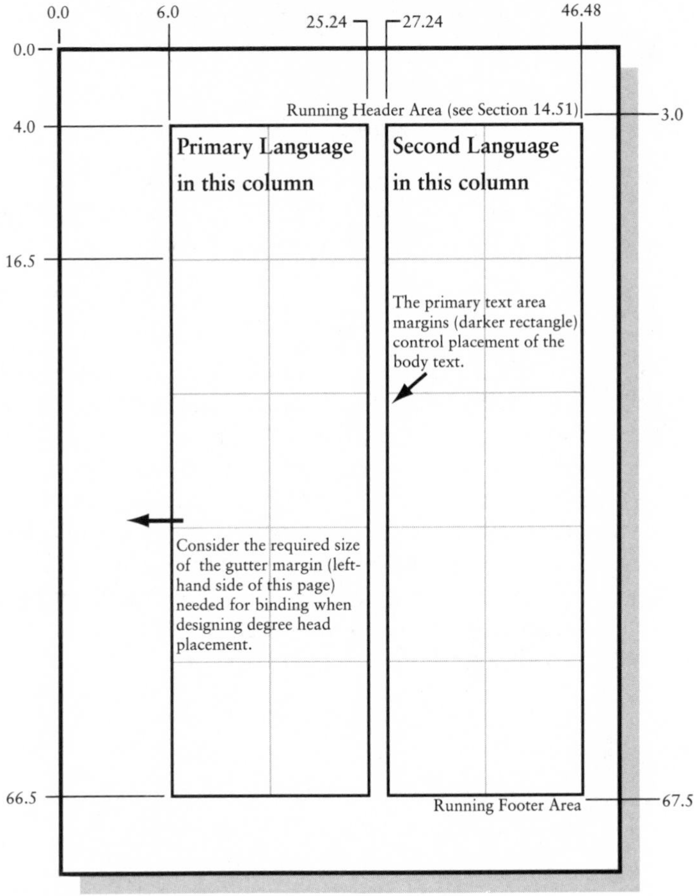
0.0
6.0
25.24
27.24
46.48
0.0
4.0
16.5
66.5
3.0
67.5
Running Header Area (see Section 14.51)
Primary Language in this column
Second Language in this column
The primary text area margins (darker rectangle) control placement of the body text.
Consider the required size of the gutter margin (left-hand side of this page) needed for binding when designing degree head placement.
Running Footer Area

Multicolumn Side-by-Side Publications

14.113 Use multicolumn side-by-side publications when many languages must be present in a document. This technique is often seen in directions for small consumer products.

A4 Multicolumn Side-by-Side Multilingual Publication
(dimensions in picas)

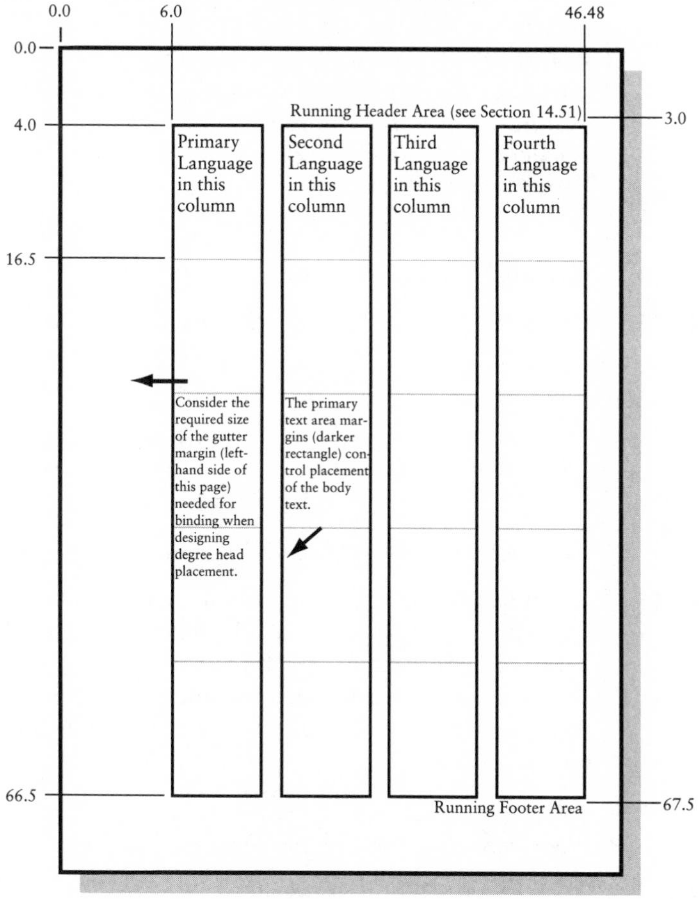

CONTROLLING LARGE DOCUMENT SETS

14.114 Organize a document and design its format to accommodate all the readers and all the information. This effort may sometimes mean creating document sets (or suites) rather than a single document. Few readers, for instance, need to understand every detail of the installation, use, operation, preventive maintenance, troubleshooting, and repair of a large product. Consider the documentation needs of all the people who play a role in flying a large jetliner. The pilots who fly the aircraft, the ground crew who check out the system before flight, the personnel who perform routine maintenance and repairs, the flight attendants who insure passenger comfort and safety—all need different types of information, which should be in separate documents.

Categorizing Information

14.115 Categorize information for a document set based on

- Audience,
- Frequency of use,
- Task, and
- System configuration.

Audience Analysis to Classify Documents

14.116 Use audience analysis to identify documents based on who needs what information; audience identification is usually combined with one of the other three approaches. For example, an aircraft document set might include a system overview for all readers and then one or more documents each for pilots only, for ground crew only, and so on.

Frequency of Use to Classify Documents

14.117 Use frequency of use to assign information to documents based on how often readers need that information. Readers generally need installation information only once and troubleshooting information only occasionally. On the other hand, they may refer every day to information about use, operation, and preventive maintenance. For example, extensive training prepares pilots to operate their aircraft without referring to a flight manual, but they do complete a documented preflight checklist before every takeoff.

Task Orientation to Classify Documents

14.118 Use task orientation to focus on procedural information that different groups of readers need to do their jobs. Answering the following questions can help identify the essential information to include in various task-oriented documents:

- Who performs the task?
- What action begins the task?
- What steps are required to perform the task?
- What action ends the task?
- Can product variations affect the task or its results? If so, what are the variations and their effects?
- How many ways can the task be performed? What are the benefits of each method? Should each method be documented?

System Configuration to Classify Documents

14.119 Use system configuration to place information in documents according to the structure or features of the product. For example, the document set for the jetliner's maintenance personnel could include separate documents about the engines, landing gear, and so on.

Making Information Accessible

14.120 To help readers find the information they need, include overviews, master indexes, and master glossaries.

14.121 An overview describes the paths readers can take through the information. For example, a novice audience might be required to read certain documents, while an expert might not need that same information. Certain documents may contain the task information most readers want most of the time, while other documents are better for first-time or reference use.

14.122 Overview design may be as simple or as complicated as needed for a specific audience and document. Simple approaches include using tables or flowcharts to list types of information and/or tasks, along with the names of the documents that contain them. A more sophisticated overview might use a graphic presentation to show how the document set is organized.

14.123 Include an index for every document in a set, unless the index has very few entries or is unlikely to be used more than once. Also include a master index to the document set to help readers find out which documents contain which information. Readers can then use the individual document indexes to look up a particular type of information in detail. Some document sets include the master index in every document, but that approach usually is not practical for large document sets, especially in technical material that is updated or revised frequently.

14.124 Similarly, include a master glossary to define terms for a topic, product, or system that uses a technical or specialized vocabulary. Individual documents could also include glossaries to define only the terms introduced in that document.

14.125 Master indexes and glossaries may duplicate some information. The convenience to readers justifies this duplication and outweighs the additional cost.

Graphics Production Considerations

14.126 Producing documentation includes

- Preparing final camera-ready pages,
- Choosing the paper,
- Choosing the binding method, and
- Arranging the printing of the document.

Preparing Camera-Ready Pages

14.127 When a document is in its final form, it can be prepared for production. Follow all the graphic design principles discussed earlier and create the final pages. These pages are called *camera ready* because, in most printing processes, the first process the printer does is to photograph them.

14.128 Camera-ready pages (also called mechanicals) use the page size, typefaces, type styles, type sizes, line spacing, line width, and illustration specifications decided during the design process. Depending on the text-processing system available, text and illustrations may be combined on the pages for the first time at this point.

14.129 Camera-ready pages can have many specialized requirements, depending on the illustrations, paper type, binding method, and printing production process chosen for the document.

Choosing Paper

14.130 The paper used for printing brochures, booklets, newsletters, and manuals varies in several ways. Carefully consider the following factors:

- Durability—the ability of the paper to withstand wear based on the ways in which readers will use the document,
- Color and/or brightness—some documents may be read under varying lighting conditions that may make text difficult to read when printed on colored paper,
- Finish—special textures and finishes may be needed for specific purposes (glossy paper can be used for printing accurate photographs; textured paper can be used to present a professional feeling for an annual report),
- Weight—a proper paper weight will help a document survive repeated use (bending, folding, and the like),
- Opacity—opaque papers are generally inappropriate for technical information because the image can be seen through the page,
- Cost—since most technical and scientific information is not sold as a separate product, the cost of its production must be justified.

14.131 Designers, printers, and paper companies are all good sources of information about paper selection. Often a printer maintains a supply of floor stock paper. Using floor stock can save money, if the paper is suitable for the printing job. (See Section 14.11.)

14.132 Among the most popular types are bond paper, coated paper, uncoated paper, and textured paper. Except for the bond paper, these papers come in book and cover weight:

- Bond paper, used primarily for letterhead and correspondence, is very strong. There are two types of bond paper: plain bond and bond with a rag or cotton content. Bond with a rag or cotton content is a higher quality paper than plain bond, and generally has a more attractive finish. It is also more expensive than plain bond.
- Coated paper, which has a glossy or dull enamel finish, is a good choice when printing photographs. However, coated paper is usually not used for documents containing large amounts of text, because it creates glare.
- Uncoated paper, which is less expensive than either coated or textured paper, has either antique or smooth finishes, and is available in various weights. Offset paper, typically used in quick-printing applications, is a type of uncoated paper.

- Textured paper—a high-quality, uncoated paper—is available in a variety of colors and finishes.

Binding the Document

14.133 Binding refers to how the pages of a document are held together. The type of binding is a particularly important decision for science and technical texts because many of these documents must be used under working conditions, and they must be updated often. There are three basic binding methods:

- Wire stitching,
- Mechanical binding, and
- Perfect binding.

14.134 Each binding method has many preprinting requirements for how the final pages should be prepared. When planning documents without a graphic designer, talk to one or more printing companies before making final decisions about page size and margins.

WIRE STITCHING

14.135 Wire stitching is the most common binding method for short documents such as pamphlets, booklets, and newsletters; it is also the simplest and least expensive. In wire (or saddle) stitching, the document is stapled along the center fold. For example, to make an 8-1/2″ × 11″ saddle-stitched document, a set of 11″ × 17″ sheets of paper are folded in half and stapled at the center.

14.136 A less common wire stitching method—side stitching—punches staples through all of the pages along the spine from front to back cover. While this is a relatively inexpensive binding method, the staple ends damage any surrounding documents on a shelf.

MECHANICAL BINDING

14.137 Mechanical binding—loose-leaf ring or coils—allows documents to lie flat when open. The primary difference is that the footprint, the total area that the open document occupies in the workplace, is usually larger for loose-leaf, ring-bound books than coil bound (which can fold back on themselves).

14.138 Loose-leaf ring binding, in which the printer punches or drills holes into the margins of the pages so that the document can be inserted into a loose-leaf binder, is one form of mechanical binding. Loose-leaf binding works better than all other binding methods for large documents because it allows for easy updating.

14.139 The other methods of mechanical binding are wire-o, spiral, and plastic comb. In wire-o and spiral binding, holes are drilled through the covers and pages, which are then bound together by a wire or plastic coil. These documents can be folded back on themselves and still lie flat, creating a small document footprint. Wire-o and spiral binding are permanent binding methods, most effective for documents less than 3⁄4-inch thick that will not need updating.

14.140 In comb binding, holes are punched in the covers and pages and a plastic binding comb is inserted. Low-cost equipment for comb binding is available for offices and copy centers. Comb-bound documents lie flat when open, but do not fold back on themselves easily to create a single-page footprint. Updating pages in comb-bound documents is possible but rarely done in practice, because removing and replacing the combs is difficult to do without damaging the document.

Perfect Binding

14.141 In perfect binding, assembled pages are trimmed along the back edge and bound with a hot, plastic glue. A preprinted cover is then glued to the spine. Paperback books are perfect bound; in large quantities, perfect binding is relatively inexpensive.

14.142 The disadvantages of perfect binding are the increased binding time and the fact that perfect-bound books do not lie flat when opened. Also, most printing companies do not have perfect-binding capabilities and must subcontract with a specialty firm, which requires additional processing time.

Bibliography

1 Audience

Barr, J. P. "Promoting the Value of Quality Technical Publications." *Proceedings of the 35th International Technical Communication Conference.* Philadelphia: Society for Technical Communication, 1988.

Bell, P. *High-Tech Writing.* New York: John Wiley & Sons, 1985.

Brockmann, R. J. *Writing Better Computer User Documentation: From Paper to Online.* New York: John Wiley & Sons, 1986.

Caernarven-Smith, P. *Audience Analysis and Response.* Pembroke, MA: Firman Technical Publications, 1983.

Cohen, G., and D. H. Cunningham. *Creating Technical Manuals.* New York: McGraw-Hill, 1984.

Huth, E. J., et al. *Council of Biology Editors Style Manual.* Arlington, VA: Council of Biology Editors, 1960.

O'Rourke, J. *Writing for the Reader.* Maynard, MA: Digital Equipment Corp., 1976.

Price, J. *How to Write a Computer Manual: A Handbook of Software Documentation.* Menlo Park, CA: Benjamin/Cummings Publishing Co., 1984.

Rosenbaum, S., and R. D. Walters. "Audience Diversity: A Major Challenge in Computer Documentation," *IEEE Transactions on Professional Communications,* PC29.4 (1986).

Tichy, H. J. *Effective Writing for Engineers, Managers, Scientists.* New York: John Wiley & Sons, 1966.

Van Buren, R., and M. F. Buehler. *The Levels of Edit,* rev. ed. Pasadena, CA: Jet Propulsion Laboratory, 1980.

2 Grammar

Aaronson, S. "Style in Scientific Writing." *Current Contents* 2 (1977): 6–15.

Alexander, P. *Sensationalism and Scientific Explanation.* London: Routledge & Kegan Paul, 1963.

Brusaw, C. T., G. J. Alred, and W. E. Oliu. *Handbook of Technical Writing.* New York: St. Martin's Press, 1982.

Dodd, J. S., ed. *The ACS Style Guide: A Manual for Authors and Editors.* Washington, DC: American Chemical Society, 1980.

Hudleston, R. D., et al. *Sentence and Clause in Scientific English.* Washington, DC: Office of Scientific and Technical Information Report #5030, 1968.

Musgrave, A. E. "Explanation, Description and Scientific Realism." *Scientia* 112 (1977): 727–741.

Schoenfeld, R. *The Chemist's English.* D-6940 Weinheim: VCH Verlags GmbH, 1986.

3 Punctuation

AIP Style Manual, 3rd ed. New York: American Institute of Physics, 1978.

Brittain, R. *Punctuation for Clarity.* New York: Barnes & Noble, 1950.

The Chicago Manual of Style, 13th ed. Chicago: University of Chicago Press, 1987.

Council of Biology Editors Style Manual, 5th ed. Bethesda, MD: Council of Biology Editors, 1983.

Government Printing Office Style Manual. Washington, DC: United States Government Printing Office, 1984.

Knuth, D. E. *T_EX and METAFONT: New Directions in Typesetting.* Bedford, MA: Digital Press, 1979.

Montgomery, M., and J. Stratton. *The Writer's Hotline Handbook.* New York: New American Library, 1981.

4 Spelling

Borror, D. J. *Dictionary of Word Roots and Combining Forms.* Palo Alto, CA: Mayfield Publishing Co., 1960.

Collocott, T. D., and A. B. Dobson, eds. *Dictionary of Science and Technology.* Edinburgh, Scotland: W & R Chambers, 1982.

The Condensed Chemical Dictionary, 10th ed. New York: Van Nostrand Reinhold, 1981.

Dictionary of Computing. New York: Oxford University Press, 1983.

Fowler, H. W. *A Dictionary of Modern English Usage,* 2nd ed. Revised and edited by E. Gowers. New York: Oxford University Press, 1965.

Hutchinson, L. I. *Standard Handbook for Secretaries,* 8th ed. New York: McGraw-Hill, 1979.

Lopedes, D. N., ed. *McGraw-Hill Dictionary of the Life Sciences.* New York: McGraw-Hill, 1976.

Parker, S. P., ed. *McGraw-Hill Dictionary of Scientific and Technical Terms,* 3rd ed. New York: McGraw-Hill Book Company, 1984.

Stein, J., ed. *The Random House Dictionary of the English Language* (unabridged). New York: Random House, 1967.

Strunk, W., Jr., and E. B. White. *The Elements of Style,* 3rd ed. New York: Macmillan, 1979.

Turabian, K. L. *A Manual for Writers of Term Papers, Theses, and Dissertations,* 4th ed. Chicago: University of Chicago Press, 1973.

12,000 Words—A Supplement to Webster's Third New International Dictionary. Springfield, MA: G. & C. Merriam, 1986.

Webster's New Collegiate Dictionary. Springfield, MA: G. & C. Merriam, 1960.

Webster's Third New International Dictionary (unabridged). Springfield, MA: G. & C. Merriam, 1961.

The Word Book II. Boston: Houghton Mifflin, 1983.

Words into Type, 3rd ed. Englewood Cliffs, NJ: Prentice-Hall, 1974.

5 Abbreviations

Abridged Index Medicus. Washington, DC: U.S. Government Printing Office, 1984.

Acronyms, Initialisms and Abbreviations Dictionary, 10th ed. Vols. 1–3. Detroit: Gale Research, 1987.

American Society of Mechanical Engineers. *Abbreviations for Use on Drawings and in Text. ANSI Y1.1-1972.* New York: United Engineering Center, 1972.

Bowden, D. D. "Abregrams or Deleweds: A Forty-Year View of Acronyms." In *International Professional Communication Conference Record: A Byte into the Future.* Winnipeg, Canada: IEEE Professional Communication Society, 1987.

Cumulated Abridged Index Medicus. Washington, DC: U.S. Government Printing Office, 1984.

Davis, N. M. *Medical Abbreviations: 5500 Conveniences at the Expense of Communications and Safety,* 4th ed. Huntington Valley, PA: Neil M. Davis Associates, 1988.

Dukes, E. "Abbreviations, Especially Symbols for Units of Measure in Technical Documentation." In *International Professional Communication Conference Record: Linking Technology and the User.* Charlotte, NC: IEEE Professional Communication Society, 1986.

———. "Quality in Terms of Symbols for Units of Measure." In *Proceedings of the 34th International Technical Communication Conference.* Washington, DC: Society for Technical Communication, 1987.

———. "Symbols for Units of Measure and Other Abbreviations: A Comparison of Some Technical Style Guides." In *Proceedings of the 34th International Technical Communication Conference.* Washington, DC: Society for Technical Communication, 1987.

Goldman, D. T., and R. J. Bell, eds. *The International System of Units (SI).* Washington, DC: National Bureau of Standards, 1986.

Hamilton, B., and B. Guidos. *MASA: Medical Acronyms, Symbols, & Abbreviations,* 2nd ed. New York: Neal-Schuman, 1988.

Hathwell, D., and A. W. K. Metzner, eds. *Style Manual: For Guidance in the Preparation of Papers for Journals Published by the American Institute of Physics and Its Member Societies,* 3rd ed. New York: American Institute of Physics, 1978.

Herb, U., and H. Keller. *The Dictionary of Engineering Acronyms and Abbreviations.* New York: Neal-Schuman, 1988.

IEEE Standard Letter Symbols for Quantities Used in Electrical Science and Electrical Engineering, ANSI/IEEE Standard 280-1985. New York: Institute of Electrical and Electronic Engineers, 1984.

IEEE Standard Letter Symbols for Units of Measurement, ANSI/IEEE Standard 260-1978. New York: Institute of Electrical and Electronic Engineers, 1985.

Index Medicus. Washington, DC: U.S. Government Printing Office, 1984 (published annually).

Iverson, C., et al. *American Medical Association Manual of Style,* 8th ed. Baltimore, MD: Williams & Wilkins, 1989.

6 Specialized Terminology

American Men and Women of Science. New York: R. R. Bowker, 1971–1976.

Geological Survey. *The National Atlas of the United States of America.* Washington, DC: U.S. Department of the Interior, 1978.

Gooch, A. and A. Garcia de Paredes, eds. *Cassell's Spanish English Dictionary.* London, England: Cassell, 1978.

IEEE Standard Dictionary of Electrical and Electronic Terms, ANSI/IEEE Standard 100-1984. New York: Institute of Electrical and Electronic Engineers, 1984.

Marks, R. W. *The New Dictionary and Handbook of Aerospace.* New York: Frederick A. Praeger, 1969.

Measures, H. *Styles of Address.* New York: St. Martin's Press, 1970.

Murdin, P., and D. A. Allen. *Catalogue of the Universe.* Cambridge, England: Cambridge University Press, 1980.

Oxford University Press. *Dictionary of Computing.* Oxford, England: Oxford Science Publications, 1986.

Rosenberg, J. M. *Dictionary of Computers, Information Processing and Telecommunications.* New York: John Wiley & Sons, 1987.

The Times Atlas of the World. London, England: John Bartholemew & Sons and Times Books, 1985.

Webster's Biographical Dictionary. Springfield, MA: G & C Merriam, 1980.

Who's Who. London, England: A & C Black (published annually).

Who's Who in Technology Today. Pittsburgh, PA: Recognition Corp., 1984.

7 Numbers and Symbols

Geological Survey. *Suggestions to Authors of the Reports of the United States Geological Survey,* 6th ed. Washington, DC: U.S. Government Printing Office, 1978.

Handbook of Chemistry and Physics, 67th ed. Cleveland, OH: Chemical Rubber Co., 1986–1987.

American National Standard Metric Practice, ANSI/IEEE Std 268-1982. New York, NY: Institute of Electrical and Electronic Engineers, 1982.

Mathematics Dictionary, 4th ed. New York: Van Nostrand Reinhold, 1976.

Murphy, P. W. *Typography for Mathematics (UCAR-10103).* Livermore, CA: Lawrence Livermore National Laboratory, 1983.

Swanson, E. *Mathematics into Type,* rev. ed. Providence, RI: American Mathematical Society, 1979.

Tinker, M. A. "How Formulae Are Read." *American Journal of Psychology* 40 (July 1928): 476–483.

———. "Legibility of Mathematical Tables." *Journal of Applied Psychology* 44.2 (1960): 83–87.

8 Nonnative Readers

Gingras, B. "Simplified English in Maintenance Manuals." *Technical Communication* 34.1 (1987): 24–28.

Gubidha, B. *Study on the Language Barrier in the Production, Dissemination and Use of Scientific and Technical Information with Special Reference to the Problems of Developing Countries.* New York: UNESCO Publication PGI-85/WS/34.

Ogden, C. K. *Basic English: A General Introduction with Rules and Grammar.* London, England: Paul Trench, Trebner & Co., 1932.

———, ed. *The General Basic English Dictionary.* New York: Norton & Co., 1942.

Peterson, D. "Approaches to Computer Assisted Translation in Canada." *Technical Communication* 35.1 (1988): 35–36.

Sanderlin, S. "Preparing Instruction Manuals for Non-English Readers." *Technical Communication,* 35.2 (1988): 96–100.

9 Quotations

See Chapters 3, 4, and 5 for relevant bibliography.

10 References

See Chapters 3, 4, and 5 for relevant bibliography.

11 Indexes

Feinberg, H., ed. *Indexing Specialized Formats and Subjects.* Metuchen, NJ: Scarecrow Press, 1983.

Hardy, P. "Computer-Aided Indexing of Technical Manuals." *Indexer* 15.1 (1986): 22–24.

Harrod, L. M., ed. *Indexers on Indexing: A Selection of Articles Published in "The Indexer."* New York: R. R. Bowker, 1978.

Jones, K. P. "Getting Started in Computerized Indexing." *Indexer* 15.1 (1986): 9–13.

Knight, G. N. *The Art of Indexing: A Guide to the Indexing of Books and Periodicals.* London, England: Allen & Unwin, 1979.

Vickers, J. A. "Index, How Not to." *Indexer* 15.1 (1987): 163–169.

Webster's Standard American Style Manual. Springfield, MA: Merriam-Webster Inc. , 1985

12 Illustrations

American Institutes for Research. "Using Icons as Communication." *Simply Stated* (September/ October 1987): 1, 3.

Bernard, R. M., C. H. Petersen, and M. Ally. "Can Images Provide Contextual Support for Prose?" *Educational Communication and Technology Journal* 29.2 (1981): 101–108.

Booher, H. R. "Relative Comprehensibility of Pictorial Information and Printed Words in Procedural Instructions." *Human Factors* 17.3 (1975): 266–277.

Dreyfuss, H. *Symbol Sourcebook.* New York: McGraw-Hill, 1972.

Dwyer, F. M. "The Effect of IQ Level on the Instructional Effectiveness of Black-and-White and Color Illustrations." *AV Communication Review* 24.1 (1976): 49–62.

———. "Exploratory Studies in the Effectiveness of Visual Illustrations." *AV Communication Review* 18.3 (1970): 235–248.

Modley, R. *Handbook of Pictorial Symbols.* New York: Dover Publications, 1976.

Peeck, J. "Retention of Pictorial and Verbal Content of a Text with Illustrations." *Journal of Educational Psychology* 66 (1974): 880–888.

Samuels, S. J. "Effects of Pictures on Learning to Read, Comprehension and Attitude." *Review of Educational Research* 40 (1970): 398–407.

13 Data Displays

Brook, J. B., and K. D. Duncan. "An Experimental Study of Flowcharts as an Aid to Identification of Procedural Faults." *Ergonomics* 23 (1980): 387–399.

Buehler, M. F. "Table Design—When the Writer/Editor Communicates Graphically." *Proceedings of 27th International Technical Communication Conference.* Washington, DC: Society for Technical Communication, 1980.

Duchastel, P. C. "Research on Illustration in Text." *Educational Communication and Technology Journal* 28 (1980): 283–287.

Duchastel, P. C., and R. Waller. "Pictorial Illustration in Instructional Text." *Educational Technology* 19.11 (1979): 20–25.

Ford, D. F. "Packaging Problem Prose." *Proceedings of 31st International Technical Communication Conference.* Washington, DC: Society for Technical Communication, 1984.

Gettys, D. "IF You Write Documentation, THEN Try a Decision Table." *IEEE Transactions On Professional Communication,* PC 29.4 (December 1986): 61–64.

Hartley, J. "On the Typing of Tables." *Applied Ergonomics* 6 (1975): 39–42.

Herdeg, W. *Graphis Diagrams.* Zurich, Switzerland: The Graphis Press, 1976.

Hill, M., and W. Cochran. *Into Print: A Practical Guide to Writing, Illustrating, and Publishing.* Los Altos, CA: William Kaufmann, 1977.

Huff, D. *How to Lie with Statistics.* New York: W. W. Norton, 1954.

Kosslyn, S. M. "Information Representation in Visual Images." *Cognitive Psychology* 7 (1957): 341–370.

———. "Scanning Visual Images: Some Structural Implications." *Perception & Psychophysics* 14.1 (1973): 90–94.

Kosslyn, S. M., et al. "Understanding Charts and Graphs: A Project in Applied Cognitive Psychology." *ERIC* ED 238 687 (1983).

Larkin, J. H., and H. A. Simon. "Why a Diagram Is (Sometimes) Worth Ten Thousand Words." *Cognitive Science* 11 (1987): 65–99.

Lefferts, R. *How to Prepare Charts and Graphs for Effective Reports.* New York: Barnes & Noble, 1982.

Levin, J. R., et al. "On Empirically Validating Functions of Pictures in Prose." In *Illustrations, Graphs and Diagrams: Psychological Theory and Educational Practice.* Edited by D. M. Willows and H. A. Houghton. New York: Springer, 1987.

Lockwood, A. *Diagrams.* New York: Watson-Guptill, 1969.

Macdonald-Ross, M. "Scientific Diagrams and the Generation of Plausible Hypotheses: An Essay in the History of Ideas." *Instructional Science* 8 (1979): 223–234.

MacGregor, A. J. *Graphics Simplified: How to Plan and Prepare Effective Charts, Graphs, Illustrations, and Other Visual Aids.* Toronto, Canada: University of Toronto Press, 1979.

Mann, Gerald A. "How to Present Tabular Information Badly." In *Proceedings of 31st International Technical Communication Conference.* Washington, DC: Society for Technical Communication, 1984.

Martin, J., and C. McClure. *Diagramming Techniques for Analysts and Programmers.* Englewood Cliffs, NJ: Prentice-Hall, 1984.

Moxley, R. "Educational Diagrams." *Instructional Science* 12 (1982): 147–160.

Murch, G. M. "Using Color Effectively: Designing to Human Specifications." *Technical Communication* 32.4 (1985): 14–20.

Neurath, M. "Isotype." *Intercultural Science* 3 (1974): 127–150.

Rankin, R. *Communicating Science Concepts Through Charts and Diagrams.* Brisbane, Australia: Griffith University, 1986.

RCA. *Guide for "RCA Engineer" Authors.* Princeton, NJ: RCA Engineer, 1984.

Robertson, B. *How to Draw Charts and Diagrams.* Cincinnati, OH: North Light Books, 1988.

Schmid, C. F., and S. E. Schmid. *Handbook of Graphic Presentation,* 2nd ed. New York: John Wiley & Sons, 1979.

Selby, P. H. *Using Graphs and Tables.* New York: John Wiley & Sons, 1979.

Tufte, E. R. *Envisioning Information.* Cheshire, CT: Graphics Press, 1990.

———. *The Visual Display of Quantitative Information.* Cheshire, CT: Graphics Press, 1983.

Vernon, M. D. "Presenting Information in Diagrams." *AV Communication Review* 1 (1953): 147–158.

Technical Writing and Style Guide. Pittsburgh, PA: Westinghouse Electric Corporation, 1987.

Willows, D. M. "A Picture Is Not Always Worth a Thousand Words: Pictures as Distractors in Reading." *Journal of Educational Psychology* 70 (1978): 255–262.

Winn, W. "The Role of Graphics in Training Documents: Toward an Explanatory Theory of How They Communicate." *IEEE Transactions on Professional Communication* 32.4 (1989): 300–309.

———. "Using Charts, Graphs and Diagrams, in Educational Materials." In *The Psychology of Illustration.* Edited by D. M. Willows and H. Houghton. New York: Springer, 1987.

Winn, W., and W. G. Holliday. "Design Principles for Charts and Diagrams." In *The Technology of Texts.* Edited by D. Jonassen. Englewood Cliffs, NJ: Educational Technology Publications, 1982. 277–299.

14 Layout

Arps, R. B., R. L. Erdmann, A. S. Neal, and C. E. Schlaipfer. "Character Legibility Versus Resolution in Image Processing of Printed Matter." *IEEE Transactions on Man-Machine Systems* 10.3 (1969): 66–71.

Ausubel, D. P. "The Use of Advance Organizers in the Learning and Retention of Meaningful Verbal Material." *Journal of Educational Psychology* 51.5 (1960): 267–272.

Bethke, F. J., et al. "Improving the Usability of Programming Publications." *IBM Systems Journal* 20 (1981): 306–320.

Breland, K., and M. K. Breland. "Legibility of Newspaper Headlines Printed in Capitals and in Lower Case." *Journal of Applied Psychology* 28.2 (1928): 117–120.

Burt, C. *A Psychological Study of Typography.* Cambridge, England: Cambridge University Press, 1959.

Carver, R. P. "Effect of a 'Chunked' Typography on Reading Rate and Comprehension." *Journal of Applied Psychology* 54.3 (1970): 288–296.

Charrow, V. R., and J. C. Redish. *A Study of Standard Headings for Warranties: Tech. Report No. 6.* Washington, DC: AIR Document Design Project, 1980.

Coffey, J. L. "A Comparison of Vertical and Horizontal Arrangements of Alpha-Numeric Material-Experiment 1." *Human Factors* 3 (1961): 86–92, 93–98.

Coles, P., and J. Foster. "Experimental Study of Typographic Cueing in Printed Text." *Ergonomics* 20 (1977): 75–66.

Cooper, M. B., et al. "Reader Preference for Report Typefaces." *Applied Ergonomics* 10.2 (1979): 66–70.

Fabrizio, M. A., I. Kaplan, and G. Teal. "Readability as a Function of the Straightness of Right-Hand Margins." *Journal of Typographic Research* 1 (1967): 90–95.

Felker, D. B., et al. *Guidelines for Document Designers.* Washington, DC: American Institute for Research, no date.

Fields, A. A. "A Test of Information Mapped Programmed Text." *NSPI Journal* (1981): 26–28.

Firman, A. H. *An Introduction to Technical Publishing.* Pembroke, MA: Firman Technical Publications, 1983.

Foster, J. J. "A Study of the Legibility of One- and Two-Column Layouts for BPS Publications." *Bulletin of the British Psychological Society* 23 (1970): 113–114.

Foster, J. J., and P. Coles. "Typographic Cueing as an Aid to Learning from Typewritten Text." *Programmed Learning and Educational Technology* 12.2 (1975): 102–108.

———. "An Experimental Study of Typographic Cueing in Printed Text." *Ergonomics* 20.1 (1977): 57–66.

Graham, N. E. "The Speed and Accuracy of Reading Horizontal, Vertical, and Circular Scales." *Journal of Applied Psychology,* 40.4 (1956): 228–232.

Hackman, R. B., and M. A. Tinker. "Effect of Variations in Color of Print and Background Upon Eye Movements in Reading." *American Journal of Optometry* 34 (1957): 354–359.

Hartley, J. *Designing Instructional Text.* New York: Nichols Publishing Co., 1977.

Hartley, J., M. Young, and P. Burnhill. "The Effects of Interline Space on Judgments of Typesize." *Programmed Learning and Educational Technology* 12.2 (1975): 115–119.

Jonassen, D., and P. Hawk. "Using Graphic Organizers in Instruction." *Information Design Journal* 4 (1984): 58–68.

Kat, A. K., and J. L. Knight, Jr. "Speed of Reading in Single Column vs. Double Column Type." *Science News* 118.19 (1980): 296.

Klare, G. R., et al. "The Relationship of Human Interest to Immediate Retention and to the Acceptability of Technical Material." *Journal of Applied Psychology* 39.1 (1955): 92–95.

———. "The Relationship of Patterning to Immediate Retention and to Acceptability of Technical Material." *Journal of Applied Psychology* 39.1 (1955): 40–42.

———. "The Relationship of Style Difficulty to Immediate Retention and to Acceptability of Technical Material." *Journal of Educational Psychology* 46 (1955): 287–295.

———. "The Relationship of Typographic Arrangements to the Learning of Technical Training Material." *Journal of Applied Psychology.* 41.1 (1957): 41–45.

Kunst, R. "Type Styles as Related to Reading Comprehension." Doctoral dissertation, Arizona State University, 1972.

Loeb, J. W. "Differences Between Cues in Effectiveness as Retrieval Aids." Doctoral dissertation, University of Southern California, 1969.

Macdonald-Ross, M., and R. Waller. "Criticism, Alternatives, and Tests: A Conceptual Framework for Improving Typography." *Programmed Learning and Educational Technology* 17 (1975): 75–83.

Mayer, R. E., and B. K. Bromage. "Different Recall Protocols for Technical Texts Due to Advance Organizers." *Journal of Educational Psychology* 72 (1980): 209–225.

McLaughlin, H. "Comparing Styles of Presenting Technical Information." *Ergonomics* 9.3: 257–259.

Paterson, D. G., and M. A. Tinker. *How to Make Type Readable: A Manual for Typographer, Printers, and Advertisers.* New York: Harper & Brothers, Publishers, 1940.

———. "Studies of Typographical Factors Influencing Speed of Reading." *Journal of Applied Psychology* 15.3 (1931): 241–247.

Phillips, R. M. "The Interacting Effects of Letter Style, Letter Stroke-width and Letter Size on the Legibility of Projected High Contrast Lettering." Doctoral dissertation, Indiana University, 1976.

Poulton, E. C. *Effects of Printing Types and Formats on the Comprehension of Scientific Journals.* Cambridge, England: Cambridge University Press, 1959.

Ronco, P. G. *Characteristics of Technical Reports that Affect Reader Behavior: A Review of the Literature.* Tufts University, Institute for Psychological Research, 1965.

Smith, S. L. "Letter Size and Legibility." *Human Factors* 221.6 (1979): 661–670.

Soar, R. S. "Height-Width Proportion and Stroke Width in Numeral Visibility." *Journal of Applied Psychology* 39.1 (1955): 43–46.

———. "Stroke Width, Illumination Level and Figure-Ground Contrast in Numeral Visibility." *Journal of Applied Psychology* 39.6 (1955): 429–436.

Tinker, M. A. *Bases for Effective Reading.* Minneapolis, MN: University of Minnesota Press, 1965.

———. *Legibility of Print.* Ames, IA: Iowa State University Press, 1963.

———. "The Relative Legibility of the Letters, the Digits, and of Certain Mathematical Signs." *Journal of General Psychology* (July–October 1928): 472–496.

Twyman, M. "The Graphic Presentation of Language." *Information Design Journal* 3.1 (1981): 2–22.

Wendt, D. "An Experimental Approach to the Improvements of the Typographic Design of Textbooks." *Visible Language* 13.2 (1979): 108–133.

White, J. V. *Graphic Design for the Electronic Age.* New York: Watson-Guptill Publications, 1988.

Wright, P. "Presenting Technical Information: A Survey of Research Findings." *Instructional Science* 6 (1977): 93–134.

Wrolstad, M. E. "Adult Preferences in Typography: Exploring the Function of Design." *Journalism Quarterly* 37.2 (1960): 211–223.

Zachrisson, B. *Studies in the Legibility of Printed Text.* Stockholm, Sweden: Almqvist & Wiksell, 1965.

Contributors

Lori K. Anschuetz
1 Audience/14 Layout

A technical writer, editor, and project manager for Tec-Ed in Ann Arbor, Michigan, Ms. Anschuetz has over 12 years of documentation experience. She has written or edited technical manuals, marketing materials, and documentation plans; written power-plant functional descriptions; and edited reports and papers on automotive electronics, aerospace electronics, and materials engineering. She has also managed a range of communication projects, and has supervised two technical writing departments.

Ms. Anschuetz has an AB in Journalism and German from the University of Michigan and is a senior member of the Society for Technical Communication.

Lilli Babits
1 Audience/14 Layout

A writer and project manager at Tec-Ed, Lilli Babits writes user and reference manuals, marketing materials, and on-line documentation. She also manages documentation projects, including technical manuals, marketing materials, and documentation plans. Before joining Tec-Ed, she worked for a CAD system startup company, where she wrote and edited software documentation and helped design the user interface.

Ms. Babits has seven years of documentation experience and is currently a student at Eastern Michigan University.

Wallace Clements
7 Numbers and Symbols

Wallace Clements has been a technical editor and writer for three decades. He worked at Lawrence Livermore National Laboratory (LLNL) for 28 years before retiring in 1988, and he is currently with KMI Energy Systems, assigned to LLNL in Livermore, California.

His special career interests are in improving the quality of technical publications and in educating writers and editors toward that end. In this connection, he has written copious commentary and has co-authored two books published by the Society for Technical Communication: *The Scientific Report—A Guide for Authors* and *Guide for Beginning Technical Editors.*

Mr. Clements has a BA in mathematics from Hobart College and has done graduate study in teaching mathematics at Harvard University and in mathematical analysis at the University of California in Berkeley.

Alberta Cox
4 Spelling

Alberta Cox is head of publications at NASA's Ames Research Center at Moffett Field, California, a position she has held for the past seven years. Before that assignment she headed the Publications Division at the Naval Weapons Center, China Lake, California. She has been a technical writer/editor/manager for the government for over 28 years.

Mrs. Cox has taught short courses in effective writing for the government and at the junior college level, and has presented numerous papers at technical writing conferences. Mrs. Cox is a fellow of the Society for Technical Communication, and past president of the society; she is also a fellow of the Institute for the Advancement of Engineering.

Eva Dukes
5 Abbreviations/7 Numbers and Symbols/10 References

Eva Dukes has worked as a writer of chemical bulletins and brochures. She has also edited chemistry and physics reference books; science textbooks; and engineering reports, proposals, and manuals. As an acquisitions editor she procured articles for two RCA journals, as well as articles on chemistry and food science and books on medical and biochemical topics. Ms. Dukes administered the technical editing department at RCA for five years.

Ms. Dukes has a BA in Chemistry from Hunter College (CUNY). She is a member of the American Medical Writers Association, a fellow of the Society for Technical Communication, and past president of the New York Chapter of the Society for Technical Communication.

John Goldie
10 References

John Goldie is a technical writer for Tech Writing Affiliates, Newtown Center, Massachusetts, and course developer. He has designed and conducted evaluations of several software products, including interactive videodisc programs.

Mr. Goldie has also designed, written, and edited computer manuals for a variety of applications. With a background in graphic arts and photography, he enjoys designing and writing online help systems as well as coordinating the visual elements of paper-based documentation.

He has an AB in Fine Arts from Harvard and an EdM from the Harvard Graduate School of Education, where he concentrated in interactive technology.

Joseph E. Harmon
3 Punctuation

Joseph E. Harmon has been a technical editor with the Chemical Technology Division, Argonne National Laboratory in Argonne, Illinois, for the past 12 years.

He has published articles in professional writing journals and proceedings. He won awards in publication competitions for the Chicago Chapter of the Society for Technical Communication in 1985 and 1987.

Mr. Harmon has a BS in Mathematics from the University of Illinois, Champaign-Urbana, and an MA in English from the University of Illinois, Chicago.

Valerie Haus
12 Illustration

Valerie Haus is a resident visitor at AT&T in Morristown, New Jersey. Formerly she was a staff manager/technical writer at Bell Communications Research. She has been responsible for software and corporate administrative documentation.

She was also a senior technical writer at Allen-Bradley Co., Industrial Computer Group. In this position she prepared hardware and software documentation for industrial minicomputers.

Ms. Haus has an MA in Professional Writing from Carnegie-Mellon University.

William Horton
13 Data Displays

William Horton, president of William Horton Consulting in Boulder, Colorado, specializes in visual communication for technical information. Before launching his own company, he managed user interface development for Intergraph.

For the past 20 years he has applied human factors and engineering principles to the design of technical manuals, on-line information, and computer interfaces.

A graduate of MIT and a Registered Professional Engineer, Mr. Horton is the author of *Designing and Writing Online Documentation, Illustrating Computer Documentation,* and *The Icon Book.*

William J. Hosier
6 Specialized Terminology

William Hosier is currently Documentation Usability Manager at Bell Northern Research (BNR) in Ottawa, Canada. In this capacity he is responsible for documentation quality, usability evaluations, and research in technical communications. Mr. Hosier is a member of the joint Northern Telecom-BNR task force on documentation issues and was project manager for the documentation standards project.

After an early career as an electronic engineer, Mr. Hosier entered technical communications in 1968. He has held a number of managerial, staff and consultancy positions with British Aerospace, Ferranti, Phillips Computers, Phillips Telecommunications, Hewlett-Packard, ITT, and Standard Telephone Labs. Mr. Hosier joined BNR in 1980.

Mr. Hosier graduated with a degree in Electrical Engineering from Liverpool Polytechnic. He carried out postgraduate work in early integrated circuits at the University of Surrey. He is a fellow of the Institute of Scientific and Technical Communication and is currently president of the Eastern Ontario Chapter of the Society For Technical Communication.

Barry Jereb
12 Illustrations

Barry Jereb is supervisor of industrial computer hardware and software documentation at Allen-Bradley Company in Highland Heights, Ohio. In this position he has been responsible for redesigning all of the corporation's standard publications.

Mr. Jereb has published in various professional communication journals, including *Visible Language,* and has spoken at the national Plain English Forum as well as numerous other international communication conferences.

He received an MA in Professional Writing from Carnegie-Mellon University.

John Kirsch
10 References

John Kirsch, president of TechWriting Affiliates, has over 23 years of experience as a consultant, educator, and manager in the field of technical and business communication. From 1968 to 1975 he managed documentation, proposal, and customer-support teams for consulting and research organizations in the Boston area. Since 1972 he has held a range of teaching and research appointments at MIT, in the School of Engineering, the Sloan School of Management, and the School of Humanities and Social Science.

Mr. Kirsch studied and taught Greek and Latin at Harvard (AB, 1960; doctoral program, 1963–65) and Beloit College (Instructor in Classics, 1961–62).

Jean L. Murphy
2 Grammar

Jean Murphy has been teaching technical writing for the past five years. Before she became a full-time faculty member at Pierce College in Puyallup, Washington, she was a technical editor and writer for Floating Point Systems, Inc., in Portland, Oregon. While there she set up training classes for writers; rewrote the company style guide; and wrote and edited software, hardware, and diagnostic manuals.

Ms. Murphy has a BA and MA in English, both from Portland State University. She has served as a judge in the Society for Technical Communications (STC) PRISM Publication Competition and presented papers at the STC's International Technical Communication Conferences. She currently teaches and does some consulting in the Pacific Northwest.

David A. T. Peterson
8 Nonnative Readers

David A. T. Peterson is a technical communications consultant in Ottawa, Canada. He has been a professional communicator in print and broadcasting since 1953. Since 1960 he has specialized in science and technical communications. In addition to a background in the aerospace and military publications fields, he has extensive experience in transportation and mining. Mr. Peterson has published articles on computer-assisted translation of technical documents and on Simplified English.

James Prekeges
9 Quotations

James Prekeges works for Microsoft in Redmond, Washington. Mr. Prekeges has a BA in English and a BS in Scientific and Technical Communication from the University of Washington. He is an active member of several professional communication societies, including the Society for Technical Communication and the National Society for Performance and Instruction.

Philip Rubens
General Editor

Before moving into academia, Philip Rubens spent over a dozen years in various writing positions, ranging from newspapers to magazines to technical writing. In academia he directed undergraduate technical writing programs at William Rainey Harper Community College and Michigan Technological University. He has also served on the graduate faculty at Rensselaer Polytechnic Institute. At present he is the senior research associate for TechWriting Affiliates.

Dr. Rubens has published numerous articles and spoken at many international communication conferences. He is a past member of the board of directors for the Society for Technical Communication and continues to serve on a variety of advisory boards for professional organizations and journals. He has a PhD in Literature from Northern Illinois University.

Richard Vacca
11 Indexes

In his 12 years in technical communication, Richard Vacca has written a variety of technical documents, ranging from engineering proposals to magazine articles to reference manuals. He has extensive experience in both the private (BBN Communications Corporation, CFI/Price-Waterhouse, Ford Motor Company) and the public sector (United Nations High Commission for Refugees, University of Wisconsin College of Engineering).

Mr. Vacca has a BA in Philosophy from the University of Wisconsin and an MS in Technical Writing from Rensselaer Polytechnic Institute. He has taught undergraduate technical writing at Salem State College (Salem, Massachusetts) and in continuing education at the University of Wisconsin.

R. Dennis Walters
1 Audience/14 Layout

A senior-level specialist in written communication and training, R. Dennis Walters has created a variety of award-winning user manuals, reference manuals, and product brochures for firms such as Convergent Technologies, Atari, and Xerox. He has also designed instructional programs, written training materials, made training presentations, and performed usability testing for Xerox, General Motors, and Control Data Corporation.

Mr. Walters has almost 20 years of documentation and training experience, and holds a PhD in English from Michigan State University.

Index

The text of *Science and Technical Writing*
was composed in the typeface Sabon
by North Market Street Graphics of Lancaster, Pa.;
printed and bound by Arcata Graphics, Fairfield, Pa.;
and designed by Katy Riegel.